Diseases of DNA Repair

ADVANCES IN EXPERIMENTAL MEDICINE AND BIOLOGY

Recent Volumes in this Series

Volume 677
PROTEINS: MEMBRANE BINDING AND PORE FORMATION
Edited by Gregor Anderluh and Jeremy Lakey

Volume 678
CHEMO FOG: CANCER CHEMOTHERAPY-RELATED COGNITIVE IMPAIRMENT
Edited by Robert B. Raffa and Ronald J. Tallarida

Volume 679
MIPS AND THEIR ROLE IN THE EXCHANGE OF METALLOIDS
Edited by Thomas P. Jahn and Gerd P. Bienert

Volume 680
ADVANCES IN COMPUTATIONAL BIOLOGY
Edited by Hamid R. Arabnia

Volume 681
MELANOCORTINS: MULTIPLE ACTIONS AND THERAPEUTIC POTENTIAL
Edited by Anna Catania

Volume 682
MUSCLE BIOPHYSICS: FROM MOLECULES TO CELLS
Edited by Dilson E. Rassier

Volume 683
INSECT NICOTINIC ACETYLCHOLINE RECEPTORS
Edited by Steeve Hervé Thany

Volume 684
MEMORY T CELLS
Edited by Maurizio Zanetti and Stephen P. Schoenberger

Volume 685
DISEASES OF DNA REPAIR
Edited by Shamim I. Ahmad

Diseases of DNA Repair

Edited by

Shamim I. Ahmad, BSc, MSc, PhD
School of Science and Technology, Nottingham Trent University
Nottingham, United Kingdom

Springer Science+Business Media, LLC

Landes Bioscience

Springer Science+Business Media, LLC
Landes Bioscience

Printed in the USA.

Springer Science+Business Media, LLC, 233 Spring Street, New York, New York 10013, USA
http://www.springer.com

Please address all inquiries to the publishers:
Landes Bioscience, 1002 West Avenue, Austin, Texas 78701, USA
Phone: 512/ 637 6050; FAX: 512/ 637 6079
http://www.landesbioscience.com

The chapters in this book are available in the Madame Curie Bioscience Database.
http://www.landesbioscience.com/curie

Diseases of DNA Repair, edited by Shamim I. Ahmad. Landes Bioscience / Springer Science+Business Media, LLC dual imprint / Springer series: Advances in Experimental Medicine and Biology.

ISBN: 978-1-4419-6447-2

Library of Congress Cataloging-in-Publication Data

Diseases of DNA repair / edited by Shamim I. Ahmad.
p. ; cm. -- (Advances in experimental medicine and biology ; v. 685)
Includes bibliographical references and index.
ISBN 978-1-4419-6447-2
1. DNA repair. 2. Genetic disorders. I. Ahmad, Shamim I. II. Series: Advances in experimental medicine and biology, v. 685. 0065-2598 ;
[DNLM: 1. Genetic Diseases, Inborn--etiology. 2. DNA Repair--physiology. 3. Genetic Diseases, Inborn--therapy. W1 AD559 v.685 2010 / QZ 50 D611 2010]
QH467.D13 2010
616'.042--dc22

2010011800

DEDICATION TO A SPECIAL PERSON—AMY

It was Wednesday, August 28th 1991, when little angel, Amy, came into this world in Liverpool Women's Hospital, UK. She was born 8-weeks prematurely weighing merely 2 lbs 11 oz. She was declared to have an intrauterine growth failure.

Here is what Amy's mum has to say about her birth: "Looking back, we think that it was apparent to specialists at her birth that there was a problem—silence when she was born, her placenta was thin and transparent. Despite the fact that a geneticist was brought to see her, nothing was said to us, and we presumed that this was a normal birth".

In the beginning Amy met all her milestones; she was tiny yet an active baby. She walked at 13 months, talked on time and health visitors told Amy's parents that she would need to wear diving boots to slow her down. At around 18 months, Amy visited the Genetics Department of the Alder Hey Children's Hospital, Liverpool, UK for follow-up, where it was advised that something was seriously wrong with her—something very rare—and that, most likely, she wouldn't grow—they gave it a name—'Amy syndrome'. One of several disorders was suspected and tests were done, but no specific diagnosis was decided upon.

Amy continued to thrive in some ways but not in stature. She was a nightmare to feed but remained happy and full of energy. She began attending a mainstream school but couldn't quite keep up, so was transferred to a Moderate Learning Difficulties School where she made lots of friends and loved her work. When Amy was 5 years-old, her parents, while researching on the internet, came across a rare genetic disorder of DNA repair deficiency called Cockayne syndrome (CS). The clinical features of this disease and photos of CS children indicated that Amy might be suffering from this syndrome. Tests were carried out in the UK but no concrete results came back. Yet her parents strongly believed that Amy had CS.

Around the age of 11, Amy began to deteriorate. Firstly she developed a tremor and then her balance began to fail. She fell over regularly and began walking with a very poor posture. She kept asking her mother "What is happening to me?" And she began to feel very sad. Despite regular visits to a number of doctors, they could give her very little help. At this stage Amy's parents took the challenge to ask "What exactly was wrong with Amy?".

Amy was 14 when her parents raised enough money to fly to the USA to meet Dr. Edward Neilan at Boston Children's Hospital, who was studying CS at that time. Dr Neilan carried out genetic analyses on Amy's chromosomes and found a mutation, which indicated (together with her symptoms and appearance) that she had atypical CS. When the diagnosis was confirmed, Amy's mother said "As a mother, I had known this for nearly 10 years and, in one way, my mind was at peace with this news but, in another, I was broken-hearted".

At this time Amy started suffering from tremors. Moreover, she was unable to eat, drink, dress or carry out her everyday tasks without help.

Subsequently Amy was taken to a neurologist, Dr. Peter Kang, also in Boston, who prescribed medications for her tremors. Within 10 days Amy's tremor had essentially disappeared and this distraught girl was back to her former happy self. Not only that, but a number of other CS children began taking the same medication and they too regained some measure of dignity. From then on Amy became desperate to help other children by participating in several invasive tests, carried out at experimental levels, and by saying "If it helps others then I want these done".

Back in England Amy's parents' thinking was that they did not want any parent or child to ever feel as lonely or isolated as they and Amy had. So in 2007 they set up a charity organization called "Amy and Friends (Cockayne Syndrome Support)", and since then have united and helped over 50 families from across the world suffering from CS. This charity can be found at the website www.amyandfriends.org (Cockayne syndrome support).

Through this organization Amy and Amy's parents have brought together a large number of children suffering from CS, provide them happiness, love and support, while sharing together heart-breaking moments when children as young as 20 months and as old as 20 years lose their life. Brave Amy has visited several of her dying friends and helped them on their way to heaven by saying "Go now, our friends are waiting to take you, don't stay here—go and run and play".

She tells everyone she meets that she is glad to have CS so that she can help others. She is a pleasure to be around, brings love to those who meet her and once anyone meets Amy, she is never forgotten. Her teacher once said that what Amy lacks in stature she makes up in spirit.

Amy is now suffering from an underactive thyroid, kidney failure, hypertension, diabetes, high cholesterol, raised liver enzymes, a low grade glioma on her thalamus, kyphosis, lordosis, scoliosis and is stiff and in regular pain. She is unable to do most things she loves doing and very often feels lonely, despite having so many people around who love her. She lives now every day wanting to live her life to the fullest and dreaming of the day when her body is free from this illness.

Never before in my life I met a person who has been suffering so much, yet very bravely faces the situation and at the same time shows her help, love and affections to others suffering as her. She must be one of the bravest persons known. She has touched my heart so much in such a short time that, not only have I decided to work for this charity organization, but also dedicate this book to her. I gave her an Indian name—DIYA—which means a 'clay lamp' that burns to give light to others.

Shamim I. Ahmad, BSc, MSc, PhD

PREFACE

DNA is an essential component of life and its integrity plays a key role in sustaining normal life functions including DNA replication, genetic recombination, transcription and DNA repair. DNA is constantly threatened by numerous environmental and intrinsic deleterious physical and chemical agents, such as UV light, ionizing radiation, reactive oxygen species (ROS) and various chemical agents, which can act as mutagens and carcinogens. Since the integrity of DNA is vital, nature has endowed living organisms with fairly intricate repair systems involving a large number of proteins and enzymes participating in a variety of DNA repair pathways. Interestingly, there exists variation in the ability of individual's to repair the damage to DNA.[1]

Since this book is geared to be used by varied groups of readers such as advanced students and instructors in the fields of biology and medicine, scientists and more importantly clinicians, it is considered important to provide brief accounts of the basics of DNA damage, repair, mutagenesis and cancer.

Studies on DNA damage and repair originated in 1935 in bacteria,[2] specifically in *Escherichia coli* and subsequently focused on lower eukaryotes such as *Saccharomyces cereviciae* and on complex organisms such as humans and plants. In the last 75 years thousands of scientific papers have been published (23,667 according to PubMed Data, October 2009), and yet we have not reached the goal of understanding DNA repair systems in humans. Interestingly in the original studies the damaging agent used was ultraviolet light of type C (180-290 nm).[2]

Studies on the bacterium *E. coli* have played important role in our understanding of DNA damage and repair in humans. A large number of DNA repair systems have been identified that operate upon different types of damage; these include **photoreactivation repair**,[3] **nucleotide excision repair** (NER) of two types, **global genomic repair** (GGR) and **transcription-coupled repair** (TCR),[4] **mismatch repair or MMR**,[5] **SOS** or **error prone repair**[6] (found in certain prokaryotes involving more than 40 genes in *E. coli* and missing in humans), **homologous recombination repair** involving RecBCD in *E. coli*,[7] **base excision repair** or BER (now considered to be of two types SP-BER or short patch BER and LP-BER or long patch BER,[8] **non-homologous DNA end joining**,[9] **alkB mechanisms** to repair methyl lesions (and also faulty or damaged RNA)[10,11] and **translesion synthesis,**[12] although the latter might be considered a damage tolerance rather than a repair mechanism. It should be noted that certain systems such as SOS repair also play roles in modulation of drug resistance and in secretion and dissemination of virulence in bacteria.[13] Likewise DNA polymerases (5 in *E. coli*, 8 in *S. cereviciae* and 15 in humans) that play critical role in DNA replication, repair, recombination and mutagenesis; certain of these are specialized polymerases that allow DNA synthesis past lesions in DNA.[14]

Although a good number of DNA repair enzymes have been conserved through evolution in species ranging from bacteria to high eukaryotes, DNA repair systems in humans have evolved more intricately, due to aging, immune systems, cell compartmentalization including nuclear-lamina,[15] presence of mitochondria with their own DNA, presence of chromatin in DNA,[16] respiratory cycle including hypoxia[17] etc, involving hundreds of genes, enzymes and proteins.[18] As an example, in humans the two BER sub-pathways are operated by different sets of proteins whereas in prokaryotes only one set exists.[8] In addition, humans have evolved sophisticated signal transduction networks to sense DNA damage, to trigger a response and to regulate the consequent gene expression. The first step of the response is cell cycle arrest followed by the activation of cellular DNA damage responses. The sensor proteins and signal transduction network are then coordinated to finally repair the DNA.[19] Cells possess the ability to sense whether normal function with DNA replication, transcription and cell progression can resume. In cases when repair cannot restore the DNA to its normal form, either due to massive damage or to impairment of repair system, the cell undergoes senescence or apoptosis.

Eukaryotic chromatin in DNA also plays important roles in maintaining genomic integrity, and multiple post-translational histone modifications via acetylation, methylation, phosphorylation and ubiquitination participate in this process.[20] The modifications also play roles in maintaining the chromatin environment, transcription and replication.

The attempt to repair damage in DNA can have three fates: (i) the DNA is repaired perfectly, for example as carried out by photoreactivation, also known as error free repair; (ii) none or incomplete repair, in which case the cell undergoes senescence or apoptosis (iii) carry out error prone repair and induce mutation, which is normally long lasting and inheritable, causing inborn errors. Although in many cases, the occurrence of mutations plays negative role in life (such as, fetal demise or induction of genetic diseases), in certain cases, it can be advantageous (e.g. survival of rare antibiotic-resistant mutant bacteria in the presence of antibiotics).

It is estimated that about 6000 single gene mutations cause human diseases (http://www.ornl.gov/sci/techresources/Human_Genome/medicine/assist.shtml). The editor of this book estimates that there are about 60-70 diseases associated with defects in DNA repair systems. This editor has recently published four books addressing the DNA repair diseases, viz, Fanconi anemia,[21] xeroderma pigmentosum,[22] ataxia telangiectasia[23] and Cockayne syndrome.[24] The purpose of this volume is to present an updated detailed account of some important additional diseases of DNA repair. It has not been possible to cover all the DNA repair deficient diseases in this work, hence diseases such as Bloom's syndrome, Werner's syndrome, Nijmegen breakage syndrome, ataxia telangiectasia-like disorder, RAD 50 deficiency, RIDDLE syndrome and others will be presented in a forthcoming volume.

In this book, the editor aimed to maintain a balance between the DNA repair diseases that have been studied exhaustively, well-defined genetic defects in DNA repair system(s) and those diseases (such as **Triple A syndrome**) that shows a tangential association with DNA repair defects and hence more studies are warranted.

DNA repair diseases can be linked to DNA repair deficiencies in chromosomes and/or in mitochondria. As the research on human neurodegenerative diseases progresses, it is becoming apparent that several of them are associated with mitochondrial dysfunction due mostly to intrinsic oxidative damage. **Amyotrophic lateral sclerosis** (ALS) is one such disease (eloquently presented in Chapter 2) caused by mutations in the gene coding for superoxide dismutase (SOD). Naturally if SOD (an important scavenger of superoxide radicals) is absent, the endogenous concentration of intrinsically produced reactive oxygen species (ROS) will increase and the amount of damage will exceed the ability of the repair complexes that remove them, and hence the disease. It is, however, still puzzling that mutations in two other genes, TDP-43 and FUS/TLS, also lead to this disease; this warrants further studies. In a recent study the editor and his colleagues have identified a number of yet unidentified enzymes in bacteria that may be responsible for scavenging two types of ROS, hydroxyl radicals and superoxide radicals (manuscript in preparation).

Alzheimer's disease is another common neurodegenerative disorder due to mitochondrial dysfunction, induced mainly by ROS, and this has been fully described in Chapter 4. In Chapter 5 another neurodegenerative disorder, **Huntington's disease**, is described as caused by unstable expansion of CAG repeats, located in the 5-prime terminal section of the *IT15* gene. In Chapter 6, Katsuno and colleagues have described another neurodegenerative disorder, **spinal and bulbar muscular atrophy**, again caused by expansion of a trinucleotide CAG repeat, which encodes a polyglutamine tract within the first exon of the androgen receptor gene.

Chapter 7 addresses another neurological disorder, **spinocerebellar ataxia with axonal neoropathy**, caused by a specific mutation in the gene coding for tyrosyl DNA phosphodiesterase. This disorder leads to multiple pathological phenotypes including neurological disease.

Another neurodegenerative disease, **early-onset ataxia with ocular motor apraxia** and **hyperalbumenimia/ataxia with oculomotor apraxia**, has been described in Chapter 3, and is caused by a mutation in the *APTX* gene and consequent lack of repair of DNA single and double strand breaks leading to neuronal cell dysfunction and death.

Cornelia de Lange syndrome (CdLS) is another disease with mental retardation; it has been described in Chapter 11. This is a more complicated disease, in which severe physiological deformities are observed, mostly facial gestalt. Cells from patients show a few specific chromosomal rearrangements. It is interesting to note that dup 3q syndrome has long been considered to be a phenocopy of

CdLS due to clinical overlap; however the several mutations that cause dup 3q do not overlap with mutations found in CdLS. How and why clinical overlaps exist between these two syndromes, while genetic studies differentiate them, is a subject of further studies.

A number of other interesting diseases associated with deficiency in DNA repair, are certain **hereditary photodermatosis** (Chapter 9), **trichothiodystrophy** (TTD) (Chapter 10) and **familial cutaneous melanoma** (Chapter 13). One prominent feature of these diseases is the sensitivity of the skin to sunlight, although in certain cases of TTD skin sensitivity is not apparent. The sun sensitivity in these diseases is the result of defective repair of damage to DNA caused by ultraviolet (UV) light. Although most laboratory work has been carried out using germicidal UV lamps (emitting UVC at 180-290 nm), in fact the patients suffer from the effects of unintentional exposure to solar UV.

Although the sun emits all kinds of UV light, only UVB (5%, 290-310 nm) and UVA (95%, 310-400 nm) reach the earth's surface. Whereas UVC causes DNA damage by inducing formation of cyclobutane pyrimidine dimers (CPD) and (6-4) pyrimidine-pyrimidone photoproducts (6-4PP), UVB retains the ability to induce CPD and 6-4PP, and also can damage DNA via ROS production. UVA rays most likely damage DNA solely by ROS production generated via the photolysis of a variety of endogenous photosensitizers.[25-28]

Another disease linked to NER is **cerebro-oculo-facio-skeletal syndrome** (COFS), which has been adequately described in Chapter 19, and in which facial dysmorphism, crophthalmia, arthrogryposis and severe developmental delays occur. Profound photosensitivity is also observed in the patients due to mutations in genes involved in the NER pathway. Molecular analysis revealed that TCR is functionally impaired in this disease. Interestingly most facial features in COFS are also observed in **ligase IV syndrome** (Chapter 16). Ligase IV is a component of a DNA repair pathway involved in non-homologous end joining of DNA strand breaks.

Deficiency in DNA repair also causes diseases of the immune systems; these include **primary immunodeficiency syndrome**, **inherited defects of immunoglobulin class switch recombination** (IDICR) and **dyskeratosis congenita**. In Chapter 14 Slatter and Gennery have elegantly described the molecular mechanisms, clinical presentation and treatment of primary immunodeficiency disorders caused by defects in MMR. Chapter 15 addresses IDICR; this disease is caused by defects in activation-induced cytidine deaminase (AID), uracil-N-glycosylase, post-meiotic segregation and in non-homologous end joining. Deficiency in immune systems naturally leads to loss of defense for infections and their consequences. Chapter 20 describes **dyskeratosis congenita** which is an inheritable bone marrow disease leading to inability to produce enough mature erythrocytes, granulocytes and platelets causing peripheral blood cytopenia along with other clinical features. Four genes are implicated in this disease, the X-linked DKC1, which codes for dyskerin protein, TERC and TERT coding for the mRNA for telomeres and telomerase respectively, and one autosomal recessive, NOP10.

Cancers of various types are by far the most important end result due to defect in DNA repair; thus hamartomas in several organs, specifically in kidney, is caused by impairment of the *TSC1* and *TSC2* genes leading to **tuberous sclerosis complex**, which has been eloquently described in Chapter 8. **Colorectal cancer** is another disease due to a deficiency in MMR. Its treatment usually involves chemoradiation therapy and/or surgery. The author of Chapter 12 discusses which kind of patients should receive what kind of treatment and why. **Muir-Torre syndrome, Wilms' tumor, retinoblastoma** and **Von Hippel Lindau syndrome**, are linked to specific cancers. Chapters 17, 18, 21 and 22 were respectively devoted to describe these diseases.

It is hoped that this book will stimulate both expert and novice researchers in the field with excellent overviews of the current status of research and pointers to future research goals. Material presented in this book should also be beneficial to the clinicians in treating their specialized cases. The insights gained will be valuable for the development of new therapeutic drugs to treat the clinical problems raised by these rare but devastating diseases as well.

Finally the editor wishes to present his most cordial thanks to all his contributors without whom it would not have been possible to produce this quality and rare volume.

Shamim I. Ahmad, BSc, MSc, PhD
School of Science and Technology, Nottingham Trent University
Nottingham, United Kingdom

ACKNOWLEDGEMENT

The editor is thankful to Professor Graciela Spivak for her valuable input in the preparation of this section of the book.

REFERENCES

1. Rihs HP, Hoffmeyer F, Bruning T. DNA repair: From the mechanism to the impact on occupational research. Pneumologie 2009; 63:319-324.
2. Hollaender A, Curtis JT. Effect of sublethal doses of monochromatic ultraviolet radiation on bacteria in liquid suspension. Proc Soc Exp Biol Med 1935; 33:61-62.
3. Lucas-Lledo JI, Lynch M. Evolution of mutation rates: Phylogenomic analysis of the photolyase/chryptochrome family. Mol Biol Evol 2009; 26:1143-1153.
4. Truglio JJ, Croteau DL, Van Houten B et al. Prokaryotic nucleotide excision repair, the UvrABC system. Chem Rev 2006; 106:233-252.
5. Li GM. Mechanism and functions of DNA mismatch repair. Cell Res 2008;18:85-98.
6. Junion C. Inducible SOS response system of DNA repair and mutagenesis in Escherichia coli. Int J Biol Sci 2008; 4:338-344.
7. Dillingham MS, Kowalczykowski SC. RecBCD enzyme and the repair of double strand DNA breaks. Microbiol Mol Biol Rev 2008; 72:642-671.

8. Krwawicz J, Arczewska KD, Speina E et al. Bacterial DNA repair genes and their eukaryotic homologues: Mutations in genes involved in base excision repair (BER) and DNA end processors and their implication in mutagenesis and human disease. Acta Biochim Pol 2007; 54:413-434.
9. Poplawski T, Stoczynska E, Blasiak J. Non-homologous DNA end joining—new proteins, new functions, new mechanisms. Postepy Biochem 2009; 55:36-45.
10. Falnes PO, Klungland A, Alseth I. Repair of methyl lesions in DNA and RNA by oxidative demethylation. Neuroscience 2007; 145:1222-1232.
11. Sundheim O, Talstad VA, Vagbo CB et al. AlkB demethylases flip out in different ways. DNA Repair (Amst) 2008; 7:1916-1923.
12. Fuchs RP, Fujii S. Translesion synthesis in Escherichia coli: Lessons from the NarI mutation hot spot. DNA Repair (Amst) 2007; 6:1032-1041.
13. Butala M, Zgur-Bertok D, Busby SJ. The bacterial LexA transcriptional repressor. Cell Mol Life Sci 2009; 66:82-93.
14. Garcia-Diaz M, Bebenek K. Multiple functions of DNA polymerases. CRC Critical Rev Plant Sci 2007; 26:105-122.
15. Chi YH, Chen ZJ, Jeang KT. The nuclear envelopathies and human diseases. J Biomed Sci 2009; 16:96.
16. Venkitaraman AR. Modifying chromatin architecture during the response to DNA breakage. Crit Rev Biochem Miol Biol 2009; Oct 30 [Epub ahead of print].
17. Yoo YG, Hayashi M, Christensen J et al. An essential role of the HIF-1alpha-c-Myc axis in malignant progression. Ann NY Acad Sci 2009; 1177:198-204.
18. Friedberg EC, Walker GC, Siede W et al. DNA Repair and Mutagenesis. Washington DC: ASM Press, 2005.
19. Stracker TH, Usui T, Petrin JH. Taking the time to make important decision: the checkpoint effector kinases Chk1 and Chk2 and the DNA damage response. DNA Repair (Amst) 2009; 8:1047-1054.
20. Yang J, Ledaki I, Turley H et al. Role of hypoxia-inducible factors in epigenetic regulation via histone demethylases. Ann NY Acad Sci 2009; 1177:185-197.
21. Ahmad SI, Kirk SH, eds. Molecular Mechanisms of Fanconi Anemia. Austin/New York: Landes Bioscience/Springer Science+Business Media, 2007.
22. Ahmad SI, Hanaoka F, eds. Molecular Mechanisms of Xeroderma Pigmentosum. Austin/New York: Landes Bioscience/Springer Science+Business Media, 2008.
23. Ahmad SI, ed. Molecular Mechanisms of Ataxia Telangiectasia. Austin: Landes Bioscience, 2009.
24. Ahmad SI, ed. Molecular Mechanisms of Cockayne Syndrome. Austin: Landes Bioscience, 2009.
25. Craggs J, Kirk SH, Ahmad SI. Synergistic action of near UV and phenylalanine, tyrosine or tryptophan on the inactivation of phage T7: Role of superoxide radicals and hydrogen peroxide. J Photochem Photobiol B 1994; 24:123-128.
26. Paretzoglou A, Stockenhuber C, Kirk S et al. Generation of reactive oxygen species from the photolysis of histidine by near ultraviolet light: Effects on T7 as a model biological system. J Photochem Photobiol B 1998; 43:101-105.
27. Ahmad SI, Hargreaves A, Taiwo FA et al. Near ultraviolet photolysis of L-mandelate, formation of reactive oxygen species, inactivation of phage T7 and implications on human health. J Photochem Photobiol B 2004; 77:55-62.
28. Hargreaves A, Taiwo FA, Duggan O et al. Near ultraviolet photolysis of β-phenyl pyruvic acid generates free radicals and results in DNA damage. J Photochem Photobiol B 2007; 89:110-116.

ABOUT THE EDITOR...

SHAMIM AHMAD after obtaining his Master's degree in Botany from Patna University, Bihar, India and his PhD in Molecular Genetics from Leicester University, England, joined Nottingham Polytechnic as Grade 1 lecturer and subsequently promoted to SL post. Nottingham Polytechnic subsequently became Nottingham Trent University where, after serving for about 35 years, he took early retirement to spend the remaining time writing books and conducting full-time research. For more than three decades he worked on different areas of biology including thymineless death in bacteria, genetic control of nucleotide catabolism, development of anti-AIDS drugs, control of microbial infection of burns, phages of thermophilic bacteria and microbial flora of Chernobyl after nuclear accident. But his primary interest, which started 25 years ago, is DNA damage and repair, particularly near UV photolysis of biological compounds, production of reactive oxygen species and their implications on human health including skin cancer and xeroderma pigmentosum. He is also investigating photolysis of non-biological compounds such as 8-methoxypsoralen+UVA, mitomycin C, and nitrogen mustard and their importance in psoriasis treatment and in Fanconi anemia. In 2003 he received a prestigious "Asian Jewel Award" in Britain for "Excellence in Education". He is also the Editor of *Molecular Mechanisms of Fanconi Anaemia, Molecular Mechanisms of Xeroderma Pigmentosum, Molecular Mechanisms of Ataxia Telangiectasia* and *Molecular Mechanisms of Cockayne Syndrome* published by Landes Bioscience.

PARTICIPANTS

Hiroaki Adachi
Department of Neurology
Nagoya University Graduate School of Medicine
Nagoya
Japan

Shamim I. Ahmad
School of Science and Technology
Nottingham Trent University
Nottingham
UK

Osamu Arisaka
Department of Pediatrics
Dokkyo Medical University
Shimotsuga-gun, Tochigi
Japan

Haruhiko Banno
Department of Neurology
Nagoya University Graduate School of Medicine
Nagoya
Japan

Gareth Baynam
Genetic Services of Western Australia
Princess Margaret Hospital for Children
School of Paediatrics and Child Health
The University of Western Australia
Perth
Australia

Sandrine Betuing
UMRS 952, INSERM
UMR 7224-CNRS
Université Pierre et Marie Curie
Paris
France

Cornelius F. Boerkoel
Department of Medical Genetics
Provincial Medical Genetics Program
Child and Family Research Institute
Children's and Women's Health Centre of British Columbia
University of British Columbia
Vancouver, British Columbia
Canada

Cecilia Bonnet
Pôle des maladies du système nerveux
Fédération de neurologie
Hopital Pitié-Salpétrière
Paris
France

Jocelyne Caboche
UMRS 952, INSERM
UMR 7224-CNRS
Université Pierre et Marie Curie
Paris
France

Valeria Calsolaro
Department of Neuroscience
Neurological Clinic
University of Pisa
Roma
Italy

Dimitry A. Chistiakov
Department of Molecular Diagnostics
National Research Center GosNIIGenetika
Moscow
Russia

Sumit Dave
Departments of Surgery & Pediatrics
Division of Urology
Schulich School of Medicine & Dentistry
The University of Western Ontario
London Health Sciences Centre-Victoria Hospital
London, Ontario
Canada

Anne Durandy
INSERM, U768
Université Paris Descartes, Faculté de Médecine René Descartes
Paris
France

Claude E. Gagna
Department of Pathology and Laboratory Medicine
UMDNJ-New Jersey Medical School
Newark, New Jersey
USA

Pauline Gardës
INSERM, U768
Université Paris Descartes, Faculté de Médecine René Descartes
Paris
France

Andrew R. Gennery
Newcastle General Hospital
Newcastle upon Tyne
UK

Vineeta Gupta
Department of Pediatrics
Institute of Medical Sciences
Banaras Hindu University
Varanasi
India

Samy L. Habib
South Texas Veterans Healthcare System, Geriatric Research, Education, and Clinical Center
and
Department of Medicine
The University of Texas Health Science Center
San Antonio, Texas
USA

Donna E. Hansel
Cleveland Clinic
Department of Anatomic Pathology
Taussig Cancer Institute and Genomic Medicine Institute
Glickman Urological and Kidney Institute
Cleveland, Ohio
USA

Johan Hansson
Department of Oncology-Pathology
Karolinska Institute
Karolinska University Hospital Solna
Stockholm
Sweden

Ryuki Hirano
Department of Neurology and Geriatrics
Graduate School of Medical and Dental Sciences
Kagoshima University
Kagoshima City
Japan

Sergio Huerta
University of Texas Southwestern Medical Center
Dallas, Texas
USA

Heidrun Interthal
Institute of Cell Biology
University of Edinburgh
Edinburgh, Scotland
UK

Jonathan Izawa
Departments of Surgery & Oncology
Divisions of Surgical Oncology & Urology
Schulich School of Medicine & Dentistry
The University of Western Ontario
London Health Sciences Centre-Victoria Hospital
London, Ontario
Canada

Masahisa Katsuno
Department of Neurology
Nagoya University Graduate School of Medicine
and
Institute for Advanced Research
Nagoya University
Nagoya
Japan

Jenny J. Kim
Cleveland Clinic
Taussig Cancer Institute
Cleveland, Ohio
USA

Sven Kracker
INSERM, U768
Université Paris Descartes, Faculté de Médecine René Descartes
Paris
France

Akash Kumar
Department of Pediatrics
Institute of Medical Sciences
Banaras Hindu University
Varanasi
India

Muriel W. Lambert
Department of Pathology and Laboratory Medicine
UMDNJ-New Jersey Medical School
Newark, New Jersey
USA

W. Clark Lambert
Department of Pathology and Laboratory Medicine
UMDNJ-New Jersey Medical School
Newark, New Jersey
USA

Jinglan Liu
Division of Human and Molecular Genetics
The Children's Hospital of Philadelphia
Philadelphia, Pennsylvania
USA

Annalisa LoGerfo
Department of Neuroscience
Neurological Clinic
University of Pisa
Roma
Italy

Dietmar Lohmann
Institut fur Humangenetik
Universitätsklinikum Essen
Essen
Germany

Takao Makifuchi
Joetsu General Hospital
Niigata
Japan

Michelangelo Mancuso
Department of Neuroscience
Neurological Clinic
University of Pisa
Roma
Italy

Carlos H. Martínez
Department of Surgery
Division of Urology
Schulich School of Medicine & Dentistry
The University of Western Ontario
London Health Sciences Centre-Victoria Hospital
London, Ontario
Canada

Pedro Mercader
Dermatology Service
Hospital General
Universitario Morales Meseguer
Murcia
Spain

Dennis H. Oh
Department of Dermatology
University of California at San Francisco
and
Dermatology Service
VA Medical Center
San Francisco, California
USA

Osamu Onodera
Department of Molecular Neuroscience
Resource Branch for Brain Disease Research
Center for Bioresource-Based Research
Brain Research Institute
Niigata University
Niigata
Japan

Daniele Orsucci
Department of Neuroscience
Neurological Clinic
University of Pisa
Roma
Italy

J. Jefferson P. Perry
Department of Molecular Biology
The Scripps Research Institute
La Jolla, California
USA
and
The School of Biotechnology
Amrita University
Kollam, Kerala
India

Brian I. Rini
Cleveland Clinic
Department of Solid Tumor Oncology
Taussig Cancer Institute
Glickman Urological and Kidney Institute
Cleveland, Ohio
USA

Emmanuel Roze
UMRS 952, INSERM
UMR 7224-CNRS
Université Pierre et Marie Curie
and
Pôle des maladies du système nerveux
Fédération de neurologie
Hopital Pitié-Salpétrière
Paris
France

Mustafa A.M. Salih
Department of Paediatrics
Division of Paediatric Neurology
College of Medicine
King Saud University
Riyadh
Saudi Arabia

Vijaya Sarathi
Department of Endocrinology
King Edward Memorial Hospital
Parel, Mumbai
India

Tatsuya Sato
Department of Neurology
Clinical Neuroscience Branch
Brain Research Institute
Niigata University
Niigata
Japan

Nalini S. Shah
Department of Endocrinology
King Edward Memorial Hospital
Parel, Mumbai
India

David S. Shin
Department of Molecular Biology
The Scripps Research Institute
La Jolla, California
USA

Gabriele Siciliano
Department of Neuroscience
Neurological Clinic
University of Pisa
Roma
Italy

Mary A. Slatter
Newcastle General Hospital
Newcastle upon Tyne
UK

Gen Sobue
Department of Neurology
Nagoya University Graduate School of Medicine
Nagoya
Japan

Graciela Spivak
Department of Biology
Stanford University
Stanford, California
USA

Keisuke Suzuki
Department of Neurology
Nagoya University Graduate School of Medicine
Nagoya
Japan

Hiroshi Suzumura
Department of Pediatrics
Dokkyo Medical University
Shimotsuga-gun, Tochigi
Japan

Masayoshi Tada
Department of Neurology
Clinical Neuroscience Branch
Brain Research Institute
Niigata University
Niigata
Japan

John A. Tainer
Department of Molecular Biology
The Scripps Research Institute
La Jolla, California
USA
and
Life Sciences Division
Lawrence Berkeley National Laboratory
Berkeley, California
USA

Hiroshi Takashima
Department of Neurology and Geriatrics
Graduate School of Medical and Dental Sciences
Kagoshima University
Kagoshima City
Japan

Fumiaki Tanaka
Department of Neurology
Nagoya University Graduate School of Medicine
Nagoya
Japan

Cheryl Walton
Department of Medical Genetics
Provincial Medical Genetics Program
Child and Family Research Institute
Children's and Women's Health Centre of British Columbia
University of British Columbia
Vancouver, British Columbia
Canada

Akio Yokoseki
Department of Neurology
Clinical Neuroscience Branch
Brain Research Institute
Niigata University
Niigata
Japan

CONTENTS

ACKNOWLEDGEMENTS

The editor of this book cordially acknowledges the various authors for their contribution of chapters with in depth knowledge and highly skilled presentations. Without them it would not have been possible to produce this unique piece of work. I would also like to acknowledge Celeste Carlton, Erin O'Brien and Cynthia Conomos of Landes Bioscience for their hard work and patience in preparation of this book. Finally, I would like to acknowledge my three daughters, Farhin Ahmad, Mahrin Ahmad and Tamsin Ahmad for their continuous love, support and encouragement to produce this book.

Chapter 1

Triple-A Syndrome

Vijaya Sarathi and Nalini S. Shah*

Abstract

Triple-A syndrome is characterized by triad of adrenocorticotrophic hormone (ACTH)—resistant adrenal insufficiency, alacrimia and achalasia cardia. It is a rare disease and inherited by autosomal recessive pattern. Allgrove syndrome is characterized by mutation(s) in AAAS gene, located on chromosome 12q13, that codes for ALADIN protein. Most mutations produce a truncated protein, although missense and point-mutations have also been reported. Some patients with Triple-A syndrome may not have mutations in AAAS gene; in those there is no specific genotype-phenotype correlation. Although alacrimia is not the usual presenting manifestation, probably it is the earliest and most consistent feature. Achalasia cardia and adrenal insufficiency are the early and usual presenting manifestations. Neurological features appear at later age and autonomic manifestations are the most common neurological disorder. Polyneuropathy, amyotrophy, optic atrophy are the other common neurological problems. Alacrimia is diagnosed by Schirmer's test while ahalasia cardia and adrenal insufficiency are best diagnosed by esophageal monometry and ACTH stimulated cortisol levels respectively. Alacrimia is treated with artificial tears while achalasia cardia with either pneumatic dilatation or Heller's myotomy. Adrenal insufficiency is treated with glucocorticoid and if necessary mineralocorticoid replacement.

Introduction

Allgrove et al, for the first time in 1978 described two pairs of siblings with glucocorticoid deficiency and achalasia cardia. Three of them also had alacrimia and hence, a new term 'Triple-A syndrome' or 'Allgrove syndrome' was coined for the combination of these three manifestations.[1] Although Triple-A syndrome is classically described as a triad of adrenal insufficiency, alacrimia and achalasia cardia, spectrum of the disease varies widely. Some times it is partially characterized and expressed by only two of the three manifestations (2A syndrome) while most often it is associated with autonomic neuropathy, deserving the term '4A syndrome'.[2,3] Some authors suggested the eponym '5A syndrome' to include amyotrophy and other neurological manifestations as fifth component of the syndrome.[4]

Epidemiology

Triple-A syndrome is a rare disorder and is usually described as case reports, often in familial clusters. It has been reported from different parts of the world. The actual incidence is difficult to estimate because of its rarity and widely variable presentation. Allgrove syndrome is characterized by autosomal recessive pattern of inheritance and most patients have consanguineous parents. Owing to its autosomal recessive inheritance pattern, the recurrence risk of Allgrove syndrome in future pregnancies of parents with an affected child is 25%. Penetrance of the identified AAAS gene (gene mutated in Triple-A syndrome) defects is close to 100% for

*Corresponding Author: Nalini S. Shah—Department of Endocrinology, King Edward Memorial Hospital, Parel, Mumbai, India 400012. Email: drshri25@rediffmail.com

Diseases of DNA Repair, edited by Shamim I. Ahmad. ©2010 Landes Bioscience and Springer Science+Business Media.

Allgrove syndrome (albeit with variable clinical expression) and is characterized by absence of any symptoms in heterozygote carriers.[5]

Mutation in AAAS gene is a relatively common. An analysis of nine patients with Triple-A, who were carriers of the IVS 14 + 1G→A mutation suggested an estimated age of the mutation of ~1000-1175 years.[6]

Etiology

Common etiologic factor affecting the three organ systems (esophagus, lacrimal glands and adrenal glands) with different function and location remained enigmatous until recently. Although initially it appeared that mutations in the ACTH receptor gene would provide the link for the association of the triad manifestations of Allgrove syndrome, it could not be proven.[8,9] Triple-A syndrome was also considered to be a variant of familial glucocorticoid deficiency (FGD) due to the presence of ACTH insensitivity in both the disorders. However, no abnormality was found in the ACTH receptor gene, located on the short arm of chromosome 18 (18p11.2), which is responsible for FGD.[9]

Several initial linkage analyses failed to localize the gene for Triple-A syndrome. In 1996 Weber et al succeeded to locate the gene for the Triple-A syndrome to chromosome 12q13 near the type II keratin gene cluster.[10] The gene was located on a chromosomal segment, flanked by the markers D12S1629 and D12S312, which are 6cM apart.[10] In 2001, Handschug et al identified the 16 exons in AAAS gene.[11]

AAAS gene shows ubiquitous expression in human tissues with particularly abundant expression in the adrenal gland, gastrointestinal structures, pituitary gland, cerebellum and fetal lung.[11] It codes for a protein called ALADIN for alacrima, achalasia, adrenal insufficiency and neurological disorder.[12]

ALADIN is a protein belonging to the WD repeat family.[13] Proteins, belonging to this family, have a wide functional diversity and are involved in protein-protein interactions, signal transduction, RNA processing, vesicular trafficking, cytoskeleton assembly and cell division control.[11] The exact function of ALADIN protein is not known but proteomic analysis has shown it to be a part of the mammalian nuclear pore complex (NPC). NPC is critical for communication between the nucleus and the cytoplasm of cells.[14] When AAAS gene is mutated, ALADIN protein mislocalizes to the cytoplasm, rather than to NPC.[15] However microscopic analysis of cells from a Triple-A syndrome patient showed no morphologic abnormalities in NPC, suggesting that mutation in AAAS results in a functional rather than a structural abnormality in NPC.[15]

Hirano et al[16] reported impaired nuclear import of DNA repair proteins, including DNA ligase I and the cerebellar ataxia causative protein 'aprataxin' in a patient with Allgrove syndrome. They also proposed that oxidative stress aggravates nuclear import failure which is already compromised in patient cells and consequent DNA damage may participate in triggering cell death. Takao et al[17] showed that nuclear import dysfunction could be overcome by fusing classical nuclear localization signal (NLS) and 137-aa downstream sequence of XRCC1, designated stretched NLS (stNLS). They found that minimum essential sequence of stNLS (mstNLS) is residues 239-276, downsized by more than 100 amino acids and mstNLS enables efficient nuclear import of DNA repair proteins in patient fibroblasts, functioned under oxidative stress and reduced oxidative-stress-induced cell death, more effectively than stNLS. They concluded that this finding may have some therapeutic implications in the management of Allgrove syndrome.

Various mutations have been reported in AAAS gene including both coding regions and intervening sequences (Table 1). Most of the reported mutations produce a truncated protein, although missense and point-mutations have also been reported.[18] Most commonly reported mutation is IVS 14 + 1 G→A which leads to premature termination of predicted protein.[19] Chromosome and/or chromatid breaks, whole chromosome arm deletions, marker chromosomes involving 9q12, a heterochromatic region known to be a fragile site, have also been demonstrated in subjects with Triple-A syndrome.[20]

Table 1. List of reported mutations in patients with AAAS

Position	Protein	cDNA Mutation	References
Ex 1	Q15K	43C>A	5,11,12,21,22
Ex 1	Q41R	122A>G	23
Ex 1	G15K	1186insC	24
Ex 2	170fsX92	210delc	22
Ex 2	H71fsX92	211delC	20
Ex 2	W84X	251G>A	11,25,26
Ex 4	R119X	355C>T	22,27
IVS 4		IVS4-2A>G	13
Ex 5	Q145X	433C>T	20
IVS 5		IVS 5+3ins T	20
IVS 5		IVS 5+1G>A	28
Ex 6	R155P	464C>G	29
Ex 6	H160R	479A>G	11
Ex 6	F157fsX171	470-471delTT	11
Ex 6	A167V	500C>T	30
Ex 7	R230X	678C>T	20,22
Ex 7	R194X	580C>T	27,31,32
Ex 8	Q237X	709C>T	33
Ex 8	S263P	787T>C	11,21,22,34
Ex 8	R258GfsX33	771delG	32,35
IVS 8		IVS8+1G>A	19
Ex 9	R286X	856C>T	11,36
Ex 9	W295X	884C>A	37
Ex 9	R312X	934C>T	13
Ex 10	S328fsX363	981-982insT	13
Ex 10	V313A	1238T>C	22
Ex 11	R342X	1024C>T	11,38
Ex 11	L356fsX362	1066-1067delCT	22
IVS 11		IVS 11+1G>A	5
Ex 12	D368fsX382	1104-1105insC	22
Ex 12	S382fsX413	1144-1147delTCTG	22
Ex 12	Q387X	1159C>T	21,32
Ex 13	397fsX27	1191insA	22
Ex 14	L430F	1288C>T	25
IVS 14		IVS 14+1G>A	5,13,20,29,39
Ex 15	Q456X	1366C>T	18
Ex 15	X492	1368-1372delGCTCA	36
Ex 15	S463fsX549	1389delC	11
Ex16	R478X	1432C>T	4,13,39
Ex 16	W474X	1421G>A	22

Mutations in AAAS gene cannot be identified in every clinically diagnosed patient. Brooks et al (three subjects), Sandrini et al (one subject), Houlden et al (two families) and Huebner et al (eight families) reported clinically diagnosed Triple-A syndrome patients with no identifiable mutations in AAAS gene.[5,19,22,40] Some of them also showed to have no linkage to 12q13 region. Mutations in regulatory or deeper intronic sequences and/or of genetic heterogeneity may be responsible for the disease in these mutation-free patients with Triple A syndrome. A deletion of all or part of the AAAS gene could also explain some cases of mutation remaining undetected. Conversely, patients with two mutations in the AAAS gene may not exhibit all the features of this syndrome, raising the possibility of modifier genes and/or environmental factors that influence the phenotype.[13,21]

There does not seem to be an obvious genotype–phenotype relationship. However, patients with the S263P mutation in the third WD repeat exhibit a similar severe phenotype with existence of all the three main symptoms of Triple-A syndrome, mental retardation and progressive peripheral neuropathy. Genotype phenotype analysis of some other families has shown that frame shift, stop codon and functionally significant mutations are likely to lead to severe phenotype, most probably occurring by a loss of functional effect on protein.[11]

Pathology

Atrophic lacrimal glands have been demonstrated in Triple-A syndrome patients.[41] Lacrimal gland biopsy has shown reduced number of serous secreting cells.[42] Consequent reduced or absent lacrimation frequently leads to alacrimia and dehydration-induced keratopathy.

Imaging has also demonstrated atrophic adrenal glands. Postmortem adrenal histology in certain cases has demonstrated atrophy of zona fasciculata and reticularis with a relative preservation of the zona glomerulosa.[43] Hormonal studies have also demonstrated the preservation of the renin-angiotensin-aldosterone axis function, suggesting that the cells of the zona glomerulosa retain appropriate response to physiological stimuli. Hence the fluid and electrolyte disturbances, typically found in primary adrenal failure, are rare in these patients.[9] Abnormalities in adrenal androgen production, characterized by low serum dehydroepiandrosterone sulfate (DHEA-S) concentration and blunted DHEA response to ACTH stimulation, have also been reported.[44]

The etiology of the neuropathy in Allgrove syndrome is obscure. Previous authors have suggested that it may result from a defect of ACTH receptors on neurones/glia with secondary demyelination, but further work did not support this theory.[45,46] Sural nerve biopsies have shown changes consistent with axonal degeneration and loss of both myelinated and unmyelinated fibres with no amyloid deposition in certain patients.[45]

The autopsy case presented by Allgrove showed muscular hypertrophy, loss of ganglion cells and a paucity of small nerves in the distal oesophagus of an affected patient.[1] Histopathological analysis has revealed fibrosis of the intermuscular plane of esophageal cardia in Allgrove syndrome patients. Absent neuronal NO synthase, decreased myenteric ganglia and lymphocytic infiltration of the myenteric plexus were seen in majority of the patients while decreased interstitial cells of Kajal were seen in some patients.[47]

Clinical Manifestations

Triple-A syndrome has wide phenotypic variability even within the same family. There is no specific genetic-phenotypic correlation and individuals with same mutation can have different phenotypes. The age of onset of each of the manifestations varies widely. Alacrimia is usually present since infancy but most often remains unrecognized. Adrenal insufficiency and achalasia cardia usually manifest during first decade of life with few exceptions.

Alacrimia is probably the earliest and most consistent feature. Most often patients are described to have absence of tears from birth. But it is unlikely to be the presenting manifestation and often elicited only on direct questioning.

Achalasia of the cardia occurs in about 75% of all cases, the age of onset ranging from infancy to 16 years. Earliest reported age of achalasia cardia is 3 months in 2 siblings with Allgrove syndrome.[48,49] Achalasia often manifests as recurrent or chronic pulmonary disease as a result of aspiration especially in infants and toddlers. In older children and adults it usually manifest as dysphagia especially for liquid diet. Although it is usually diagnosed during first or second decade of life, it may remain asymptomatic and may be diagnosed when specifically investigated for or may become apparent only during adulthood.

Adrenal insufficiency is also an early manifestation although rarely it may remain undiagnosed till fifth decade of life.[45] It usually manifests as hypoglycemic episodes or shock during childhood. It is often associated with noticeable hyperpigmentation. Majority of patients have isolated glucocorticoid deficiency, but mineralocorticoid production may also be impaired in about 15% of patients.[50] Isolated aldosterone impairment has also been reported in few cases.[51,52] Adrenal insufficiency does not occur immediately in postnatal period but results from a progressive disorder leading to hypofunction of the adrenal gland at a variable time after birth.[27]

Neurological manifestations appear at a later age when compared to other manifestations of Triple-A syndrome and it may involve both central and peripheral nervous systems. Symptoms may be static or progressive. Autonomic manifestations are the most common neurological manifestations, predominantly involving the pupillary reaction. The most frequent among these is pupillotonia, characterized by pupillary reaction to 0.125% of pilocarpine. Absent or reduced and very slow light reflexes and accommodative spasms are the other pupillary abnormalities. Other autonomic manifestations include orthostatic hypotension, abnormalities of heart rate variability, abnormal sweating and abnormal reaction to intra dermal histamine.[19]

Distal motor and sensory polyneuropathy is a common manifestation.[4] Reflexes are usually brisk; however, ankle jerks are usually absent. Mental retardation is also a common feature and it appears to be an intricate part of the disease rather than secondary to recurrent episodes of hypoglycemia. Other manifestations include optic atrophy, bulbospinal amyotrophy, ataxia, dysarthria, hypernasal speech, dementia, Parkinsonian features, dystonia, chorea and sensory impairment.[19]

Other features include microcephaly, short stature, dysmorphic facies with long narrow face, long philtrum, down-turned mouth, thin upper lip and lack of eyelashes, poor wound healing, palmar and plantar hyperkeratosis, scoliosis, osteoporosis, long QT syndrome, hyperlipoproteinemia Type IIb.[53,55]

Diagnosis

Alacrimia can be diagnosed by using Schirmer's test. It can be performed by placing a filter paper strip under the eyelid, preferably the lower one. Both eyes are tested at the same time. Most often, the eyes are closed gently for about 5 minutes. After 5 minutes, paper is removed and examined. Less than 10 mm of moisture on the filter paper in 5 minutes is considered abnormal.

The accurate diagnosis of adrenal insufficiency in patients with Allgrove syndrome is important because they may need either daily or intermittent (during periods of stress) glucocorticoid replacement therapy. Inadequate rise in serum cortisol concentration after 250 μg ACTH stimulation test proves the diagnosis of adrenal insufficiency. However, if the response to 250 μg ACTH is normal, insulin-induced hypoglycemia test should be performed.[29]

Barium meal examination of the esophagus is a commonly used method which is very effective in early determination of achalasia cardia. Gastroesophagoscopy may reveal esophagitis. At present, esophageal monometry has been widely used in the diagnosis of esophageal achalasia because of its simple manipulation and high accuracy. It is very accurate and sensitive in detecting achalasia. It could be useful for assessment of severity of the disease, for selection of proper treatment and for evaluation of treatment.[56]

Differential Diagnosis

Patients who present with adrenal insufficiency need to be differentiated from other disorders of adrenal insufficiency like Adrenoleukodystrophy (ALD) and familial glucocorticoid deficiency. ALD is an X-linked recessively inherited neurological disease that can also cause impaired glucocorticoid production with minimal or no disturbance of mineralocorticoid production. Similar to Triple A syndrome, adrenal failure may precede neurological dysfunction in ALD also. Measurement of plasma concentration of very-long-chain fatty acids would exclude ALD. Triple-A syndrome patients that present with dominant neurological manifestations may be confused with disorders like familial autonomic neuropathies. Allgrove syndrome can be easily differentiated from familial glucocorticoid deficiency.

Allgrove syndrome may rarely mimic Sjogren syndrome with features of dry eyes and mouth and often needs to be differentiated from the latter. False positive anti Ro and anti La antibodies are also reported with Allgrove syndrome and may not be helpful to differentiate the two conditions. Salivary gland biopsy may differentiate these two conditions, especially in the experienced hands.[2]

Treatment

Adrenal insufficiency is a life threatening manifestation which needs life long treatment with glucocorticoids. Hydrocortisone is the preferred drug for glucocorticoid replacement especially during childhood and maintenance dose is 10-15 $mg/m^2/day$. Dose of glucocorticoids should be increased (2-3 times) during periods of stress. Unlike other causes of primary adrenal insufficiency, only 15% of the Triple-A patients suffer from mineralocorticoid deficiency and hence it is advisable to add fludrocortisone to only those with proven mineralocorticoid deficiency.

Reduction of the pressure of the lower esophageal sphincter is the primary objective in the management of achalasia cardia. As a nonsurgical treatment, pneumatic dilatation is a traditional and effective treatment. Compared with esophagomyotomy, this treatment is cheap, effective and is easily manipulated but effect is short lasting. Heller's esophagocardiomyotomy has a high degree of safety, effective results and a long effective period. It is often combined with an antireflux procedure to minimize the incidence of postoperative esophageal reflux.[56]

Artificial tears are used as the primary treatment modality as often as necessary to relieve ocular discomfort. Viscous preparations, such as gels and ointments may be used for severe cases and during night time. Progressive symptoms may require sustained-release ocular inserts. Surgical treatment like punctal occlusion and tarsorrhaphy may be considered for cases not responding to medical therapy.

Neurological manifestations are the most untreatable among the manifestations of triple-A syndrome.

Conclusion

Allgrove syndrome is characterized by triad of adreno corticotrophic hormone (ACTH) - resistant adrenal insufficiency, alacrimia and achalasia cardia. It is characterized by mutation(s) in AAAS gene, located on chromosome 12q13. Most mutations produce a truncated protein, although missense and point-mutations have also been reported. Some patients with Triple-A syndrome may not have mutations in AAAS gene. Hence, absence of mutations in AAAS gene does not negate the diagnosis of Allgrove syndrome. AAAS codes a protein called ALADIN for alacrima, achalasia, adrenal insufficiency and neurological disorder. ALADIN protein belongs to WD repeat protein family, which are essential for normal cellular function. It is a part of nuclear pore complex and mutations in AAAS gene lead to mislocalisation of the protein to cytoplasm. Recently chromosome/chromatid breaks have been described in Allgrove patients. Lately, impaired nuclear import of DNA repair proteins and in vitro techniques to normalize nuclear import failure have been demonstrated. This has provided a ray of hope for possibility of gene therapy in Allgrove syndrome patients.

No specific genotype-phenotype correlation is found among Allgrove syndrome patients. Alacrimia is the earliest and most consistent feature while achalasia cardia and adrenal insufficiency are the usual presenting manifestations. Neurological features appear at later age with autonomic abnormalities, polyneuropathy, amyotrophy, optic atrophy. Alacrimia is diagnosed by Schirmer's test while ahalasia cardia and adrenal insufficiency are best diagnosed by esophageal monometry and ACTH stimulated cortisol levels respectively. Alacrimia is treated with artificial tears while achalasia cardia with either pneumatic dilatation or Heller's myotomy. Adrenal insufficiency is treated with glucocorticoid and if necessary mineralocorticoid replacement. However, currently there is no effective therapy for neurological manifestations.

References

1. Allgrove J, Clayden GS, Grant DB et al. Familial glucocorticoid deficiency with achalasia of the cardia and deficient tear production. Lancet 1978; 1:1284-1286.
2. Onat AM, Pehlivan Y, Buyukhatipoglu H et al. Unusual presentation of triple A syndrome mimicking Sjögren's syndrome. Clin Rheumatol 2007; 26:1749-1751.
3. Gazarian M, Cowell CT, Bonney M et al. The "4 A" syndrome: adrenocortical insufficiency associated with achalasia, alacrimia, autonomic and other neurological abnormalities. Eur J Pediatr 1995; 154:18-23.
4. Goizet C, Catargi B, Tison F et al. Progressive bulbospinal amyotrophy in triple A syndrome with AAAS gene mutation. Neurology 2002; 26(58):962-965.
5. Sandrini F, Farmakidis C, Kirschner LS et al. Spectrum of mutations of the AAAS gene in Allgrove syndrome: lack of mutations in six kindreds with isolated resistance to corticotropin. J Clin Endocrinol Metab 2001; 86:5433-5437.
6. Genin E, Tullio-Pelet A, Begeot F et al. Estimating the age of rare disease mutations: the example of Triple-A syndrome. J Med Genet 2004; 41:445-449.
7. De Wied D. Neurotropic effects of ACTH/MSH neuropeptides. Acta Neurobiol Exp 1990; 50:353-366.
8. Jahn R, Padel U, Porsch P-H et al. Adrenocorticotropin hormone and a-melanocyte-stimulating hormone induce secretion and protein phosphorylation in the rat lacrimal gland by activating cAMP dependent pathway. Euro J Biochem 1982; 126:623-629.
9. Onyechi Modebe. Achalasia-alacrima-acth insensitivity syndrome (Triple a syndrome) in a saudi family. Annals of Saudi Medicine 2000; 20:3-4.
10. Weber A, Wienker TF, Jung M et al. Linkage of the gene for the triple A syndrome to chromosome 12q13 near the type II keratin gene cluster. Hum Mol Genet 1996; 5:2061-2066.
11. Handschug K, Sperling S, Yoon S-JK et al. Triple A syndrome is caused by mutations in AAAS, a new WD-repeat protein gene. Hum Mol Genet 2001; 10:283-290.
12. Ashis Mukhyopadhya, Sumita Danda, Angela Huebner et al. Mutations of the AAAS gene in an Indian family with Allgrove's syndrome. World J Gastroenterol 2006; 12:4764-4766.
13. Tullio-Pelet A, Salomon R, Hadj-Rabia S et al. Mutant WD-repeat protein in triple-A syndrome. Nat Genet 2000; 26:332-335.
14. Cronshaw JM, Krutchinsky AN, Zhang W et al. Proteomic analysis of the mammalian nuclear pore complex. J Cell Biol 2002; 158:915-927.
15. Cronshaw JM, Matunis MJ. The nuclear pore complex protein ALADIN is mislocalized in triple A syndrome. Proc Natl Acad Sci USA 2003; 100:5823-5827.
16. Hirano M, Furiya Y, Asai H et al. ALADINI482S causes selective failure of nuclear protein import and hypersensitivity to oxidative stress in triple A syndrome. Proc Natl Acad Sci USA 2006; 103:2298-2303.
17. Kiriyama T, Hirano M, Asai H et al. Restoration of nuclear-import failure caused by triple A syndrome and oxidative stress. Biochem Biophys Res Commun 2008; 374:631-634.
18. Soltani A, Arab Ameri M, Hasani Ranjbar SH. Allgrove Syndrome: A Case Report: Int J Endocrinol Metab 2007; 4:214-217.
19. Brooks BP, Kleta R, Stuart C et al. Genotypic heterogeneity and clinical phenotype in triple A syndrome: a review of the NIH experience 2000-2005. Clin Genet 2005; 68:215-221.
20. Reshmi-Skarja S, Huebner A, Handschug K et al. Chromosomal fragility in patients with triple A syndrome. Am J Med Genet 2003; 117A:30-36.
21. Prpic I, Huebner A, Persic M et al. Triple A syndrome: genotype—phenotype assessment. Clin Genet 2003; 63:415-417.
22. Houlden H, Smith S, de Carvalho M et al. Clinical and genetic characterization of families with triple A (Allgrove) syndrome. Brain 2002; 125:2681-2690.
23. Villanueva-Mendoza C, artínez-Guzmán O, Rivera-Parra D et al. Triple A or Allgrove syndrome. A case report with ophthalmic abnormalities and a novel mutation in the AAAS gene. Ophthalmic Genet 2009; 30:45-49.
24. Houlden H. The small, spastic and furrowed tongue of Allgrove syndrome. Neurology 2009; 72:1366.
25. Koehler K, Brockmann K, Krumbholz M et al. Axonal neuropathy with unusual pattern of amyotrophy and alacrima associated with a novel AAAS mutation p.Leu430Phe. Eur J Hum Genet 2008; 16:1499-1506. Epub 2008.
26. Toromanovic A, Tahirovic H, Milenkovic T et al. Clinical and molecular genetic findings in a 6-year-old Bosnian boy with triple A syndrome. Eur J Pediatr 2009; 168:317-320. Epub 2008.
27. Kinjo S, Takemoto M, Miyako K et al. Two cases of Allgrove syndrome with mutations in the AAAS gene. Endocr J 2004; 51:473-477.

28. Ismail EA, Tulliot-Pelet A, Mohsen AM et al. Allgrove syndrome with features of familial dysautonomia: a novel mutation in the AAAS gene. Acta Paediatr 2006; 95:1140-1143.
29. Salehi M, Houlden H, Sheikh A et al. The diagnosis of adrenal insufficiency in a patient with Allgrove syndrome and a novel mutation in the ALADIN gene. Metabolism 2005; 54:200-205.
30. Salmaggi A, Zirilli L, Pantaleoni C et al. Late-onset triple A syndrome: a risk of overlooked or delayed diagnosis and management. Horm Res 2008; 70:364-372. Epub 2008.
31. Dusek T, Korsic M, Koehler K et al. A novel AAAS gene mutation (p.R194X) in a patient with triple A syndrome. Horm Res 2006; 65:171-176. Epub 2006.
32. Lam YY, Lo IF, Shek CC et al. Triple-A syndrome—the first Chinese patient with novel mutations in the AAAS gene. J Pediatr Endocrinol Metab 2006; 19:765-770.
33. Katsumata M, Hirose H, Kagami M et al. Analysis of the AAAS gene in a Japanese patient with triple A syndrome. Endocr J 2002; 49:49-53.
34. Milenković T, Koehler K, Krumbholz M et al. Three siblings with triple A syndrome with a novel frameshift mutation in the AAAS gene and a review of 17 independent patients with the frequent p.Ser263Pro mutation. Eur J Pediatr 2008; 167:1049-1055. Epub 2008.
35. Gong CX, Wen YR, Zhao XL et al. Allgrove syndrome in the mainland of China: clinical report and mutation analysis. Zhonghua Er Ke Za Zhi 2007; 45:422-425.
36. Brooks BP, Kleta R, Caruso RC et al. Triple-A syndrome with prominent ophthalmic features and a novel mutation in the AAAS gene: a case report. BMC Ophthalmol 2004; 4:7.
37. Schmittmann-Ohters K, Huebner A, Richter-Unruh A et al. Clinical and novel molecular findings in a 6.8-year old Turkish boy with triple A syndrome. Horm Res 2001; 56:67-72.
38. Buderus S, Utsch B, Huebner A et al. Dysphagia due to triple A syndrome: successful treatment of achalasia by balloon dilatation. Exp Clin Endocrinol Diabetes 2007; 115:533-536.
39. Roubergue A, Apartis E, Vidailhet M et al. Myoclonus and generalized digestive dysmotility in triple A syndrome with AAAS gene mutation. Mov Disord 2004; 19:344-346.
40. Huebner A, Kaindl AM, Knobeloch KP et al. The triple A syndrome is due to mutations in ALADIN, a novel member of the nuclear pore complex. Endocr Res 2004; 30:891-899.
41. Mullaney PB, Weatherhead R, Millar L et al. Keratoconjunctivitis sicca associated with achalasia of the cardia, adrenocortical insufficiency and lacrimal gland degeneration: Keratoconjunctivitis sicca secondary to lacrimal gland degeneration may parallel degenerative changes in esophageal and adrenocortical function. Ophthalmology 1998; 105:643-650.
42. Babu K, Murthy KR, Babu N et al. Triple A syndrome with ophthalmic manifestations in two siblings. Indian J Ophthalmol 2007; 55:304-306.
43. Clark AJL, Weber A. Adrenocorticotropin insensitivity syndrome. Endocr Rev 1998; 19:828-843.
44. Elise T, Loeper S, Jung R et al. A patient with the Triple A syndrome presenting with isolated partial adrenal androgen deficiency. 85th annual meeting, Endocrine Society, Philadelphia, PA. 2003;622.
45. Kimber J, McLean BN, Prevett M et al. Allgrove or 4 "A" syndrome: an autosomal recessive syndrome causing multisystem neurological disease. J Neurol Neurosurg Psychiatry 2003; 74:654-657.
46. Stuckey BG, Mastaglia FL, Reed WD et al. Glucocorticoid insufficiency, achalasia, alacrima with autonomic motor neuropathy. Ann Intern Med 1987; 10:61-63.
47. Khelif K. Achalasia of the cardia in Allgrove's (triple A) syndrome: histopathologic study of 10 cases. Am J Surg Pathol 2003; 27:667-672.
48. Singh A, Shah A. Esophageal achalasia and alacrima in siblings. Indian Pediatr 2006; 43:161-163.
49. Thomas RJ, Sen S, Zachariah N et al. Achalasia cardia in infancy and childhood: an Indian experience. J R Coll Surg Edinb 1998; 43:103-104.
50. Grant DB, Barnes ND, Dumic M et al. Neurological and adrenal dysfunction in the adrenal insufficiency/alacrima/achalasia (3A) syndrome. Arch Dis Child 1993; 68:779-782.
51. Grant D, Dunger D, Smith I et al. Familial glucocorticoid deficiency with achalsia of the cardia associated with a mixed neuropathy, long tract degeneration and mild dementia. Eur J Pediatr 1992; 151:85-89.
52. Pombo M, Deuesa J, Taborda A et al. Glucocorticoid deficiency with achalasia of the cardia and lack of lacrimation. Clin Endocrinol 1985; 23:237-243.
53. Ozgen AG, Ercan E, Ozutemiz O et al. The 4A syndrome association with osteoporosis. Endocr J 1999; 46:227-230.
54. Dumic M, Mravak-Stipetic M, Kaic Z et al. Xerostomia in patients with triple A syndrome—a newly recognized finding. Eur J Pediatr 2000; 159:885-888.
55. Khong PL, Peh WC, Low LC et al. Variant of the Triple A syndrome. Australas Radiol 1994; 38:222-224.
56. Zhang Y, Chun-Di Xu, Zaouche A et al. Diagnosis and management of esophageal achalasia in children: analysis of 13 cases. World J Pediatr 2009; 5:56-59.

Chapter 2

Amyotrophic Lateral Sclerosis

J. Jefferson P. Perry,* David S. Shin and John A. Tainer

Abstract

Amyotrophic lateral sclerosis (ALS) is a common neurological disorder that results in loss of motor neurons, leading to a rapidly progressive form of muscle paralysis that is fatal. There is no available cure and current therapies only provide minimal benefit at best. The disease is predominantly sporadic and until very recently only the Cu,Zn superoxide dismutase (Cu,ZnSOD), which is involved in a small number of sporadic cases and a larger component of familial ones, have been analyzed in any detail. Here we describe the clinical aspects of ALS and highlight the genetics and molecular mechanisms behind the disease. We discuss the current understanding and controversies of how mutations in Cu,ZnSOD may cause the disease. We also focus on the recent discovery that mutations in either TDP-43 or FUS/TLS, which are both involved in DNA/RNA synthesis, are likely the cause behind many cases of ALS that are not linked to Cu,ZnSOD.

Introduction

Amyotrophic lateral sclerosis (ALS) is a rapidly progressive and fatal neurodegenerative disorder, for which there is no known available cure. ALS is also known as Lou Gehrig's disease, Motor Neuron disease or Charcot's disease and it is one of the most common human neurological disorders, affecting 1 in 200,000 people.[1] The disease is defined by a loss of control on voluntary muscle movement and increased muscular paralysis, occurring through the degeneration of motor neurons in the spinal cord, brainstem and motor cortex. As the disease is selective for motor neurons, the intellect normally remains intact during the disease pathology. The incidence of ALS is reasonably uniform in Western nations, though clusters of increased incidence occur in the Western Pacific. Most of the ALS cases are sporadic (SALS), with only five to ten percent of ALS cases being familial (FALS). The clinical symptoms of ALS, which are described in detail below, are muscle weakness initially arising in the limbs or involving the face and tongue. However, due the progressive nature of this devastating disease, general paralysis develops and this is followed by respiratory failure, typically within a couple of years. Despite intense efforts, much remains to be defined regarding the molecular mechanisms of the disease. ALS-linked mutations have been determined in multiple genes/proteins. Here we describe the molecular basis of the disease including the most characterized Cu,ZnSOD and two new important additions, TDP-43 and FUS/TLS.

Clinical Features

The founder of modern neurology, Jean Martin Charcot, first described the clinical and pathological features of ALS in 1869.[2] ALS was named as such because Amyotrophic refers to the muscle fiber atrophy observed in the disease and Lateral Sclerosis relates to the hardening of

*Corresponding Author: J. Jefferson P. Perry—Department of Molecular Biology, The Scripps Research Institute, La Jolla, California 92037, USA, and The School of Biotechnology, Amrita University, Kollam, Kerala 690525, India. Email: jjperry@scripps.edu

Diseases of DNA Repair, edited by Shamim I. Ahmad. ©2010 Landes Bioscience and Springer Science+Business Media.

the anterior and lateral corticospinal tracts that occurs during motor neuron degeneration. ALS affects 1 in 200,000 people[1] and the lifetime risk of the SALS is estimated to be at 1 in 400 to 1 in 1000, by age 70.[3,4] The incidence of ALS in the parts of the Western Pacific has been observed to occur with a rate of 50-100 times greater than average.[5] The cause of this increase is not known, but the accumulation of methylaminoalanine in the diet of the people of this region has been under study.[6] SALS amounts to 90-95% of cases and usually occurs in mid-life, with a mean age of onset at 55-65 years;[7] however approximately 5% of the SALS cases have a juvenile onset, below the age 30.[7] FALS accounts for 5-10% of the total ALS cases and in this inherited form of the disease the onset of symptoms typically appears a decade earlier than in SALS and the survival period is also shorter.[8-10] Additionally, the inheritance in FALS is usually Mendelian, autosomal dominant and has a Gausian distribution for the age of onset that differs from the age-dependent distribution for SALS.

Initial studies had found a small prevalence for ALS occurrence in males, although subsequent analyses suggest a more equal ratio between the genders.[11-13] Due to the selective killing of motor neurons, early symptoms rapidly lead to progressive paralysis, which is followed by death within 3 years for 50% of patients,[14,15] although approximately 4% of individuals can survive for 10 years or longer.[16,17] Some of the genetic risk factors for developing ALS have been defined, as discussed below, while any specific environmental risks have not be confirmed. Potential increased risks have been suggested, including risks to military personnel, professional Italian football players and smokers.[18-22] However, a complex genetic/environmental interaction is likely to be at play in causing disease occurrence.

Two-thirds of patients have the limb onset, spinal form of the disease. The first symptoms of this form are usually focal muscle weakness in either the upper limbs (cervical onset) or lower limbs (lumbar onset) that is either proximal or distal. Early examination typically finds focal atrophy in the muscles of the hands, forearms, shoulders or proximal thigh or distal foot. Onset is usually asymmetrical, though the other limbs will go on to develop weakness and wasting and this increases into spasticity affecting both the gait and dexterity of the individual. Most spinal ALS individuals will progress to bulbar symptoms and respiratory failure within 3-5 years. Sensory and cognitive symptoms, dementia and Parkinson disease may rarely occur with the spinal onset ALS.

Almost a third of cases have a bulbar onset and this form occurs more commonly in women and older individuals.[7,23,24] Bulbar onset can be either upper motor neuron based (pseudobublar palsy) or lower motor neuron based (bulbar palsy). In bulbar palsy, initial onset leads to slurred speech (dysarthria) and a wasting of the tongue and difficulties in swallowing (dysphagia). The lower part of the face becomes weakened and individuals develop difficulties in closing the lips and can have excessive drooling. Pseudobulbar symptoms cause uncontrollable laughing or crying, known as emotional liability and excessive yawning, brisk jaw jerk and dysarthria. In both of these forms, limb-weakening symptoms will occur mostly by 1-2 years after the initial onset and progressive paralysis results in respiratory failure within 2-3 years, slightly quicker than the spinal onset form.

Artificial ventilation of ALS patients can extend their survival and enhance their quality of life. However, patients maintained with ventilation may degrade to a 'totally locked in state', which can even include oculomotor impairment.[25,26] Unfortunately, the only approved drug so far treating ALS is riluzole, which can only modestly extend life for about 2-3 months.[27] The mechanism of action for riluzole includes inhibition of glutamate release from presynaptic terminals and also increase extracelluar glutamate uptake,[28] but its full actions are only partially known. Therefore, care for individuals with ALS is primarily supportive and includes nutritional management to ensure adequate intake of foods and water, due to difficulties in swallowing, in addition to ventilation management.

Genetic Basis and the Molecular Mechanism of the Disease

Basic research defining the mechanisms behind this disease is of much significance and will likely provide new avenues for the successful generation of therapeutics combating ALS.

However, it is not yet clear which cellular pathway(s) are disrupted in ALS. Moreover, like many other neurodegenerative diseases, ALS pathology may arise through a complex interplay between varying cellular mechanisms. However, common to all forms of ALS are intra-neuronal inclusions, which can be seen in the degenerating neurons. Bunina bodies, which are small eosinophilic inclusions, that are positively stained by cystatin, are present in 70-100% of cases.[29] Also, Lewis-body like inclusions, bearing similarity to the Parkinson inclusions are also observed in 95% of cases; they stain positive for ubiquitin and have TDP-43 protein as the major constituent. In FALS, hyaline conglomerates are observed, which are rarely seen in SALS and these form neurofilament inclusions.[30]

The different factors that may cause ALS are being actively researched. One contribution to ALS may be excitotoxcity, where injury to neurons is the result of excessive stimulation by amino acid modulators, such as glutamate, kainite, or AMPA (α-amino-hydroxy-5-methylisoxasole-4 propionic acid). This over stimulation can lead to excessive calcium influx into the motor neuron, resulting in cell death. Supporting this theory is, that certain ALS patients have been defined as having elevated levels of glutamate in the cerebrospinal fluid[31,32] and, it was the excitotoxic theory that led to the identification of the glutamate release inhibitor, riluzole. This drug subsequently was licensed to become the first therapeutic agent to be used to reduce (albeit moderately) ALS progression.[27]

Mitochondrial dysfunction is linked to a variety of neuropathologies[33] and in ALS individuals, a defect in mitochondrial energy metabolism has been implicated[34] along with mutation in mitochondrial DNA.[35-37] Alterations in mitochondrial biochemistry and morphology are also observed in both SALS patients and in SOD1 transgenic mice.[38,39] Additionally, Cu,ZnSOD occurs in the mitochondrial membrane space and proteomic data have suggested that Cu,ZnSOD mutants, potentially causing mitochondrial defects, lead to caspase mediated cell death.[40] Deficits in neurotrophic factors may be another contributing factor in ALS, because motor neuron loss has biochemical markers and key elements in common with apoptosis.[41-45] This has led to nerve growth factors (neurotrophins) being analyzed to see if they limit disease progression substances, but preliminary studies have failed to show any benefits from these agents.[46-49] Other potential causes of ALS include neurofilament disorganization, leading to axonal strangulation.[50] Also, axonal transport defects may be deleterious, as the postmitotic motor neurons are greater than a meter in humans and their neurofilament-based transport system is vulnerable to cellular stress due to extended protein lifetimes. Cellular stresses include oxidative damage, which is known to contribute to many neurodegenerative diseases and inflammatory processes may also contribute to ALS.[32,51-54]

Perhaps the largest strides in delineating factors that contribute to ALS have been taken in determining the genetic factors involved, but even progression of this work is slow. It is now known that ALS mutations in the Cu,ZnSOD gene (SOD1) cause about 20% of the cases of FALS, which are inherited in an autosomal dominant fashion[55,56] and also 2% of patients with SALS. Mutations in other genes, that result in FALS, include Alsin (ALS2) and Senataxin (ALS4), both of which are linked to a juvenile onset form of FALS that occurs before 25 years of age.[57-59] Several other genes are of significant interest, including those with DNA metabolism/repair functions. A particular active field of research is on the TDP-43 and FUS/TLS DNA/RNA binding proteins, which have very recently been determined to have ALS-linked mutations.[60-69]

Cu,ZnSOD

Cu,ZnSOD protects cells from oxidative damage by converting superoxide anions (a reactive oxygen species or ROS) into hydrogen peroxide and this peroxide product is then safely removed by catalase and peroxidase. Since cellular damage, induced by ROS, have been linked to multiple diseases, hence a mutant SOD most likely will increase damage on the motor neurons too leading to neurodegenerative diseases. Remarkably, over 100 isolated cases of single-residue mutations, occurring in more than 70 SOD amino acid positions of this 153 residue protein, have been found among ALS patients.[70] These FALS mutation sites occur with varying frequency in all 5 of the exons of the SOD1 gene that encodes Cu,ZnSOD. Several theories have been put forward for how these mutations affect Cu,ZnSOD action, including increased stability or a gain of function.[55,71] However, the structural destabilization concept that was proposed, when the human Cu,ZnSOD structure was determined, has stood the test of time and it can explain how one disease is caused by a considerable number of distinct Cu,ZnSOD mutations.[56] Moreover, a clear gain of function for the FALS Cu,ZnSOD mutants has not been clearly defined. The structural destabilization theory implies that protein aggregation may be a significant contributing factor to ALS. This is supported by intra-neuronal inclusions that are observed in the degenerating neurons, especially the neurofilament hyaline conglomerates that are observed in FALS.[30] Moreover, in many other neurodegenerative diseases, such as Alzheimer, Huntington and Parkinson disease, protein aggregates are observed in the brain and are linked to their pathology.[72-74]

The crystal structure of human Cu,ZnSOD has been determined and it forms a structure that is highly conserved between various eukaryotic Cu,ZnSODs[75,76] (Fig. 1). Human Cu,ZnSOD is composed of two identical subunits and each subunit consists of a main β-barrel,[77] composed of eight antiparallel β-strands arranged in a 'Greek-key motif' (Fig. 1). This Greek-key motif and the hydrophobic dimer interface provide structural stability to Cu,ZnSOD, a requirement for its in vivo function. Also needed is an optimally fast catalysis, which occurs through electrostatic guidance of the superoxide through a channel from the protein surface to the active site and this channel is formed by two external loops. The first, 'zinc loop', tethers the dimer interface with the active site zinc and contains a disulfide bond that aids structural stability.[75,76,78-80] The second, 'electrostatic loop', primarily functions to guide the superoxide substrate to the active site.[80,81] In each of the subunits, the active site contains two metal sites that are closely spaced, a Cu-binding site and a Zn-binding site and these are required to mediate the disproportionate reaction. The Cu ion is the more solvent exposed metal and this is where the substrate/product/reaction intermediate is likely to bind during the enzymatic reaction.[82]

Mapping ALS mutations onto the protein structure reveals that the mutations are dispersed throughout its structure (Fig. 1). This further supports the theory that mutations give rise to structural destabilization, because a gain of function would likely result in mutations clustering in and around certain sites on the protein. Structural destabilization likely leads to protein aggregation and interestingly, protein aggregates have been observed in both mouse and cell culture models of ALS and they are immunoreactive to Cu,ZnSOD antibodies.[83,84] Also, the destabilization mechanism for FALS is consistent with suggestion that mutant SOD-mediated toxicity may be through the coprecipitation of destabilized SOD with other key cellular components.[1,83,85] Additionally, ALS mutations have been determined to promote fibril formation,[86] which is likely toxic to the motor neuron.

TDP-43

More recently, a break through in ALS research has occurred by the discovery of the TAR DNA-binding protein, TDP-43. It is a major component of the ubiquitinated protein aggregates found in many cases of SALS and also in FALS.[87] These aggregates appear unique to previously defined aggregates, in that they do not form amyloid deposits and do not stain for tau, α-synuclein, β-amyloid and expanded polyglutamines. Further studies indicate that the TDP-aggregates are common to most ALS individuals, with the exception of patients with SOD1 mutations, who have hyaline conglomerates instead.[87,88] Notably, TDP-43 is also directly linked to frontotemporal lobar degeneration (FTLD)[89,90] and in cystic fibrosis.[91] In ALS and FTLD, TDP-43 protein observed in the brains and spinal cords is abnormally hyperphosphorylated and ubiquitinated and the protein is also missing its C-terminal region.[87,88] TDP-43 is partly cleared from the nuclei of cells containing aggregates, which has led to the hypothesis that the phenotype is in part due to loss of normal TDP-43 functions.[66,88] However, it is not yet clear if this is the case or if toxicity is caused through aggregation and/or a gain of function. TDP-43 was initially identified as being involved in transcriptional processes, through its transcriptional repression of *HIV-1* genes,[92] mouse *SP-10* gene[93] and the expression of human cyclin-dependent kinase 6.[94] TDP-43 is now known to also

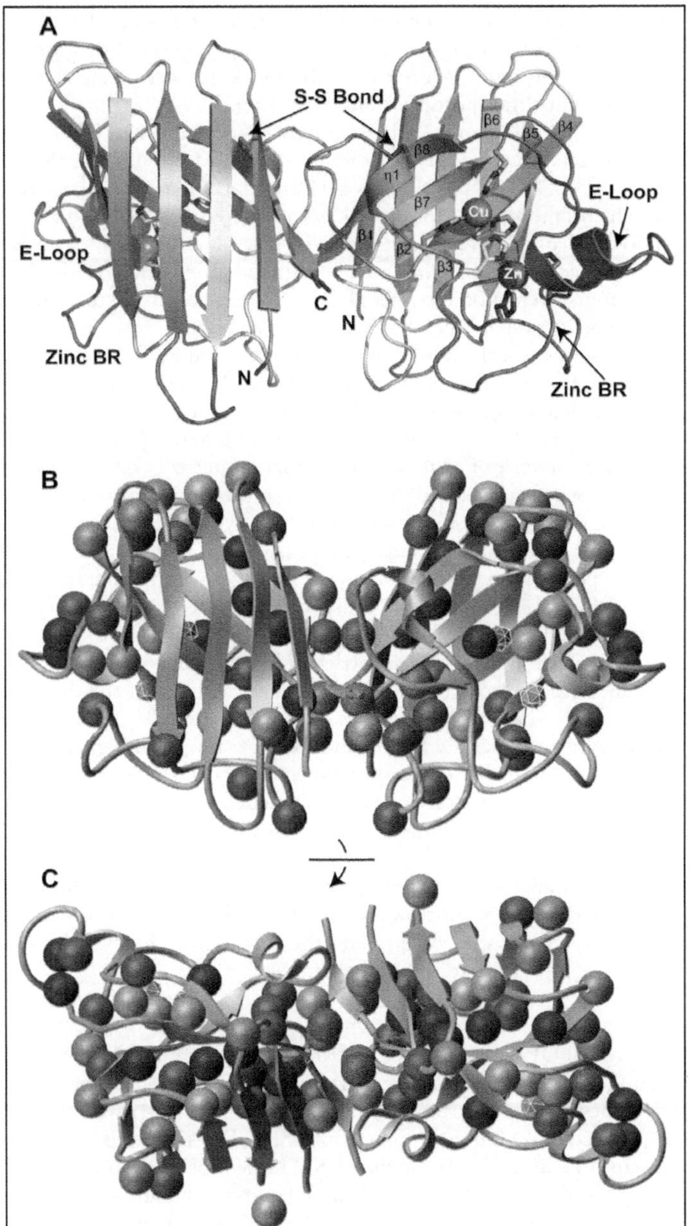

Figure 1. Cu,ZnSOD structure and ALS-inducing mutations. A) Crystal structure of Cu,ZnSOD with the bound metals ions depicted as spheres and metal chelating side chains as sticks. Also highlighted are key structural elements of each subunit, including the electrostatic loop (E-loop), the Zinc binding loop (BR), the stabilizing disulphide bond (S-S). B) Spheres depict mutations of Cu,ZnSOD that cause FALS, which are observed throughout the structure and are expected to perturb the integrity of the β-barrel. Increased darkness in the grey scale sphere coloration indicates increased severity and an earlier age of onset linked to the specific mutation. The active site metal ions are depicted as spheres within cages. C) A 90 degrees rotation of (B).

function as a splicing factor. In cystic fibrosis the binding of TDP-43 to the UG-repeats, located at the 3′-splice site of CFTR intron 8, leads to exon 9 skipping, resulting in the expression of a shorter, inactive CFTR protein.[95] Moreover, TDP-43 has further defined roles in neurons in regulating mRNA stability, transportation and translation.[96,97]

More recently, a direct link to ALS has strongly been established where dominant mutations have been described in the TDP-43 protein, coded by TDP-43 gene (*TARDBP*) that are the primary cause of ALS.[60-63,65,66] TDP-43 is a 414-residue polypeptide, encoded by a six exon containing gene on chromosome 1. The protein is known to have DNA and RNA binding functions and has two defined RNA-recognition motifs (RRM1 and RRM2), a nuclear localization signal, a predicted nuclear export signal and a C-terminal glycine rich region that is thought to act as a protein-protein interaction interface (Fig. 2). The crystal structure of the RRM2 domain in complex with single-stranded DNA has been reported (Fig. 2).[98] These structural studies determined that TDP-43 uses both RRM domains for DNA and RNA binding and that TDP-43 forms a homo-dimer. The crystal structure of RRM2 revealed that it has an atypical RRM-fold, containing a five-stranded β-sheet, instead of the usual 4-stranded sheet, which is packed with two α-helices. A single strand of DNA has been observed bound to this β-sheet and this provided an understanding of the preference of TDP-43/RRM2 for binding to TG/UG nucleic acids. Towards the beginning of the oligonucleotide, a thymine base at position 3 forms three hydrogen bonds with RRM2 and the G base at position four forms four hydrogen bonds with the protein (Fig. 2). Substituting these bases with others may reduce hydrogen bond formation and hence lower affinity: a U base at position 3 would maintain the three hydrogen bonds, but replacement by a C would reduce the hydrogen bonds to two. Also, replacing the G-4 base with an A would result in only a single hydrogen bond being retained.

Notably, the dimerization of RRM2 is mediated by the additional strand, β4, of the atypical RRM2 domain, with β4 from one subunit interacts with β4 of the dimerization partner. Interestingly, this creates a homodimer that has a high thermostablity, at least 35°C more stable than RRM1 domain or full-length TDP-43 and that can also form higher order thermostable complexes. Interestingly, about 30 distinct mutations are now known in 22 unrelated families that cause FALS and a further 29 mutations have been defined in SALS. Nearly all of these map to the C-terminus and they are missense mutations, with the exception of a single truncation mutant (Fig. 2).[61] Every individual with these TDP-43 mutations developed ALS, but there are some variability within the families for both, the age and the site of onset. Thus, this recently identified thermostability of the RRM2 domain now provides a testable model, where loss of the C-terminal glycine rich region through mutation can promote the formation of thermostable RRM2-mediated aggregates of TDP-43, giving rise to the ALS phenotype.

FUS/TLS

Following on from TDP-43, another DNA/RNA binding protein, FUS or TLS (fused in sarcoma/translocation in liposarcoma) has been very recently found to have functions in ALS.[69] FUS/TLS was identified because a linkage between ALS and chromosome 16 was known. Sequencing of genes, encoding DNA/RNA binding proteins (similar to TDP-43) in this chromosome, identified a dominant missense mutation, R521C, in the FUS/TLS gene of a large British family with cases of FALS.[69] Surveying a further 197 FALS cases identified the same mutation in a further four families, in addition to two missense mutations in four families.[69] A second, independent study, used linkage analysis on an autosomal recessive inheritance of ALS in a family from Cape Verde Islands.[68] This study identified a H517Q mutation in the FUS/TLS gene. Subsequent screening elucidated a further 12 dominant mutations in 16 families, when screening 292 FALS cases.[68] However, no FUS/TLS mutations were identified in a survey of 293 SALS patients.[68] Thus, together these two studies revealed FUS/TLS mutations in approximately 4% of FALS cases, which have a classical form of the disease without cognitive defects.

The FUS/TLS protein has 526 amino acid residues with several distinct domains (Fig. 3). In the N-terminus contains a glutamine, glycine, serine and tyrosine rich (QGSY) region, this is

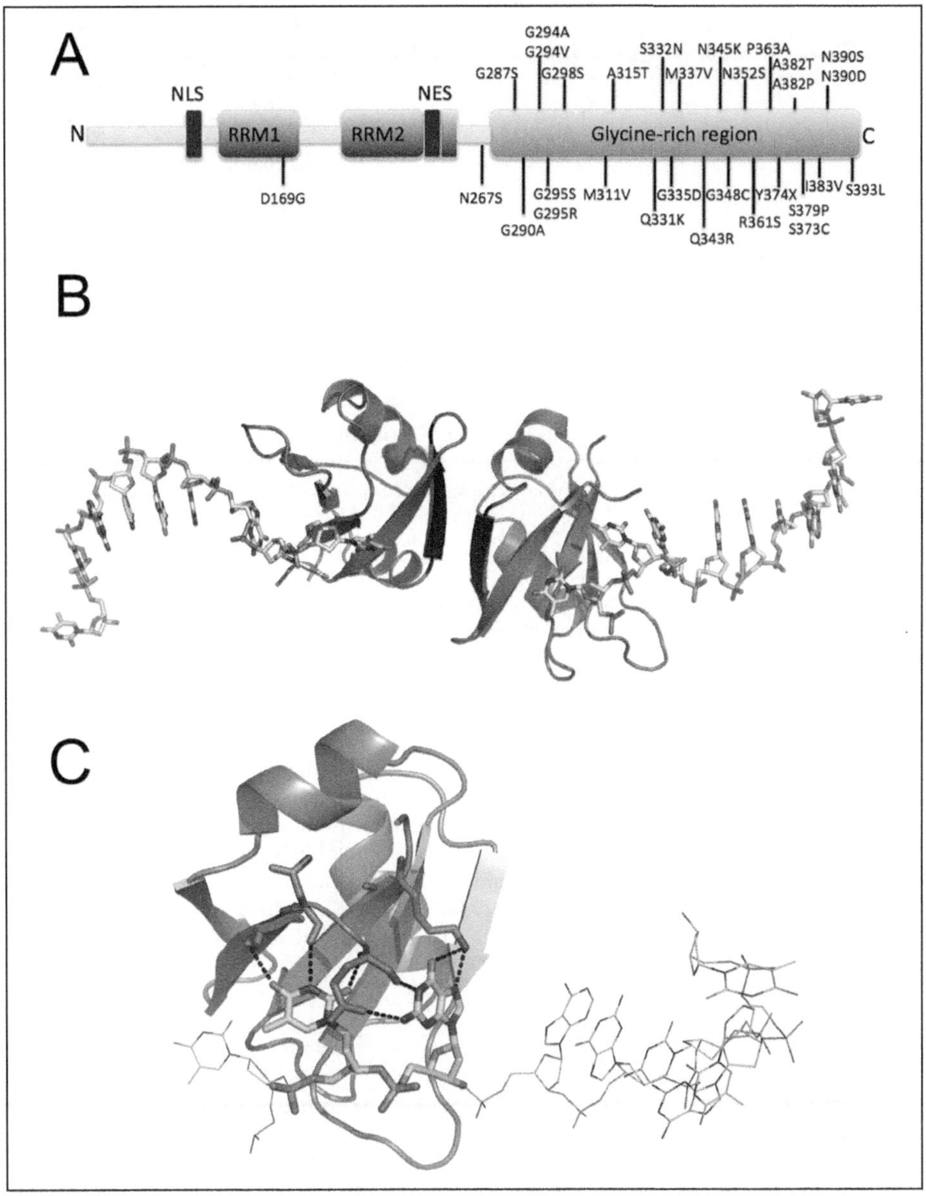

Figure 2. Schematic of the TDP-43 and Structure of TDP-43 RRM2 domain in complex with single-stranded DNA. A) TDP-43 contains a nuclear localization signal (NLS) a predicted nuclear export signal (NES), two DNA/RNA recognition motifs, RRM1 and RMM2 and a C-terminal glycine-rich region. Most of the mutations resulting in TDP-43 that result in ALS occur in the C-terminal glycine-rich region and are missense mutations, with the exception of a truncation at Y374. B) TDP-43 RRM2 forms an abnormal RRM domain fold, where an extra strand β4 (black) is utilized for homodimerization. C) TDP-43 has a preference for U/T:G binding and the complex with ssDNA reveals multiple, stabilizing hydrogen bonds of T3 and G4 with the β-sheet of TDP-43 (bases and side chains depicted in sticks, with black dashed lines showing the hydrogen bonds), many of which would be lost if other bases were at this position.

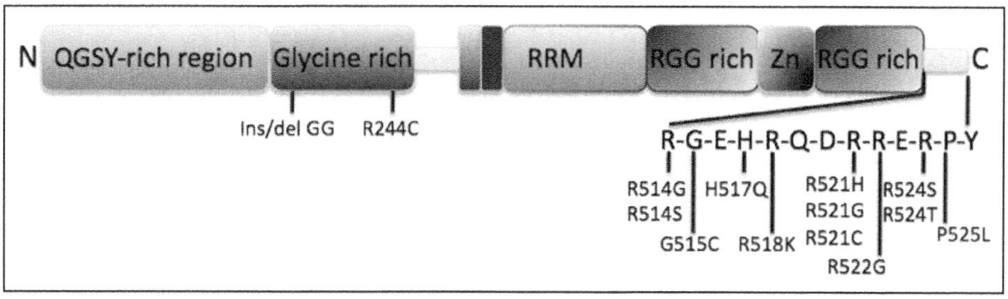

Figure 3. Schematic of the FUS/TLS protein. FUS/TLS contains several motifs, an N-terminal QGSY-rich region, a glycine-rich region, a nuclear export signal, RNA recognition motif, two RGG motifs and a Zn finger motif. ALS causing mutations mostly map to the C-terminus of FUS/TLS, similar to TDP-43.

followed by a glycine rich region, an RRM region, RNA binding RGG repeats, C-terminal zinc finger motif and a highly conserved C-terminal region. A significant number of ALS-causing mutations is in the conserved C-terminal region and they are mostly missense mutations, with the exception of either a two glycine residue insertion or a 2 glycine residue deletion of the glycine rich region. FUS/TLS is ubiquitously expressed, similar to TDP-43 and is mainly localized in the nucleus. It, however, also occur in the cytoplasm in most cells. Individuals with FUS/TLS-linked ALS have aggregates of the protein in the cytoplasm of the neurons and have normal staining for the protein in their nuclei. These cytoplasmic inclusions are not present in normal individuals or those with mutations in *SOD1* or *TARDBP*; also TDP-43 aggregates are not present in the neurons of FUS/TLS mutants, indicating a separate mechanism for the disease.

It is recently that the roles of FUS/TLS are being uncovered, which is similar to TDP-43. Also in common with TDP-43 is that FUS/TLS is linked to other human disease states, where chromosomal translocations result in a FUS/TLS fusion protein that promotes carcinogenesis.[99-102] One normal cellular function of FUS/TLS is the regulation of transcription in response to DNA damage.[103] This occurs through FUS/TLS recruitment by sense and anti-sense noncoding RNAs, which are transcribed in the 5′ regulatory region of the cell-cycle kinase *cyclin D1* gene. This recruits FUS/TLS to bind and inhibit the CREB-binding protein and p300 histone acetyltransferase activities, which leads to the repression of *cyclin D1* transcription. Other potential roles include a more general influence on transcription, as FUS/TLS interacts with the transcriptional machinery, including RNA polymerase II and the TFIID complex, in addition to several nuclear hormone receptors (reviewed in ref. 104-106).

Conclusion

Despite the exciting progress that has been made in determining the genetic and molecular mechanisms of development of ALS, much still remains to be learned. Furthermore, it is important to explore the treatments for this most common form of motor neuron disease. The breakthrough in the last couple of years has been the discovery of DNA/RNA metabolizing proteins TDP-43 and FUS/TLS being involved in many cases of ALS, with the latter having defined functions in the DNA repair response and genomic integrity. However, key molecular questions include characterizing the cellular roles of these two proteins further and determining whether aggregation or loss of normal cellular function of these proteins is behind their pathogenicity. Questions also remain on the mechanisms behind mutant Cu,ZnSOD mediated pathogenicity, but the concept of mutations causing structural instability and aggregation appears to be the most likely reason.

Endeavors focusing on treatment for ALS could try several different approaches, including a small molecule-based method to bind to and perturb the aggregation of SOD1, TDP-43 or FUS/TLS. A vaccine approach against epitopes of misfolded/truncated aggregates forms another potentially important direction, as results form immunization studies in transgenic mouse models of the somewhat analogous Alzheimer's disease resulted in reduced amyloid load and increased cognition. Furthermore, a T-cell protective autoimmune response can be boosted by Cop-1 (Teva Pharmaceuticals) and in SOD1 mutant mouse model of ALS, Cop-1 vaccination provided a significant increase in life span. Stem cell therapy could also be explored, with the caveat that the neurons affected cover a large area of the nervous system and therefore therapy would differ from the more targeted approaches currently being explored. Overall, the recent findings in ALS research have been occurring at a rapid pace and we hope this continues, to ultimately provide novel therapies that help fight against this relatively common and fatal disease.

References

1. Cleveland DW, Rothstein JD. From Charcot to Lou Gehrig: deciphering selective motor neuron death in ALS. Nat Rev Neurosci 2001; 2(11):806-19.
2. Charcot JM. Deux cas d'atrophie musculaire progressive avec lesions de la substance grise et des faisceaux antero-lateraux de la moelle epiniere. Arch Physiol Neurol Pathol 1869: 2:744-54.
3. Traynor BJ, Codd MB, Corr B et al. Incidence and prevalence of ALS in Ireland, 1995-1997: a population-based study. Neurology 1999; 52(3):504-9.
4. Johnston CA, Stanton BR, Turner MR et al. Amyotrophic lateral sclerosis in an urban setting: a population based study of inner city London. J Neurol 2006; 253(12):1642-3.
5. Steele JC, McGeer PL. The ALS/PDC syndrome of Guam and the cycad hypothesis. Neurology 2008; 70(21):1984-90.
6. Ince PG, Codd GA. Return of the cycad hypothesis—does the amyotrophic lateral sclerosis/parkinsonism dementia complex (ALS/PDC) of Guam have new implications for global health? Neuropathol Appl Neurobiol 2005; 31(4):345-53.
7. Haverkamp LJ, Appel V, Appel SH. Natural history of amyotrophic lateral sclerosis in a database population. Validation of a scoring system and a model for survival prediction. Brain 1995; 118(Pt 3):707-19.
8. Mulder DW, Kurland LT, Offord KP et al. Familial adult motor neuron disease: amyotrophic lateral sclerosis. Neurology 1986; 36(4):511-7.
9. Li TM, Alberman E, Swash M. Comparison of sporadic and familial disease amongst 580 cases of motor neuron disease. J Neurol Neurosurg Psychiatry 1988; 51(6):778-84.
10. Veltema AN, Roos RA, Bruyn GW. Autosomal dominant adult amyotrophic lateral sclerosis. A six generation Dutch family. J Neurol Sci 1990; 97(1):93-115.
11. Worms PM. The epidemiology of motor neuron diseases: a review of recent studies. J Neurol Sci 2001; 191(1-2):3-9.
12. Abhinav K, Stanton B, Johnston C et al. Amyotrophic lateral sclerosis in South-East England: a population-based study. The South-East England register for amyotrophic lateral sclerosis (SEALS Registry). Neuroepidemiology 2007; 29(1-2):44-8.
13. Zoccolella S, Beghi E, Palagano G et al. Analysis of survival and prognostic factors in amyotrophic lateral sclerosis: a population based study. J Neurol Neurosurg Psychiatry 2008; 79(1):33-7.
14. Norris F, Shepherd R, Denys E et al. Onset, natural history and outcome in idiopathic adult motor neuron disease. J Neurol Sci 1993; 118(1):48-55.
15. Logroscino G, Traynor BJ, Hardiman O et al. Descriptive epidemiology of amyotrophic lateral sclerosis: new evidence and unsolved issues. J Neurol Neurosurg Psychiatry 2008; 79(1):6-11.
16. Testa D, Lovati R, Ferrarini M et al. Survival of 793 patients with amyotrophic lateral sclerosis diagnosed over a 28-year period. Amyotroph Lateral Scler Other Motor Neuron Disord 2004; 5(4):208-12.
17. Turner MR, Parton MJ, Shaw CE et al. Prolonged survival in motor neuron disease: a descriptive study of the King's database 1990-2002. J Neurol Neurosurg Psychiatry 2003; 74(7):995-7.
18. Chio A, Benzi G, Dossena M et al. Severely increased risk of amyotrophic lateral sclerosis among Italian professional football players. Brain 2005; 128(Pt 3):472-6.
19. Weisskopf MG, McCullough ML, Calle EE et al. Prospective study of cigarette smoking and amyotrophic lateral sclerosis. Am J Epidemiol 2004; 160(1):26-33.
20. Fang F, Bellocco R, Hernan MA et al. Smoking, snuff dipping and the risk of amyotrophic lateral sclerosis—a prospective cohort study. Neuroepidemiology 2006; 27(4):217-21.
21. Horner RD, Grambow SC, Coffman CJ et al. Amyotrophic lateral sclerosis among 1991 Gulf War veterans: evidence for a time-limited outbreak. Neuroepidemiology 2008; 31(1):28-32.
22. Miranda ML, Alicia Overstreet Galeano M, Tassone E et al. Spatial analysis of the etiology of amyotrophic lateral sclerosis among 1991 Gulf War veterans. Neurotoxicology 2008; 29(6):964-70.
23. Beghi E, Millul A, Micheli A et al. Incidence of ALS in Lombardy, Italy. Neurology 2007; 68(2):141-5.

24. Forbes RB, Colville S, Swingler RJ. The epidemiology of amyotrophic lateral sclerosis (ALS/MND) in people aged 80 or over. Age Ageing 2004; 33(2):131-4.
25. Hayashi H, Kato S. Total manifestations of amyotrophic lateral sclerosis. ALS in the totally locked-in state. J Neurol Sci 1989; 93(1):19-35.
26. Sasaki S, Tsutsumi Y, Yamane K et al. Sporadic amyotrophic lateral sclerosis with extensive neurological involvement. Acta Neuropathol 1992; 84(2):211-5.
27. Miller RG, Mitchell JD, Lyon M et al. Riluzole for amyotrophic lateral sclerosis (ALS)/motor neuron disease (MND). Cochrane Database Syst Rev 2007(1):CD001447.
28. Distad BJ, Meekins GD, Liou LL et al. Drug therapy in amyotrophic lateral sclerosis. Phys Med Rehabil Clin N Am 2008; 19(3):633-51, xi-xii.
29. Piao YS, Wakabayashi K, Kakita A et al. Neuropathology with clinical correlations of sporadic amyotrophic lateral sclerosis: 102 autopsy cases examined between 1962 and 2000. Brain Pathol 2003; 13(1):10-22.
30. Leigh PN, Dodson A, Swash M et al. Cytoskeletal abnormalities in motor neuron disease. An immunocytochemical study. Brain 1989; 112(Pt 2):521-35.
31. Rothstein JD, Tsai G, Kuncl RW et al. Abnormal excitatory amino acid metabolism in amyotrophic lateral sclerosis. Ann Neurol 1990; 28(1):18-25.
32. Shaw PJ, Forrest V, Ince PG et al. CSF and plasma amino acid levels in motor neuron disease: elevation of CSF glutamate in a subset of patients. Neurodegeneration 1995; 4(2):209-16.
33. Nicholls DG, Budd SL. Mitochondria and neuronal survival. Physiol Rev 2000; 80(1):315-60.
34. Wiedemann FR, Winkler K, Kuznetsov AV et al. Impairment of mitochondrial function in skeletal muscle of patients with amyotrophic lateral sclerosis. J Neurol Sci 1998; 156(1):65-72.
35. Dhaliwal GK, Grewal RP. Mitochondrial DNA deletion mutation levels are elevated in ALS brains. Neuroreport 2000; 11(11):2507-9.
36. Wiedemann FR, Manfredi G, Mawrin C et al. Mitochondrial DNA and respiratory chain function in spinal cords of ALS patients. J Neurochem 2002; 80(4):616-25.
37. Ro LS, Lai SL, Chen CM et al. Deleted 4977-bp mitochondrial DNA mutation is associated with sporadic amyotrophic lateral sclerosis: a hospital-based case-control study. Muscle Nerve 2003; 28(6):737-43.
38. Kong J, Xu Z. Massive mitochondrial degeneration in motor neurons triggers the onset of amyotrophic lateral sclerosis in mice expressing a mutant SOD1. J Neurosci 1998; 18(9):3241-50.
39. Krasnianski A, Deschauer M, Neudecker S et al. Mitochondrial changes in skeletal muscle in amyotrophic lateral sclerosis and other neurogenic atrophies. Brain 2005; 128(Pt 8):1870-6.
40. Ferri KF, Kroemer G. Mitochondria—the suicide organelles. Bioessays 2001; 23(2):111-5.
41. Guegan C, Przedborski S. Programmed cell death in amyotrophic lateral sclerosis. J Clin Invest 2003; 111(2):153-61.
42. Pasinelli P, Houseweart MK, Brown RH Jr et al. Caspase-1 and -3 are sequentially activated in motor neuron death in Cu,Zn superoxide dismutase-mediated familial amyotrophic lateral sclerosis. Proc Natl Acad Sci USA 2000; 97(25):13901-6.
43. Pasinelli P, Borchelt DR, Houseweart MK et al. Caspase-1 is activated in neural cells and tissue with amyotrophic lateral sclerosis-associated mutations in copper-zinc superoxide dismutase. Proc Natl Acad Sci USA 1998; 95(26):15763-8.
44. Vukosavic S, Dubois-Dauphin M, Romero N et al. Bax and Bcl-2 interaction in a transgenic mouse model of familial amyotrophic lateral sclerosis. J Neurochem 1999; 73(6):2460-8.
45. Li M, Ona VO, Guegan C et al. Functional role of caspase-1 and caspase-3 in an ALS transgenic mouse model. Science 2000; 288(5464):335-9.
46. Mitchell JD, Wokke JH, Borasio GD. Recombinant human insulin-like growth factor I (rhIGF-I) for amyotrophic lateral sclerosis/motor neuron disease. Cochrane Database Syst Rev 2002(3):CD002064.
47. Bongioanni P, Reali C, Sogos V. Ciliary neurotrophic factor (CNTF) for amyotrophic lateral sclerosis/motor neuron disease. Cochrane Database Syst Rev 2004(3):CD004302.
48. Ochs G, Penn RD, York M et al. A phase I/II trial of recombinant methionyl human brain derived neurotrophic factor administered by intrathecal infusion to patients with amyotrophic lateral sclerosis. Amyotroph Lateral Scler Other Motor Neuron Disord 2000; 1(3):201-6.
49. Meininger V, Bensimon G, Bradley WR et al. Efficacy and safety of xaliproden in amyotrophic lateral sclerosis: results of two phase III trials. Amyotroph Lateral Scler Other Motor Neuron Disord 2004; 5(2):107-17.
50. Hirano A, Donnenfeld H, Sasaki S et al. Fine structural observations of neurofilamentous changes in amyotrophic lateral sclerosis. J Neuropathol Exp Neurol 1984; 43(5):461-70.
51. Ferrante RJ, Browne SE, Shinobu LA et al. Evidence of increased oxidative damage in both sporadic and familial amyotrophic lateral sclerosis. J Neurochem 1997; 69(5):2064-74.

52. Smith RG, Henry YK, Mattson MP et al. Presence of 4-hydroxynonenal in cerebrospinal fluid of patients with sporadic amyotrophic lateral sclerosis. Ann Neurol 1998; 44(4):696-9.
53. Tohgi H, Abe T, Yamazaki K et al. Remarkable increase in cerebrospinal fluid 3-nitrotyrosine in patients with sporadic amyotrophic lateral sclerosis. Ann Neurol 1999; 46(1):129-31.
54. Tohgi H, Abe T, Yamazaki K et al. Increase in oxidized NO products and reduction in oxidized glutathione in cerebrospinal fluid from patients with sporadic form of amyotrophic lateral sclerosis. Neurosci Lett 1999; 260(3):204-6.
55. Rosen DR. Mutations in Cu/Zn superoxide dismutase gene are associated with familial amyotrophic lateral sclerosis. Nature 1993; 364(6435):362.
56. Deng HX, Hentati A, Tainer JA et al. Amyotrophic lateral sclerosis and structural defects in Cu,Zn superoxide dismutase. Science 1993; 261(5124):1047-51.
57. Ben Hamida M, Hentati F, Ben Hamida C. Hereditary motor system diseases (chronic juvenile amyotrophic lateral sclerosis). Conditions combining a bilateral pyramidal syndrome with limb and bulbar amyotrophy. Brain 1990; 113(Pt 2):347-63.
58. Chance PF, Rabin BA, Ryan SG et al. Linkage of the gene for an autosomal dominant form of juvenile amyotrophic lateral sclerosis to chromosome 9q34. Am J Hum Genet 1998; 62(3):633-40.
59. Hentati A, Bejaoui K, Pericak-Vance MA et al. Linkage of recessive familial amyotrophic lateral sclerosis to chromosome 2q33-q35. Nat Genet 1994; 7(3):425-8.
60. Corrado L, Ratti A, Gellera C et al. High frequency of TARDBP gene mutations in Italian patients with amyotrophic lateral sclerosis. Hum Mutat 2009; 30(4):688-94.
61. Daoud H, Valdmanis PN, Kabashi E et al. Contribution of TARDBP mutations to sporadic amyotrophic lateral sclerosis. J Med Genet 2009; 46(2):112-4.
62. Gitcho MA, Baloh RH, Chakraverty S et al. TDP-43 A315T mutation in familial motor neuron disease. Ann Neurol 2008; 63(4):535-8.
63. Kabashi E, Valdmanis PN, Dion P et al. TARDBP mutations in individuals with sporadic and familial amyotrophic lateral sclerosis. Nat Genet 2008; 40(5):572-4.
64. Rutherford NJ, Zhang YJ, Baker M et al. Novel mutations in TARDBP (TDP-43) in patients with familial amyotrophic lateral sclerosis. PLoS Genet 2008; 4(9):e1000193.
65. Sreedharan J, Blair IP, Tripathi VB et al. TDP-43 mutations in familial and sporadic amyotrophic lateral sclerosis. Science 2008; 319(5870):1668-72.
66. Van Deerlin VM, Leverenz JB, Bekris LM et al. TARDBP mutations in amyotrophic lateral sclerosis with TDP-43 neuropathology: a genetic and histopathological analysis. Lancet Neurol 2008; 7(5):409-16.
67. Kuhnlein P, Sperfeld AD, Vanmassenhove B et al. Two German kindreds with familial amyotrophic lateral sclerosis due to TARDBP mutations. Arch Neurol 2008; 65(9):1185-9.
68. Kwiatkowski TJ Jr, Bosco DA, Leclerc AL et al. Mutations in the FUS/TLS gene on chromosome 16 cause familial amyotrophic lateral sclerosis. Science 2009; 323(5918):1205-8.
69. Vance C, Rogelj B, Hortobagyi T et al. Mutations in FUS, an RNA processing protein, cause familial amyotrophic lateral sclerosis type 6. Science 2009; 323(5918):1208-11.
70. Gaudette M, Hirano M, Siddique T. Current status of SOD1 mutations in familial amyotrophic lateral sclerosis. Amyotroph Lateral Scler Other Motor Neuron Disord 2000; 1(2):83-9.
71. Wiedau-Pazos M, Goto JJ, Rabizadeh S et al. Altered reactivity of superoxide dismutase in familial amyotrophic lateral sclerosis. Science 1996; 271(5248):515-8.
72. Wanker EE. Protein aggregation in Huntington's and Parkinson's disease: implications for therapy. Mol Med Today 2000; 6(10):387-91.
73. Soto C. Protein misfolding and disease; protein refolding and therapy. FEBS Lett 2001; 498(2-3):204-7.
74. Dobson CM. Protein folding and its links with human disease. Biochem Soc Symp 2001(68):1-26.
75. Parge HE, Hallewell RA, Tainer JA. Atomic structures of wild-type and thermostable mutant recombinant human Cu,Zn superoxide dismutase. Proc Natl Acad Sci USA 1992; 89(13):6109-13.
76. Tainer JA, Getzoff ED, Beem KM et al. Determination and analysis of the 2 A-structure of copper, zinc superoxide dismutase. J Mol Biol 1982; 160(2):181-217.
77. Richardson JS. Beta-Sheet topology and the relatedness of proteins. Nature 1977; 268(5620): 495-500.
78. Hallewell RA, Imlay KC, Lee P et al. Thermostabilization of recombinant human and bovine CuZn superoxide dismutases by replacement of free cysteines. Biochem Biophys Res Commun 1991; 181(1):474-80.
79. Getzoff ED, Tainer JA, Stempien MM et al. Evolution of CuZn superoxide dismutase and the Greek key beta-barrel structural motif. Proteins 1989; 5(4):322-36.
80. Getzoff ED, Tainer JA, Weiner PK et al. Electrostatic recognition between superoxide and copper, zinc superoxide dismutase. Nature 1983; 306(5940):287-90.

81. Getzoff ED, Cabelli DE, Fisher CL et al. Faster superoxide dismutase mutants designed by enhancing electrostatic guidance. Nature 1992; 358(6384):347-51.
82. Shin DS, Didonato M, Barondeau DP et al. Superoxide dismutase from the eukaryotic thermophile Alvinella pompejana: structures, stability, mechanism and insights into amyotrophic lateral sclerosis. J Mol Biol 2009; 385(5):1534-55.
83. Bruijn LI, Houseweart MK, Kato S et al. Aggregation and motor neuron toxicity of an ALS-linked SOD1 mutant independent from wild-type SOD1. Science 1998; 281(5384):1851-4.
84. Durham HD, Roy J, Dong L et al. Aggregation of mutant Cu/Zn superoxide dismutase proteins in a culture model of ALS. J Neuropathol Exp Neurol 1997; 56(5):523-30.
85. Johnston RB Jr, Godzik CA, Cohn ZA. Increased superoxide anion production by immunologically activated and chemically elicited macrophages. J Exp Med 1978; 148(1):115-27.
86. DiDonato M, Craig L, Huff ME et al. ALS mutants of human superoxide dismutase form fibrous aggregates via framework destabilization. J Mol Biol 2003; 332(3):601-15.
87. Arai T, Hasegawa M, Akiyama H et al. TDP-43 is a component of ubiquitin-positive tau-negative inclusions in frontotemporal lobar degeneration and amyotrophic lateral sclerosis. Biochem Biophys Res Commun 2006; 351(3):602-11.
88. Neumann M, Sampathu DM, Kwong LK et al. Ubiquitinated TDP-43 in frontotemporal lobar degeneration and amyotrophic lateral sclerosis. Science 2006; 314(5796):130-3.
89. Talbot K, Ansorge O. Recent advances in the genetics of amyotrophic lateral sclerosis and frontotemporal dementia: common pathways in neurodegenerative disease. Hum Mol Genet 2006; 15 Spec No 2:R182-7.
90. Kwong LK, Uryu K, Trojanowski JQ et al. TDP-43 proteinopathies: neurodegenerative protein misfolding diseases without amyloidosis. Neurosignals 2008; 16(1):41-51.
91. Buratti E, Baralle FE. Multiple roles of TDP-43 in gene expression, splicing regulation and human disease. Front Biosci 2008; 13:867-78.
92. Ou SH, Wu F, Harrich D et al. Cloning and characterization of a novel cellular protein, TDP-43, that binds to human immunodeficiency virus type 1 TAR DNA sequence motifs. J Virol 1995; 69(6):3584-96.
93. Abhyankar MM, Urekar C, Reddi PP. A novel CpG-free vertebrate insulator silences the testis-specific SP-10 gene in somatic tissues: role for TDP-43 in insulator function. J Biol Chem 2007; 282(50):36143-54.
94. Ayala YM, Misteli T, Baralle FE. TDP-43 regulates retinoblastoma protein phosphorylation through the repression of cyclin-dependent kinase 6 expression. Proc Natl Acad Sci USA 2008; 105(10):3785-9.
95. Mantovani V, Garagnani P, Selva P et al. Simple method for haplotyping the poly(TG) repeat in individuals carrying the IVS8 5T allele in the CFTR gene. Clin Chem 2007; 53(3):531-3.
96. Strong MJ, Volkening K, Hammond R et al. TDP43 is a human low molecular weight neurofilament (hNFL) mRNA-binding protein. Mol Cell Neurosci 2007; 35(2):320-7.
97. Wang IF, Wu LS, Chang HY et al. TDP-43, the signature protein of FTLD-U, is a neuronal activity-responsive factor. J Neurochem 2008; 105(3):797-806.
98. Kuo PH, Doudeva LG, Wang YT et al. Structural insights into TDP-43 in nucleic-acid binding and domain interactions. Nucleic Acids Res 2009; 37(6):1799-808.
99. Yang X, Nagasaki K, Egawa S et al. FUS/TLS-CHOP chimeric transcripts in liposarcoma tissues. Jpn J Clin Oncol 1995; 25(6):234-9.
100. Mills KI, Walsh V, Gilkes AF et al. High FUS/TLS expression in acute myeloid leukaemia samples. Br J Haematol 2000; 108(2):316-21.
101. Perez-Losada J, Pintado B, Gutierrez-Adan A et al. The chimeric FUS/TLS-CHOP fusion protein specifically induces liposarcomas in transgenic mice. Oncogene 2000; 19(20):2413-22.
102. Perez-Losada J, Sanchez-Martin M, Rodriguez-Garcia MA et al. Liposarcoma initiated by FUS/TLS-CHOP: the FUS/TLS domain plays a critical role in the pathogenesis of liposarcoma. Oncogene 2000; 19(52):6015-22.
103. Wang X, Arai S, Song X et al. Induced ncRNAs allosterically modify RNA-binding proteins in cis to inhibit transcription. Nature 2008; 454(7200):126-30.
104. Law WJ, Cann KL, Hicks GG. TLS, EWS and TAF15: a model for transcriptional integration of gene expression. Brief Funct Genomic Proteomic 2006; 5(1):8-14.
105. Cassiday LA, Maher LJ, 3rd. Having it both ways: transcription factors that bind DNA and RNA. Nucleic Acids Res 2002; 30(19):4118-26.
106. Lagier-Tourenne C, Cleveland DW. Rethinking ALS: the FUS about TDP-43. Cell 2009; 136(6):1001-4.

Chapter 3

Early-Onset Ataxia with Ocular Motor Apraxia and Hypoalbuminemia/Ataxia with Oculomotor Apraxia 1

Masayoshi Tada, Akio Yokoseki, Tatsuya Sato, Takao Makifuchi and Osamu Onodera*

Abstract

DNA single-strand breaks (SSBs) are non-overlapping discontinuities in strands of a DNA duplex. Significant attention has been given on the DNA SSB repair (SSBR) system in neurons, because the impairment of the SSBR causes human neurodegenerative disorders, including early-onset ataxia with ocular motor apraxia and hypoalbuminemia (EAOH), also known as ataxia-oculomotor apraxia Type 1 (AOA1). EAOH/AOA1 is characterized by early-onset slowly progressive ataxia, ocular motor apraxia, peripheral neuropathy and hypoalbuminemia. Neuropathological examination reveals severe loss of Purkinje cells and moderate neuronal loss in the anterior horn and dorsal root ganglia. EAOH/AOA1 is caused by the mutation in the *APTX* gene encoding the aprataxin (APTX) protein. APTX interacts with X-ray repair cross-complementing group 1 protein, which is a scaffold protein in SSBR. In addition, *APTX*-defective cells show increased sensitivity to genotoxic agents, which result in SSBs. These results indicate an important role of APTX in SSBR. SSBs are usually accompanied by modified or damaged 5′- and 3′-ends at the break site. Because these modified or damaged ends are not suitable for DNA ligation, they need to be restored to conventional ends prior to subsequent repair processes. APTX restores the 5′-adenylate monophosphate, 3′-phosphates and 3′-phosphoglycolate ends. The loss of function of APTX results in the accumulation of SSBs, consequently leading to neuronal cell dysfunction and death.

Introduction

DNA is continuously damaged by endogenous reactive oxygen species (ROS) or exogenous environmental genotoxins, with an estimated frequency of more than 10,000 DNA lesions/cell/day.[1] In neurons, ROS pose particularly great threat to DNA integrity and cause DNA single-strand breaks (SSBs) much more frequently than double-strand breaks (DSBs).[1] SSBs are discontinuities in the sugar-phosphate backbone at non-overlapping sites of strands of a DNA duplex and result directly from sugar damage or indirectly from base damage via the enzymatic excision of DNA during DNA repair processes.[2-5] SSBs are also generated by the activity of topoisomerase I (TOP1), when TOP1 cleaves DNA to relax a DNA helix during transcription or DNA replication and covalently binds to the 3′-end of the break.[2-5] The DNA single-strand break-repair (SSBR) system, which includes the base

*Corresponding Author: Osamu Onodera—Department of Molecular Neuroscience, Resource Branch for Brain Disease Research, Center for Bioresource-Based Research, Brain Research Institute, Niigata University, 1-757 Asahimachi, Niigata 951-8122, Japan. Email: onodera@bri.niigata-u.ac.jp

Diseases of DNA Repair, edited by Shamim I. Ahmad. ©2010 Landes Bioscience and Springer Science+Business Media.

excision repair (BER), plays an important role in repairing SSBs.[2-5] Recent studies have revealed that the impairment of SSBR causes at least two distinct human neurodegenerative diseases, early-onset ataxia with ocular motor apraxia and hypoalbuminemia/ataxia with oculomotor apraxia Type 1 (EAOH/AOA1) and spinocerebellar ataxia with axonal neuropathy Type 1 (SCAN1).[4-7]

EAOH/AOA1 is caused by the loss of function of aprataxin (APTX),[8,9] a protein that contributes to repair of SSBs by restoring modified or damaged ends of SSBs.[5,6,10] SCAN1 is caused by the dysfunction of Tyrosil-DNA phosphodiesterase 1 (TDP1),[11] which removes the TOP1 residue from the 3′-end of TOP1-mediated SSBs.[5,7,12,13] Thus both, APTX and TDP1, are involved in processing the modified or damaged ends of SSBs in SSBR and the dysfunction of either protein can cause the accumulation of SSBs.[7,14] These findings suggest that the accumulation of SSBs or the impairment of SSBR might cause selective neuronal cell death in these disorders.[4,5,12,13] In this chapter, we present the clinical and pathological features, genetic basis and molecular mechanisms of EAOH/AOA1.

Definition

EAOH/AOA1 is a recessively inherited progressive neurodegenerative disorder characterized by early-onset ataxia accompanied with sensory and motor peripheral neuronopathy caused by mutations in the *APTX*, the gene which encodes aprataxin (APTX) protein.[8,9] Patients with EAOH/AOA1 show childhood onset ocular motor apraxia (OMA), which is a unique eye movement characterized by the impairment of initiation of voluntary saccadic eye movements.[15-17] In adulthood, OMA disappears and ocular movements are severely limited instead. Hypoalbuminemia and hypercholesterolemia develop in the middle to late stages of the disease.

Synonyms and Historical Annotations

EAOH/AOA1 was initially described as a variant form of Friedreich's ataxia (FRDA) in Japan.[18-26] In contrast to FRDA, patients with EAOH/AOA1 never develop cardiomyopathy, but do develop severe sensory and motor neuronopathy, hypoalbuminemia and moderate cognitive impairment. Based on the clinical and biochemical signature, the disease has been designated as early-onset cerebellar ataxia with hypoalbuminemia (EOCA-HA),[21,23] hereditary motor and sensory neuropathy associated with cerebellar atrophy (HMSNCA),[22] or early-onset ataxia associated with hypoalbuminemia (EOAHA).[26] Early onset ataxia with the unique eye movement abnormality, OMA, has been reported in Portuguese and Japanese families[27] and designated as AOA1.[28] In 2001, the gene mutated in EOAHA and AOA1 was identified independently.[8,9] It was subsequently revealed that both AOA1 and EOAHA are caused by an identical mutation in the *APTX* gene. The differences in clinical features between AOA1 and EOAHA depend on the age at diagnosis. Thus we proposed these disorders be designated EAOH.[8] The syndrome of ataxia with OMA is sometimes referred to as Aicardi syndrome.[29] This name is not proper, however, because it may lead to confusion with the other Aicardi syndrome, agenesis of the corpus callosum with infantile spasms and chorioretinal abnormalities.[30]

Epidemiology

Incidence and Prevalence

The incidence and prevalence of the spinocerebellar ataxias (SCAs) differ markedly among different ethnic groups and geographic areas. In Caucasians, FRDA is the most frequently observed hereditary ataxia, with an estimated prevalence of 1-2 per 50,000 population.[27,31,32] In the Portuguese population, EAOH/AOA1 accounts for ~21% of recessive inherited ataxias (RIAs), whereas FRDA accounts for 38% of RIAs.[27] The estimated prevalence of EAOH/AOA1 in the Portuguese population is 0.2-0.3 per 50,000 population.[27] Patients with EAOH/AOA1 have been reported from other parts of the world.[17,33-41]

In the Japanese population, the estimated prevalence of hereditary SCAs is 2-3 per 50,000 population, whereas the estimated prevalence of recessive inherited ataxias is ~0.2 per 50,000 population.[42] In contrast to Caucasians, FRDA has not been reported in Japan and EAOH/AOA1 seems to be the most frequently observed form of RIAs in adults.[31,32]

Sex and Age Distribution

EAOH/AOA1 affects men and women equally.[43] Mean age at onset is ~5 years, ranging typically from 1-18 years.[8,9,17,27,28,33-41,43] The onset after 20 years is extremely rare.[33] The oldest reported age of onset is in a 40 years old patient with a homozygous 6668T→C mutation (exon 5), resulting in the substitution of proline for leucine at codon 223 (L223P).[44]

Risk Factors

EAOH/AOA1 is an inherited disorder with no known environmental risk factors.

Genetics of EAOH/AOA1

The gene for AOA1 has been linked to a locus on chromosome 9p13 in Portuguese and Japanese families.[28] From the clinical similarities between AOA1 and EOAHA, we found that the gene for EOAHA was also linked to the same locus as AOA1. In late 2001, we and others independently identified the gene mutated in EOAHA and AOA1 and designated it as aprataxin (*APTX*).[8,9]

The *APTX* gene consists of nine exons, which include two additional exons in the 5′ region of the *APTX*.[5] At least five transcript variants encoding distinct isoforms have been identified for the gene, although the nature of some variants has not been determined. The longest form of APTX (aprataxin isoform a: NP-778243.1) is the major isoform consisting of 342 amino acids. The isoform (NP-778243.1) has three domains including an N-terminal forkhead-associated (FHA) domain, a central histidine triad (HIT) domain and a C-terminal zinc-finger (ZNF) domain[2,5,8,9,45] (Fig. 1). To date, 18 pathogenic mutations have been identified, which include two frameshift, two splice site, two nonsense and 12 missense mutations[5,8,9,17] (Fig. 1). All of the missense mutations in the *APTX* are located in Exon 7 and 8,[5,8,9] which encode the HIT domain, suggesting that the domain is important for the function or stability of APTX.

A founder effect in the *APTX* gene is evident among some populations. The homozygous mutation of 689insT (V230fs) is in 78% of 42 Japanese patients with EAOH/AOA1 of 27 families,[8,39,43] whereas W279X mutation is most frequently observed in the European population. The mRNA with 689insT mutation is degradated by means of nonsense-mediated mRNA decay, which is elicited when a premature stop codon is located at least 50 to 55 nucleotides upstream of the exon-exon junction close to the 3′ end.[46] Heterozygous K153E or S242N mutations in *APTX*

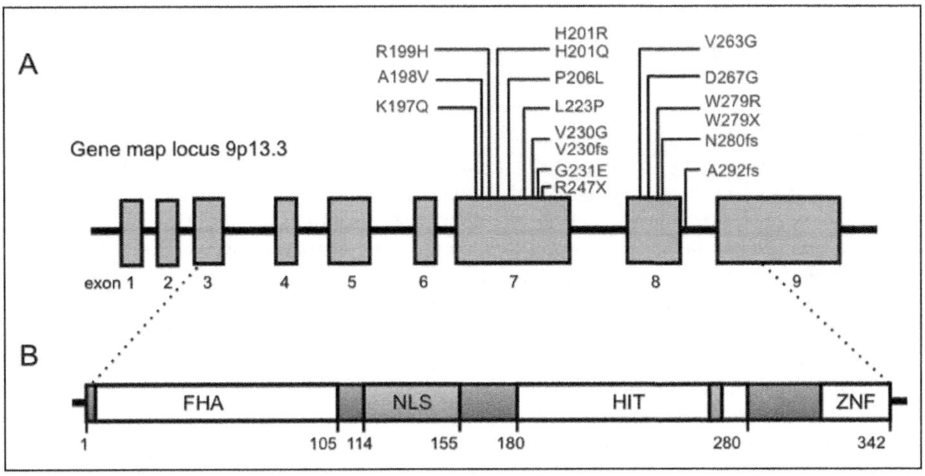

Figure 1. Aprataxin and disease-associated mutations. A) Genomic organization of *APTX*. B) The structure of the long-form aprataxin (aprataxin isoform a: NP-778243.1). The isoform consists of 342 amino acids and have a forkhead-associated (FHA) domain, a histidine triad (HIT) domain and a zinc-finger (ZNF) domain. NLS, nuclear localization signal.

were reported in patients with clinical features resembling the cerebellar type of Multiple System Atrophy-C (MSA-C);[47] however, it is still controversial whether these heterozygous mutations are pathogenic.

The amount of APTX in the autopsied brain tissue from patients carrying compound-heterozygous 689insT/840delT mutations, lymphoblastoid cells with compound-heterozygous 689insT/840delT, homozygous P206L, or compound-heterozygous P206L/V263G mutations and fibroblasts with homozygous V230G is reduced compared to healthy individuals.[40,48] Consistent with these findings, the disease-associated missense mutations decrease the stability of APTX in culture cells.[49-51] These results indicate that decreased levels of APTX may underlie the molecular pathogenesis of EAOH/AOA1.

Genotype and Phenotype Correlation

The correlation between genotype and clinical phenotype has not been demonstrated clearly. Patients with homozygous frameshift mutations, however, tend to show a more severe phenotype, including earlier onset and more severe cognitive impairment, than patients with homozygous missense mutations.[17,33,39,43,44,52,53]

Clinical Features

Signs and Symptoms

EAOH/AOA1 is characterized by three clinical features: (i) cerebellar and sensory ataxia and involuntary movements; (ii) OMA, defined as the impaired initiation of saccadic eye movement, accompanied with head thrusts; and (iii) sensory and motor neuronopathy. These clinical features are accompanied by two biochemical markers, hypoalbuminemia and hypercholesterolemia[17,43,54,55] (Fig. 2A). Gait disturbance typically first becomes apparent between the ages of two and eighteen years. In early childhood, OMA and ataxic gait are prominent features, frequently accompanied with involuntary movements including tremor, chorea and athetosis. Then vibration sense and deep tendon reflexes progressively decrease. Subsequently, muscular weakness with muscle wasting and the loss of superficial and deep sensation develop (Fig. 2B,C). In adulthood, most patients develop ophthalmoplegia and OMA becomes more difficult to recognise. They frequently have mild cognitive impairment. Pes cavus, scoliosis, dystonia and hypotonia are also observed. Most patients are in a wheelchair-bound state by ~20 years of age. The life span is comparable to that of normal individuals.

OMA in EAOH/AOA1

OMA is an ocular motor disturbance, initially defined by Cogan as the impaired initiation of horizontal saccades. To overcome this impairment, patients with OMA often blink or use involuntary head movements to initiate their saccadic eye movements and it is often these compensatory mechanisms that are most striking on the clinical examination.[15,57-59] The oculographic pattern is characterized by increased latencies and decreased amplitude of horizontal saccades. When patients move their gaze toward a lateral target, they first turn their head overshooting the target then return their head slightly in the opposite direction. The loss of cancellation of vestibuloocular reflex (VOR) is another characteristic feature in EAOH/AOA1. Therefore, by the head rotation, the remaining VOR helps patients to move their gaze towards a target.[17,55]

The precise mechanism underlying the OMA is unclear. The activation of excitatory burst neurons at the caudal pons initiates the saccadic eye movement[15] (Fig. 3A). The activation of excitatory burst neurons is continuously inhibited by the activation of ominipause neurons. The activation of ominipause neurons is inhibited by the fastigial neurons in the cerebellum and is stimulated by the fixation neurons at the superior colliculus. The fastigial neurons also inhibit the fixation neurons. Thus, the impairment of the fastigial nucleus has been proposed as one potential mechanism for OMA in EAOH/AOA1. In the autopsied brains of patients with EAOH/AOA1, however, pathological involvement of the fastigial nucleus is still controversial.[55,60,61] Purkinje cells in the dorsal vermis (lobules VI and VII), designated as the oculomotor vermis, stimulate the fastigial nucleus and may initiate the saccadic eye movement.[62-66] The pharmacological inhibition of Purkinje cells results in

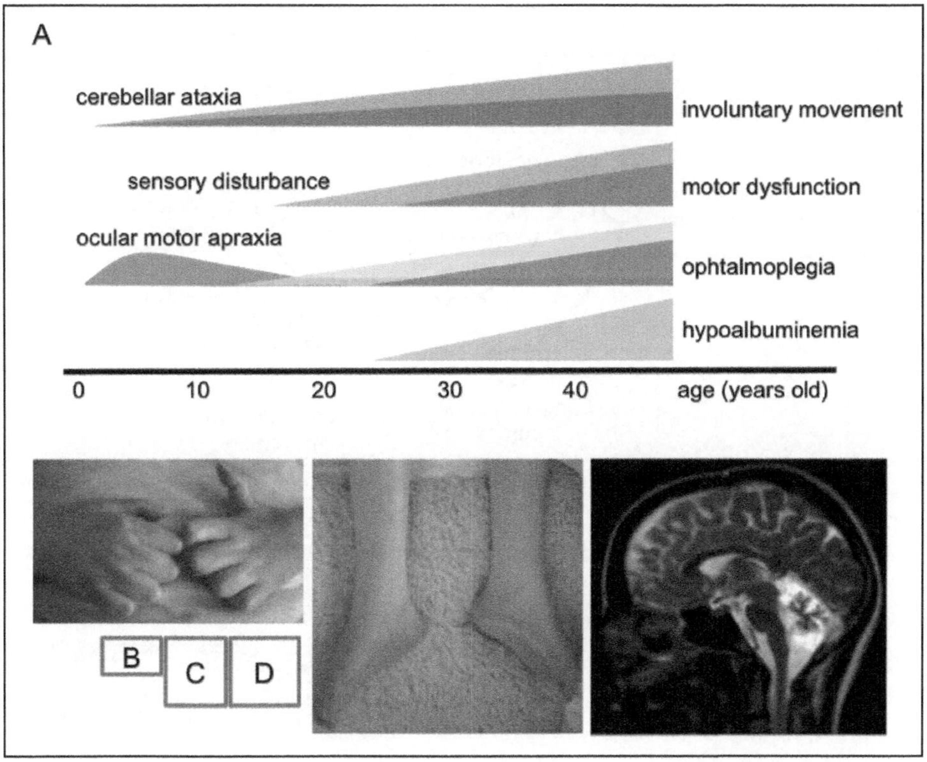

Figure 2. Clinical characteristic features of EAOH/AOA1. A) Clinical spectrum of EAOH/AOA1. The clinical spectrum of EAOH/AOA1 is divided into four axes; motor symptoms, peripheral neuropathy, eye symptoms and biochemical markers. The frequency of each phenotype depends on age. B,C) Images of upper and lower extremities of patients with EAOH/AOA1, showing deformity of distal extremities caused by severe peripheral neuropathy. D) Marked atrophy of the cerebellum on MRI. (Modified from ref. 55.)

marked inhibition of saccadic eye movement.[66] Therefore, the selective involvement of Purkinje cells in the oculomotor vermis might also be responsible for OMA in EAOH/AOA1.

Imaging and Laboratory Findings

Atrophy of the cerebellum, particularly of the cerebellar vermis, is obvious on MRI imaging even early in the course of the disease[17,38,43,60] (Fig. 2D). Cerebellar hypoperfusion is also seen in patients who undergo brain ECD-SPECT.[17] Motor nerve conduction studies are characterized by the decreased amplitude of compound muscle action potentials with decreased nerve conduction velocities, indicating axonal degeneration.[38,43] Sensory nerve action potentials are decreased in early childhood and are not detected until 20 years of age.[17,38,43]

Although hypoalbuminemia and hypercholesterolemia are characteristic biochemical markers of EAOH/AOA1, they become prominent only after 20 years of age.[43,55] Hypercholesterolemia is corrected by infusion of albumin, whereas hypoalbuminemia is not corrected by reducing the total level of serum cholesterol, indicating that the hypoalbuminemia causes hypercholesterolemia.[22,26] The degradation rate of serum albumin is normal in EAOH/AOA1,[22,55] suggesting that synthesis of albumin in the liver decreases in EAOH/AOA1. The serum alpha-fetoprotein, which is increased in ataxia-telangiectasia (AT) and ataxia with oculomotor apraxia Type 2 (AOA2), is normal in EAOH/AOA1.

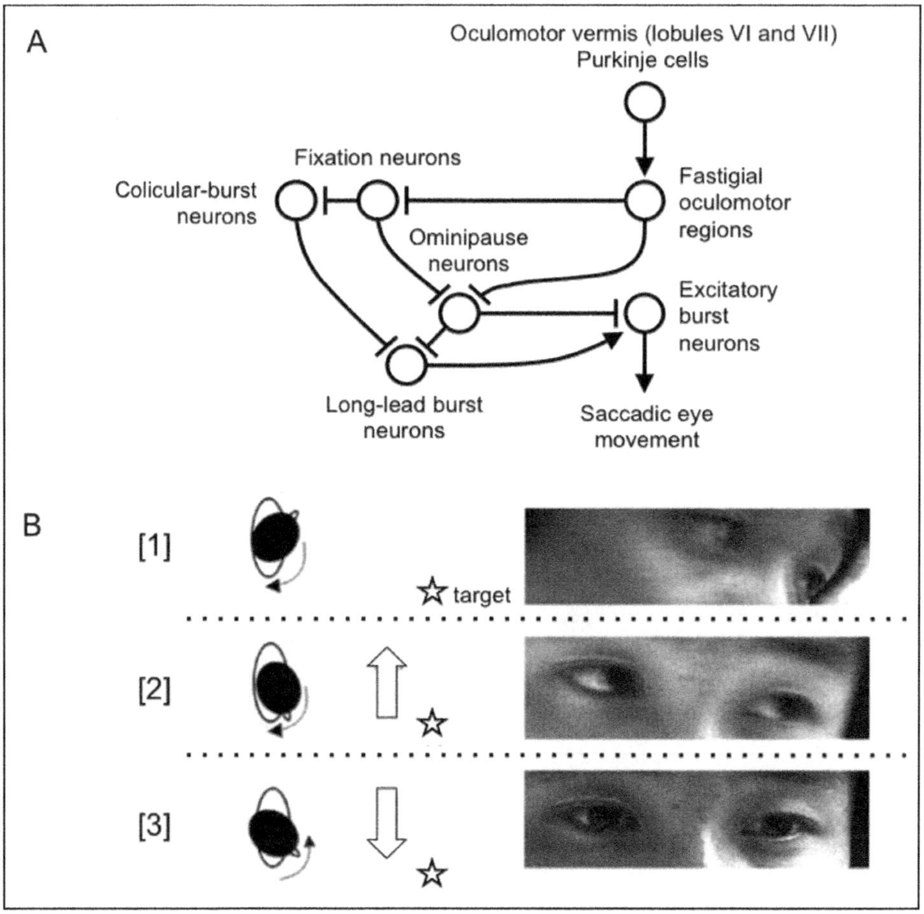

Figure 3. A model for regulation of saccadic eye movement and head thrust in EAOH/AOA1. A) A model for regulation of the initiation of saccadic eye movement. O, neuron; →, excitatory projection; ⊣, inhibitory projection. B) Head thrust and vestibuloocular reflex (VOR). The patient was given instructions. [1] Ask the patient to gaze toward the right lateral target. [2] The patient turns his head overshooting the target. His gaze moves to the left owing to the remaining VOR. [3] He then turns his head to the left. His gaze moves to the right owing to the remaining VOR. (Modified from ref. 55.)

Neuropathological Features

The most characteristic neuropathological finding in EAOH/AOA1 is severe loss of Purkinje cells without prominent grumose degeneration[22,55,60,61,67] (Fig. 4). The loss of Purkinje cells is likely an early event in the disease because cerebellar ataxia is an initial symptom and marked atrophy of the cerebellar vermis is seen on MRI even in early childhood. Other neuropathological findings include severe degeneration of the posterior columns and spinocerebellar tracts and moderate neuronal loss in the anterior horn and dorsal root ganglia of the spinal cord.[22,55,60,61] Sural nerve biopsies of patients with EAOH/AOA1 reveal the loss of large myelinated nerve fibers without demyelination.[17,19,22] Although the patients with EAOH/AOA1 frequently have cognitive impairment and involuntary movements, no obvious neuropathological findings are reported in the cerebral cortex or basal ganglia. There are no neuronal and glial inclusions as are seen in many other cerebellar ataxias.

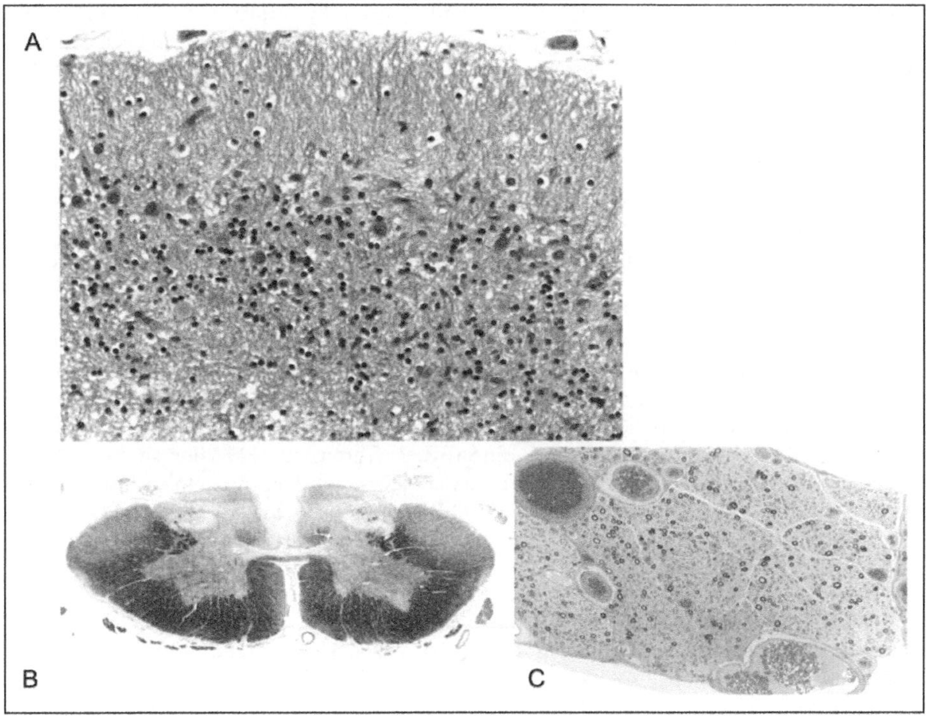

Figure 4. Neuropathological characteristic features of EAOH/AOA1. A) Severe loss of Purkinje cells in the cerebellum. H.E. stain. B) Severe degeneration of the posterior columns and spinocerebellar tracts. Kluver-Barrera stain. C) The loss of myelinated fibers of the sural nerve. Toluidine blue stain.

Differential Diagnosis

The early development of most patients with EAOH/AOA1 is almost normal. However, in rare cases resulting from complete loss of APTX by a total deletion, a frameshift, or a nonsense mutation, such as 689insT (V230fs) or W279X, ataxia can be present at birth or in the early infancy. In these cases, other congenital ataxias or metabolic disorders should be considered as part of the differential diagnoses. The key to the diagnosis is the presence of OMA, which is most easily diagnosed by monitoring for abnormal head movements or head thrusts in early childhood. Congenital OMA, which is characterized by OMA in the absence of other remarkable neurological abnormalities, should be also considered as a differential diagnosis. Although patients with congenital OMA sometimes show mild truncal ataxia, they never show marked cerebellar atrophy, progressive ataxia and peripheral neuropathy.

There are at least three other disorders with clinical similarities to EAOH/AOA1, including ataxia telangiectasia (AT), an ataxia-telangiectasia-like disorder (ATLD) and ataxia with oculomotor apraxia Type 2 (AOA2).[16,29,68-71] Based on the unique clinical features of these diseases, we can clinically distinguish them from EAOH/AOA1. AT is characterized by telangiectasia, immunodeficiency and malignancy predisposition, particularly, leukemia and lymphoma.[16,68,71] A predisposition to malignancy is not seen in patients with EAOH/AOA1.[32,55,56] The serum alpha fetoprotein levels are elevated in patients with AT and AOA2,[68,70,71] whereas the serum albumin levels are decreased in patients with EAOH/AOA1.[20,22,26,43,72] The frequency of OMA in AOA2 is much lower than those in EAOH/AOA1.[70] SCAN1 is another neurodegenerative disorder with clinical similarities to EAOH/AOA1.[11,73] Only one family with SCAN1, however, has been reported[11] and the clinical phenotype of this family is not fully evaluated. We have to wait for further reports to evaluate the clinical phenotype of SCAN1. EAOH/AOA1 has been described as a variant form of FRDA in Japanese families in 1970s.[3-13] In contrast to FRDA, EAOH/AOA1 patients never develop cardiomyopathy, but do develop marked cerebellar atrophy and severe motor neuropathy and hypoalbuminemia.[32,55,56]

Pathogenesis

Function of Aprataxin

APTX is widely expressed in most tissues including the nervous system, heart, liver, kidneys and lymph nodes.[48] In the nervous system, APTX is expressed in the cerebellum, basal ganglia, cerebral cortex and spinal cord.[48] In situ hybridization analysis of mouse brains reveals that *APTX* mRNA is expressed at high levels in both Purkinje cells and cerebellar granular cells.[74]

Based on amino acid sequence homology, APTX is a member of the HIT superfamily of nucleotide hydrolases and transferases.[8,9,75,76] In its N-terminus, APTX has a forkhead-associated (FHA) domain, which is homologous to the FHA domain of polynucleotide kinase 3′-phosphatase (PNKP). PNKP is an end-processing enzyme in SSBR.[3,5,77] Through the FHA domain, both, APTX and PNKP, competitively interact with phosphorylated X-ray repair cross-complementing group 1 protein (XRCC1), which is a scaffold protein in SSBR.[5,78-80] Consistent with this, *APTX*-defective lymphoblastoid cells and fibroblasts show increased sensitivity to genotoxic agents or oxidative stresses which result in SSBs.[41,78,81] Taken together these results suggest that APTX plays an important role in SSBR.

An SSB is usually accompanied by loss of a single nucleotide and by modified or damaged 5′- and 3′-ends of the break site. These modified or damaged 5′- and 3′-ends include 5′-adenylate monophosphate (5′-AMP), 5′-hydroxyl, 5′-aldehyde, 5′-deoxyribophosphate (5′-dRP), 3′-α, β-unsaturated aldehyde, 3′-phosphate, 3′-phosphoglycolate (PG) and 3′-TOP1 peptide ends.[5,45] Because these modified or damaged ends are not suitable for DNA ligation, they are normally restored to conventional 5′-phosphate and 3′-hydroxyl ends to reseal the breaks. In the process of DNA ligation, AMP transiently binds to 5′-phosphate through a pyrophosphate bond and is subsequently removed by the attack of 3′-hydroxyl ends. If the 3′-end is modified or damaged, 5′-AMP remains as an abortive ligation intermediate. APTX removes the AMP from 5′-ends and results in restoration of abortive DNA ligation intermediates to conventional 5′-phosphate ends.[6,14,82-84] In addition, APTX restores 3′-phosphates and 3′-phosphoglycolates (PG) ends, which are frequently observed specifically at the SSB sites arising from sugar damage induced by ROS.[10] Thus, APTX restores the modified or damaged ends at SSBs to suitable ends for the subsequent repair processes (Fig. 5).

Defects in SSBR and Neurodegeneration

Impairment of DNA repair systems is associated with many human diseases. Whereas the impairment of the DNA double-strand break repair (DSBR) system results in systemic involvement, including immunodeficiency and malignancy predisposition, the impairment of the SSBR system results in early onset neurodegenerative disorders, including EAOH/AOA1 and SCAN1. In fact, the pathological findings of EAOH/AOA1 are largely restricted to the nervous system, particularly Purkinje cells and the dorsal root ganglia and anterior horn cells.[22,55,60,61] Considerable attention has thus been given on the association between SSBR system and neuronal survival. Interestingly, TDP1, the causative gene product for SCAN1, removes the TOP1 peptide at the 3′-end of TOP1-mediated SSBs and also removes 3′-PG ends induced by oxidative stress.[7,13,85] The similarity of function between APTX and TDP1 suggests that the restoration of the modified or damaged ends of SSBs might be important to maintain the function and viability of neurons.

The most likely contributing factor to the selective vulnerability of the nervous system seems to be the ability of its cells to compensate for defective SSBR. In both, nonproliferating and proliferating cells, SSBs are rapidly repaired by SSBR. If SSBR is defective in proliferating cells, unrepaired SSBs can cause potentially lethal DSBs during DNA replication. Proliferating cells

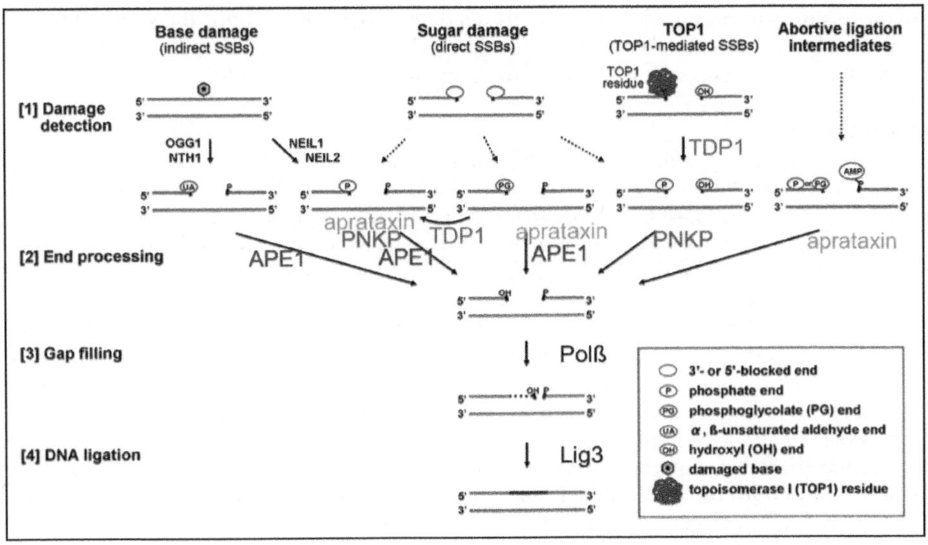

Figure 5. A model for aprataxin-dependent SSBR. SSBR pathways defined by the types of enzymes that remove damaged 3′-, or 5′-ends are shown. SSBs can arise indirectly from base damage, or directly from sugar damage or TOP1 cleavage, or as an abortive ligation intermediate. Red circles denote the damaged ends, the specific types of which depend on the source of the break. [1] Damage detection. [2] The processing of damaged 3′- and 5′-ends is mediated by either APE1, APTX, PNKP, or TDP1, depending on the type of modified or damaged ends. These modified or damaged ends should be converted to 3′-hydroxyl and 5′-phosphate ends for subsequent repair processes. In the pathway for repairing indirectly induced SSBs, damaged 3′-α, β unsaturated aldehyde ends are removed by APE1. In the pathway for repairing directly induced SSBs, 3′-PG ends might be removed by APTX or APE1 and 3′-phosphate ends by APTX, PNKP or APE1. In the pathway for repairing TOP1-mediated SSBs, TOP1 covalent complexes at the 3′-ends are restored to 3′-phosphate ends by TDP1. In the repairing abortive ligation intermediates, 5′-adenylate monophosphate (AMP) ends are removed by APTX. [3] After removing damaged 3′- and 5′-ends, DNA polymerase β (Pol β) fills the gap. [4] DNA ligase III (Lig3) seals the single-strand nick. A color version of this image is available at www.landesbioscience.com/curie.

therefore possess alternative repair systems such as the homologous recombination repair as a back-up for repairing SSBs during DNA replication.[5,45] If SSBR is defective in nonproliferating cells such as terminally differentiated neurons, unrepaired SSBs persist and potentially block transcription by stalling RNA polymerases at the SSB sites, leading to cell dysfunction and consequent cell death.[12,13] Consistent with this scenario, homologous recombination is selectively elevated in proliferating TDP1-defective lymphoblastoid cells as indicated by an increased frequency of sister chromatid exchanges.[7]

However, it is still unclear why neurons in general and purkinje neurons in particular are so exquisitely sensitive to the impairment of SSBR, compared with other nonproliferating cells including hepatocytes, myocytes and adipocytes. Likely factors for the vulnerability of neurons include the high amount of oxidative lesions in it, due to high levels of oxidative stress and low levels of antioxidant enzymes; also can be a high transcriptional demand and the limited regenerative capacity of neurons, as compared with other nonproliferating cells.[12,13]

Experimental Models

Mice lacking APTX have shown no distinct clinical phenotype or pathological findings in their life time.[6] Their too short lifespan may be the reason not to allow the progressive accumulation of the lesions to be measured. More recently, a mouse model lacking both APTX and TDP1 has been developed.[86] Neuronal cells of the mice show the decreased rate of repair of SSBs induced by oxidative stress, suggestive of the synergistic action of APTX and TDP1.[86]

Conclusion

Whereas increasing evidence indicates the APTX functions in SSBR, it remains to be elucidated which function of APTX, 3′-end processing or 5′-end processing, plays key roles in the molecular pathogenesis of EAOH/AOA1. Moreover, it remains unclear whether restoring the modified or damaged ends of SSBs is the only function of APTX. It might be expected that the 5′-AMP removal activity is a major activity of APTX under physiological conditions, because in vitro studies show that the 5′-AMP removal activity is stronger than the restoring activity of the 3′-modified ends.[6,10,14] However, the modified or damaged 3′-ends still remain to be restored, even though APTX restores the abortive ligation 5′-ends. If the modified or damaged 3′-end is restored to a hydroxyl end, the 3′-hydroxyl end itself can remove the 5′-AMP end through the intermediation of DNA ligase without APTX. Therefore, a question arises whether the 5′-AMP removal activity of APTX is necessary in neurons. Similarly, the function of APTX to restore the modified or damaged 3′-ends might be compensated by other end-processing enzymes, including PNKP and APE1. These findings raise the possibility that other functions of APTX may contribute to the pathogenesis of EAOH/AOA1.

Several findings suggest other functions of APTX. APTX interacts with XRCC4, a scaffold protein in DNA DSBR, suggesting that APTX might be involved in DSBR as well as SSBR.[80] To date, however, impairment of DSBR has not been detected in APTX-defective cells.[41,78,86] The role of APTX in its interaction with XRCC4 should be worked out. APTX also interacts with the nucleolar proteins nucleolin, nucleophosmin and upstream-binding factor 1 (UBF1) and APTX localizes to the nucleolus.[48,87] The nucleolus is a place for transcription and maturation of ribosomal RNAs. Ribosomal RNAs is transcribed from ribosomal DNAs (rDNAs) at nucleolus. Therefore, there is a possibility that APTX acts on the repair of rDNAs in the nucleolus. The loss of APTX function might cause the accumulation of SSBs on rDNAs, consequently leading to a defect in ribosomal mRNAs transcription. However, the depletion of APTX does not significantly affect ribosomal RNA transcription.[87] Thus, further studies are necessary to clarify the role of APTX in the nucleolus.

Despite the identification and understanding of the function of the causative gene for EAOH/AOA1, we have no effective treatment to improve or to prevent a progression of the disease. If we argue that the accumulation of SSBs may be responsible for the molecular pathogenesis of EAOH/AOA1, antioxidant treatment with coenzyme Q10, vitamin E derivates or idebenone (which protect DNA from oxidative stress) might have the potential to ameliorate the disease progression. However, to evaluate the effectiveness of such treatments, there is a pressing need to develop molecular markers of progression that can be followed over time and which represent the disease progression more sensitively than does serum albumin.

In conclusion, APTX plays a role in SSBR by restoring the modified or damaged ends of SSBs and the loss of function of APTX results in the accumulation of SSBs, consequently leading to neuronal cell dysfunction and death. These findings raise the idea that protecting DNA against oxidative damage may be a potential therapeutic strategy to prevent the disease progression in EAOH/AOA1. Moreover, further analysis of the APTX's physiological functions may open new avenues for understanding DNA quality control system in neurons.

Acknowledgements

This study was supported in part by a Grant-in-Aid for Scientific Research on Priority Areas (C) -Advanced Brain Science Project- and a Grant-in-Aid for Scientific Research (A) and (B) from the Ministry of Education, Culture, Sports, Science and Technology, Japan, a grant for the Research for the Future Program from the Japan Society for the Promotion of Science, a grant for Surveys and Research on Specific Diseases, the Ministry of Health, Labor and Welfare, Japan, a grant from the Tsubaki Memorial Neuroscience Research Foundation. We are grateful to Drs. Shakkottai VG and Todd PK for providing helpful comments.

References

1. Brooks PJ. DNA repair in neural cells: basic science and clinical implications. Mutat Res 2002; 509:93-108.
2. Caldecott KW. Mammalian DNA single-strand break repair: an X-ra(y)ted affair. Bioessays 2001; 23:447-455.
3. Caldecott KW. XRCC1 and DNA strand break repair. DNA Repair (Amst) 2003; 2:955-969.
4. Caldecott KW. DNA single-strand breaks and neurodegeneration. DNA Repair (Amst) 2004; 3:875-882.
5. Caldecott KW. Single-strand break repair and genetic disease. Nat Rev Genet 2008; 9:619-631.
6. Ahel I, Rass U, El-Khamisy SF et al. The neurodegenerative disease protein aprataxin resolves abortive DNA ligation intermediates. Nature 2006; 443:713-716.
7. El-Khamisy SF, Saifi GM, Weinfeld M et al. Defective DNA single-strand break repair in spinocerebellar ataxia with axonal neuropathy-1. Nature 2005; 434:108-113.
8. Date H, Onodera O, Tanaka H et al. Early-onset ataxia with ocular motor apraxia and hypoalbuminemia is caused by mutations in a new HIT superfamily gene. Nat Genet 2001; 29:184-188.
9. Moreira MC, Barbot C, Tachi N et al. The gene mutated in ataxia-ocular apraxia 1 encodes the new HIT/Zn-finger protein aprataxin. Nat Genet 2001; 29:189-193.
10. Takahashi T, Tada M, Igarashi S et al. Aprataxin, causative gene product for EAOH/AOA1, repairs DNA single-strand breaks with damaged 3′-phosphate and 3′-phosphoglycolate ends. Nucleic Acids Res 2007; 35:3797-3809.
11. Takashima H, Boerkoel CF, John J et al. Mutation of TDP1, encoding a topoisomerase I-dependent DNA damage repair enzyme, in spinocerebellar ataxia with axonal neuropathy. Nat Genet 2002; 32:267-272.
12. El-Khamisy SF, Caldecott KW. TDP1-dependent DNA single-strand break repair and neurodegeneration. Mutagenesis 2006; 21:219-224.
13. El-Khamisy SF, Caldecott KW. DNA single-strand break repair and spinocerebellar ataxia with axonal neuropathy-1. Neuroscience 2007; 145:1260-1266.
14. Reynolds JJ, El-Khamisy SF, Katyal S et al. Defective DNA ligation during short-patch single-strand break repair in ataxia oculomotor apraxia 1. Mol Cell Biol 2009; 29:1354-1362.
15. Leigh RJ, Zee DS. The neurology of eye movements. Philadelphia 1999; xi, 281.
16. Lewis RF, Lederman HM, Crawford TO. Ocular motor abnormalities in ataxia telangiectasia. Ann Neurol 1999; 46:287-295.
17. Le Ber I, Moreira MC, Rivaud-Pechoux S et al. Cerebellar ataxia with oculomotor apraxia type 1: clinical and genetic studies. Brain 2003; 126:2761-2772.
18. Inoue N, Izumi K, Mawatari S et al. Congenital ocular motor apraxia and cerebellar degeneration—report of two cases. Rinsho Shinkeigaku 1971; 11:855-861.
19. Kawasaki S, Ideta T, Ueno H et al. [Three cases of autosomal recessively inherited neuropathy with cerebellar ataxia, optic atrophy and hyperlipidemia (author's transl)]. Rinsho Shinkeigaku 1982; 22:15-23.
20. Uekawa K, Yuasa T, Kawasaki S et al. [A hereditary ataxia associated with hypoalbuminemia and hyperlipidemia—a variant form of Friedreich's disease or a new clinical entity?]. Rinsho Shinkeigaku 1992; 32:1067-1074.
21. Kubota H, Sunohara N, Iwabuchi K et al. [Familial early onset cerebellar ataxia with hypoalbuminemia]. No To Shinkei 1995; 47:289-294.
22. Fukuhara N, Nakajima T, Sakajiri K et al. Hereditary motor and sensory neuropathy associated with cerebellar atrophy (HMSNCA): a new disease. J Neurol Sci 1995; 133:140-151.
23. Hanihara T, Kubota H, Amano N et al. [Siblings of early onset cerebellar ataxia with hypoalbuminemia]. Rinsho Shinkeigaku 1995; 35:83-86.
24. Sekijima Y, Ohara S, Nakagawa S et al. Hereditary motor and sensory neuropathy associated with cerebellar atrophy (HMSNCA): clinical and neuropathological features of a Japanese family. J Neurol Sci 1998; 158:30-37.
25. Tachi N, Kozuka N, Ohya K et al. Hereditary cerebellar ataxia with peripheral neuropathy and mental retardation. Eur Neurol 2000; 43:82-87.
26. Koike R, Tanaka H, Tsuji S. Early onset ataxia associated with hypoalbuminemia. Neurol Med 1998; 48:237-242.
27. Barbot C, Coutinho P, Chorao R et al. Recessive ataxia with ocular apraxia: review of 22 Portuguese patients. Arch Neurol 2001; 58:201-205.
28. Moreira MC, Barbot C, Tachi N et al. Homozygosity mapping of Portuguese and Japanese forms of ataxia-oculomotor apraxia to 9p13 and evidence for genetic heterogeneity. Am J Hum Genet 2001; 68:501-508.

29. Aicardi J, Barbosa C, Andermann E et al. Ataxia-ocular motor apraxia: a syndrome mimicking ataxia-telangiectasia. Ann Neurol 1988; 24:497-502.
30. Aicardi J. Aicardi syndrome. Brain Dev 2005; 27:164-171.
31. Di Donato S, Gellera C, Mariotti C. The complex clinical and genetic classification of inherited ataxias. II. Autosomal recessive ataxias. Neurol Sci 2001; 22:219-228.
32. Fogel BL, Perlman S. Clinical features and molecular genetics of autosomal recessive cerebellar ataxias. Lancet Neurol 2007; 6:245-257.
33. Criscuolo C, Mancini P, Sacca F et al. Ataxia with oculomotor apraxia type 1 in Southern Italy: late onset and variable phenotype. Neurology 2004; 63:2173-2175.
34. Habeck M, Zuhlke C, Bentele KH et al. Aprataxin mutations are a rare cause of early onset ataxia in Germany. J Neurol 2004; 251:591-594.
35. Amouri R, Moreira MC, Zouari M et al. Aprataxin gene mutations in Tunisian families. Neurology 2004; 63:928-929.
36. Tsao CY, Paulson G. Type 1 ataxia with oculomotor apraxia with aprataxin gene mutations in two American children. J Child Neurol 2005; 20:619-620.
37. Quinzii CM, Kattah AG, Naini A et al. Coenzyme Q deficiency and cerebellar ataxia associated with an aprataxin mutation. Neurology 2005; 64:539-541.
38. Tranchant C, Fleury M, Moreira MC et al. Phenotypic variability of aprataxin gene mutations. Neurology 2003; 60:868-870.
39. Shimazaki H, Takiyama Y, Sakoe K et al. Early-onset ataxia with ocular motor apraxia and hypoalbuminemia: the aprataxin gene mutations. Neurology 2002; 59:590-595.
40. Ferrarini M, Squintani G, Cavallaro T et al. A novel mutation of aprataxin associated with ataxia ocular apraxia type 1: phenotypical and genotypical characterization. J Neurol Sci 2007; 260:219-224.
41. Mosesso P, Piane M, Palitti F et al. The novel human gene aprataxin is directly involved in DNA single-strand-break repair. Cell Mol Life Sci 2005; 62:485-491.
42. Tsuji S, Onodera O, Goto J et al. Sporadic ataxias in Japan—a population-based epidemiological study. Cerebellum 2008; 7:189-197.
43. Yokoseki A, Date H, Onodera O. Clinical features of early onset ataxia with ocular motor apraxia and hypoalbuminemia (EAOH). Neurol Med 2002; 57:108-112.
44. Criscuolo C, Mancini P, Menchise V et al. Very late onset in ataxia oculomotor apraxia type I. Ann Neurol 2005; 57:777.
45. Caldecott KW. DNA single-strand break repair and spinocerebellar ataxia. Cell 2003; 112:7-10.
46. Kuzmiak HA, Maquat LE. Applying nonsense-mediated mRNA decay research to the clinic: progress and challenges. Trends Mol Med 2006; 12:306-316.
47. Baba Y, Uitti RJ, Boylan KB et al. Aprataxin (APTX) gene mutations resembling multiple system atrophy. Parkinsonism Relat Disord 2007; 13:139-142.
48. Sano Y, Date H, Igarashi S et al. Aprataxin, the causative protein for EAOH is a nuclear protein with a potential role as a DNA repair protein. Ann Neurol 2004; 55:241-249.
49. Seidle HF, Bieganowski P, Brenner C. Disease-associated mutations inactivate AMP-lysine hydrolase activity of Aprataxin. J Biol Chem 2005; 280:20927-20931.
50. Hirano M, Nishiwaki T, Kariya S et al. Novel splice variants increase molecular diversity of aprataxin, the gene responsible for early-onset ataxia with ocular motor apraxia and hypoalbuminemia. Neurosci Lett 2004; 366:120-125.
51. Hirano M, Asai H, Kiriyama T et al. Short half-lives of ataxia-associated aprataxin proteins in neuronal cells. Neurosci Lett 2007; 419:184-187.
52. Sekijima Y, Hashimoto T, Onodera O et al. Severe generalized dystonia as a presentation of a patient with aprataxin gene mutation. Mov Disord 2003; 18:1198-1200.
53. Yoon G, Westmacott R, MacMillan L et al. Complete deletion of the aprataxin gene: ataxia with oculomotor apraxia type 1 with severe phenotype and cognitive deficit. J Neurol Neurosurg Psychiatry 2008; 79:234-236.
54. Onodera O. [DNA repair and neurodegeneration]. Rinsho Shinkeigaku 2005; 45:979-981.
55. Onodera O. Spinocerebellar ataxia with ocular motor apraxia and DNA repair. Neuropathology 2006; 26:361-367.
56. Pandolfo M, Koenig M. Friereich's Ataxia. London 1998;373-398.
57. Cogan DG. A type of congenital ocular motor apraxia presenting jerky head movements. Trans Am Acad Ophthalmol Otolaryngol 1952; 56:853-862.
58. Cogan DG. A type of congenital ocular motor apraxia presenting jerky head movements. Am J Ophthalmol 1953; 36:433-441.
59. Zee DS, Yee RD, Singer HS. Congenital ocular motor apraxia. Brain 1977; 100:581-599.

60. Sugawara M, Wada C, Okawa S et al. Purkinje cell loss in the cerebellar flocculus in patients with ataxia with ocular motor apraxia type 1/early-onset ataxia with ocular motor apraxia and hypoalbuminemia. Eur Neurol 2008; 59:18-23.
61. Makifuchi T, Fukuhara N. Pathology of early onset ataxia with ocular motor apraxia and hypoalbuminemia (EAOH). Neurol Med 2002; 57.
62. Shinmei Y, Yamanobe T, Fukushima J et al. Purkinje cells of the cerebellar dorsal vermis: simple-spike activity during pursuit and passive whole-body rotation. J Neurophysiol 2002; 87:1836-1849.
63. Fukushima K. Roles of the cerebellum in pursuit-vestibular interactions. Cerebellum 2003; 2:223-232.
64. Belton T, McCrea RA. Role of the cerebellar flocculus region in cancellation of the VOR during passive whole body rotation. J Neurophysiol 2000; 84:1599-1613.
65. Belton T, McCrea RA. Role of the cerebellar flocculus region in the coordination of eye and head movements during gaze pursuit. J Neurophysiol 2000; 84:1614-1626.
66. Noda H. Cerebellar control of saccadic eye movements: its neural mechanisms and pathways. Jpn J Physiol 1991; 41:351-368.
67. Nagata T, Aoki M, Hasegawa T et al. [An autopsy case of atypical Friedreich's ataxia with chronic idiopathic intestinal pseudo-obstruction]. Rinsho Shinkeigaku 2001; 41:412-417.
68. Chun HH, Gatti RA. Ataxia-telangiectasia, an evolving phenotype. DNA Repair (Amst) 2004; 3:1187-1196.
69. Taylor AM, Groom A, Byrd PJ. Ataxia-telangiectasia-like disorder (ATLD)-its clinical presentation and molecular basis. DNA Repair (Amst) 2004; 3:1219-1225.
70. Le Ber I, Bouslam N, Rivaud-Pechoux S et al. Frequency and phenotypic spectrum of ataxia with oculomotor apraxia 2: a clinical and genetic study in 18 patients. Brain 2004; 127:759-767.
71. Sedgwick RP, Boder E. Ataxia-Telangiectasia. 1991; 60:347-423.
72. Watanabe M, Sugai Y, Concannon P et al. Familial spinocerebellar ataxia with cerebellar atrophy, peripheral neuropathy and elevated level of serum creatine kinase, gamma-globulin and alpha-fetoprotein. Ann Neurol 1998; 44:265-269.
73. Takashima H. [Molecular genetics of inherited neuropathies]. Rinsho Shinkeigaku 2006; 46:1-18.
74. http://mouse.brain-map.org/welcome.do.
75. Brenner C. Hint, Fhit, Gal T. Function, structure, evolution and mechanism of three branches of the histidine triad superfamily of nucleotide hydrolases and transferases. Biochemistry 2002; 41:9003-9014.
76. Kijas AW, Harris JL, Harris JM et al. Aprataxin forms a discrete branch in the HIT (histidine triad) superfamily of proteins with both DNA/RNA binding and nucleotide hydrolase activities. J Biol Chem 2006; 281:13939-13948.
77. Rass U, Ahel I, West SC. Defective DNA repair and neurodegenerative disease. Cell 2007; 130:991-1004.
78. Gueven N, Becherel OJ, Kijas AW et al. Aprataxin, a novel protein that protects against genotoxic stress. Hum Mol Genet 2004; 13:1081-1093.
79. Date H, Igarashi S, Sano Y et al. The FHA domain of aprataxin interacts with the C-terminal region of XRCC1. Biochem Biophys Res Commun 2004; 325:1279-1285.
80. Clements PM, Breslin C, Deeks ED et al. The ataxia-oculomotor apraxia 1 gene product has a role distinct from ATM and interacts with the DNA strand break repair proteins XRCC1 and XRCC4. DNA Repair (Amst) 2004; 3:1493-1502.
81. Hirano M, Yamamoto A, Mori T et al. DNA single-strand break repair is impaired in aprataxin-related ataxia. Ann Neurol 2007; 61:162-174.
82. Rass U, Ahel I, West SC. Actions of aprataxin in multiple DNA repair pathways. J Biol Chem 2007; 282:9469-9474.
83. Rass U, Ahel I, West SC. Molecular mechanism of DNA deadenylation by the neurological disease protein aprataxin. J Biol Chem 2008; 283:33994-34001.
84. Reynolds JJ, El-Khamisy SF, Caldecott KW. Short-patch single-strand break repair in ataxia oculomotor apraxia-1. Biochem Soc Trans 2009; 37:577-581.
85. Zhou T, Lee JW, Tatavarthi H et al. Deficiency in 3′-phosphoglycolate processing in human cells with a hereditary mutation in tyrosyl-DNA phosphodiesterase (TDP1). Nucleic Acids Res 2005; 33:289-297.
86. El-Khamisy SF, Katyal S, Patel P et al. Synergistic decrease of DNA single-strand break repair rates in mouse neural cells lacking both Tdp1 and aprataxin. DNA Repair (Amst) 2009.
87. Becherel OJ, Gueven N, Birrell GW et al. Nucleolar localization of aprataxin is dependent on interaction with nucleolin and on active ribosomal DNA transcription. Hum Mol Genet 2006; 15:2239-2249.

Chapter 4

Clinical Features and Pathogenesis of Alzheimer's Disease:
Involvement of Mitochondria and Mitochondrial DNA

Michelangelo Mancuso,* Daniele Orsucci, Annalisa LoGerfo, Valeria Calsolaro and Gabriele Siciliano

Abstract

Alzheimer's disease (AD) is a neurodegenerative disorder which results in the irreversible loss of cortical neurons, particularly in the associative neocortex and hippocampus. AD is the most common form of dementia in the elderly people. Apart from the neuronal loss, the pathological hallmarks are extracellular senile plaques containing the peptide beta-amyloid (Aβ) and neurofibrillary tangles. The Aβ cascade hypothesis remains the main pathogenetic model, as suggested by familiar AD, mainly associated to mutation in amyloid precursor protein and presenilin genes. The remaining 95% of AD patients are mostly sporadic late-onset cases, with a complex aetiology due to interactions between environmental conditions and genetic features of the individual.

Mitochondria play a central role in the bioenergetics of the cell and apoptotic cell death. Morphological, biochemical and genetic abnormalities of the mitochondria in several AD tissues have been reported. Impaired mitochondrial respiration, particularly COX deficiency, has been observed in brain, platelets and fibroblasts of AD patients. Somatic mutations in mitochondrial DNA (mtDNA) could cause energy failure and increased oxidative stress. No causative mutations in the mtDNA have been detected and studies on mtDNA polymorphisms are controversial, but the "mitochondrial cascade hypothesis", here revised, could explain many of the biochemical, genetic and pathological features of sporadic AD.

Alzheimer's Disease

Alzheimer's disease (AD) is the most common cause of dementia in the elderly people, accounting for 65-70% of all cases.[1] It affects approximately 7% of population older than 65 and up to 40% of people over the age of 80. It is a progressive neurodegenerative disorder resulting in the irreversible loss of cortical neurons, mainly in the associative neocortex and hippocampus. AD usually begins with episodic memory impairment and encompasses language, visuospatial and behavioural dysfunction.[2] Amnesia is the most common symptom of AD and typically the earliest to be noticed, but the disease is soon accompanied by a range of cognitive disruptions such as in linguistic and visuospatial abilities and in overall executive function.[3] Sporadic AD is a slowly progressive disorder with insidious onset and a mean duration of 8-12 years. AD is usually classified according to the age of onset. When the disease starts before 65 years of age it is defined as early-onset AD form. Late-onset AD occurs in the subjects over 65 years of age and accounts

*Corresponding Author: Michelangelo Mancuso—Department of Neuroscience, Neurological Clinic, University of Pisa, Via Roma 67, 56126, Italy. Email: mmancuso@inwind.it

Diseases of DNA Repair, edited by Shamim I. Ahmad. ©2010 Landes Bioscience and Springer Science+Business Media.

for 90-95% of AD cases. The former type is familial, inherited as an autosomal dominant trait, whereas the latter is more frequently sporadic. Only a minority of late-onset cases show a clear family history with autosomal dominant inheritance. Mild cognitive impairment (MCI) is the phase during which subjects have measurable cognitive deficits, but not sufficient to fulfil criteria for any specific dementing disease.[4] MCI could constitute a transitional stage between normal aging and AD. While some of these patients may remain stable, many of them convert to AD with an annual conversion rate to clinical dementia of 10-20%.[5]

The pathological hallmarks of AD include neuronal loss, extracellular senile plaques containing the peptide beta amyloid (Aβ) and neurofibrillary tangles, composed of a hyperphosphorylated form of the microtubular protein tau.[2] Recently, an overall reduction in the abundance of factors associated with haematopoiesis and inflammation has been observed in blood plasma of AD patients.[6] To date, the Aβ cascade hypothesis remains the main pathogenetic model of AD.[2] However, although this cascade is viable in familial AD cases with mutation in amyloid precursor protein (*APP*) and presenilins (*PSEN*) genes, its role in the majority of sporadic AD cases is still unclear. Sporadic AD has probably a complex aetiology due to environmental and genetic factors which taken alone are not sufficient to develop the disease. Presently the major risk factor in sporadic AD is recognized in the allele *ε4* of apolipoprotein E (*ApoE4*). However, only less than 50% of nonfamilial AD cases are carriers of the *ApoE4* allele. Therefore, other factors must be involved in the pathogenesis of the disease.

Age is a major risk factor for AD and small numbers of plaques and tangles form in most elderly individuals.[7] The cognitive changes in AD begin with memory impairment and subsequently spread to language and visuo-spatial deficits. Word finding difficulties and circumlocution are often present. In the middle stages of AD, patients are unable to work, are easily lost and confused and require daily supervision. In late stages, they develop akinetic mutism, with rigidity and incontinence. Myoclonus, hyperactive tendon reflexes and seizures may be also present. The death is usually caused by infection, pulmonary emboli, malnutrition or heart disease.

Diagnosis of AD requires a combination of psychological testing, imaging and exclusion of other neurological disorders, but the definitive diagnosis can be made only at autopsy. To date, no validated peripheral markers for the early diagnosis of AD are available.[4] In disease stages with evident cognitive disturbances, the clinical diagnosis of probable AD is made with around 90% accuracy using modern clinical, neuropsychological and imaging methods. Diagnostic sensitivity and specificity even in early disease stages are improved by cerebrospinal fluid markers, in particular combined tau and Aβ and plasma markers (e.g., Aβ-42/Aβ-40 ratio).[4]

Other causes of dementia include Lewy body dementia, Parkinson's disease, frontotemporal dementia with parkinsonism, alcoholism, drug intoxications, infections such as AIDS and syphilis, brain tumours, vitamin deficiencies, thyroid disease and others. It is extremely important to exclude other causes of dementia from AD, because certain dementias respond to specific treatments. CT or MRI scans are performed on most patients with dementia, in order to identify potentially treatable conditions or patterns consistent with AD. As AD progresses, diffuse cortical atrophy becomes apparent and MRI scans show atrophy of the hippocampus. Nuclear medicine techniques such as PET and SPECT may show decreased metabolism and decreased regional blood flow in the parietal and temporal lobes and also in other cortical areas at later stages. In AD patients, fluorodeoxyglucose (FDG)-PET measurements show regional reduction of the cerebral metabolic rate for glucose, which is related to the clinical disabilities of AD patients. Reduced cerebral metabolism has been observed in the temporo-parietal cortices and it preceded both the neuropsychological impairment and the cortical atrophy detected with conventional neuroimaging.[8] Increasing evidence suggests that reduction of the cerebral metabolic rate for glucose occur at the preclinical stages of AD. It has been observed on FDG-PET before the onset of disease in several groups of at-risk individuals, including patients with MCI and presymptomatic individuals carrying mutations responsible for early-onset familial AD.[9] The causes of the early metabolic dysfunction forerunning the onset of AD are not known, but mitochondrial dysfunction could have a role.

Treatment options for AD are still disappointing. Therapies may focus on associated symptoms, such as depression, agitation and hallucinations. Currently approved therapies (the cholinesterase inhibitors and the N-methyl-D-aspartate receptor antagonist, memantine) offer only modest symptomatic relief, which can be enhanced using combination therapy with both classes of drugs.[10] Alternative therapies such as nonsteroidal anti-inflammatory drugs, vitamin E, selegiline, Ginkgo biloba extracts, estrogens and statins, as well as behavioural and lifestyle changes, have been explored as therapeutic options, with uncertain results.[10] The long presymptomatic phase of AD augurs well for the development of preventive strategies. To test their effectiveness, it will be necessary to identify neuropathological abnormalities before the development of cognitive changes.[7]

As already discussed above, the causes of the much more frequently occurring sporadic AD remain unknown: patients with sporadic AD generally lack mutations of *APP* gene, presenilin 1 and presenilin 2; therefore, it is unclear what initiates plaque formation in such cases. Furthermore, plaques are a relatively common finding in the nondemented elderly. Several studies suggest that abnormalities in oxidative metabolism and specifically in mitochondria may play an important role in late-onset neurodegenerative disorders including AD.[11,12] We will now focus on the role of the oxidative stress, of the mitochondria and its metabolism in AD.

Mitochondrial Structure and Function

1.5 billions years ago primitive aerobic bacteria established a symbiotic relationship with a single-cell anaerobic organism.[13] Mitochondria, the descendants of the original symbionts, are organelles sized 1–10 μm. The word "mitochondrion" comes from the Greek μίτος (mitos), thread + χονδρίον (khondrion), granule. Because of their double-membraned organization, there are five distinct compartments within the mitochondrion: the outer mitochondrial membrane, the intermembrane space, the inner mitochondrial membrane, the cristae space (formed by infoldings of the inner membrane) and the matrix. In the matrix there are 2-10 copies of the mitochondrial DNA (mtDNA).[14] The human mtDNA is a circular molecule of about 16 kilobases. Out of 37 genes, 13 are for subunits of respiratory complexes I, III, IV and V, 22 for mitochondrial transfer RNAs (tRNA) and 2 for ribosomal RNAs (rRNA).[15,16] Mitochondria are highly dynamic organelles that continuously divide and fuse, thus regulating their size, shape and distribution.[17] In humans, mitochondria replicate mainly in response to the energy needs of the cell, rather than in phase with the cell cycle.

An important role for the mitochondria is the production of adenosine triphosphate (ATP) by oxidizing glucose, pyruvate and NADH.[14] This process ("aerobic respiration") needs the presence of oxygen (O_2). When O_2 is limited, the glycolytic products are metabolized by anaerobic respiration. The production of ATP from glucose has an approximately 13-fold higher yield during aerobic respiration compared to anaerobic metabolism. Each pyruvate molecule produced by glycolysis is actively transported across the inner mitochondrial membrane and into the matrix where it is oxidized and combined with coenzyme A (CoA) to form CO_2, acetyl-CoA and NADH. The main pathway for oxidation of glucose in brain is the tricarboxylic acid (TCA) cycle (the "Krebs' cycle"), which takes place in the mitochondria. The oxidative decarboxylation of pyruvate, the product of glycolysis, by the pyruvate dehydrogenase complex (PDHC), provides acetyl CoA to initiate the TCA cycle, which includes eight different enzymes. The acetyl-CoA is the primary substrate to enter the TCA cycle. The enzymes of the TCA cycle are located in the mitochondrial matrix, with the exception of succinate dehydrogenase, which is bound to the inner mitochondrial membrane as part of Complex II. The TCA cycle oxidizes the acetyl-CoA to CO_2 and, in the process, produces reduced cofactors (three molecules of NADH and one molecule of $FADH_2$) which are a source of electrons for the electron transport chain (ETC) and a molecule of GTP (which is readily converted to ATP). The redox energy from NADH and $FADH_2$ is transferred to O_2 in several steps via the ETC. Protein complexes in the inner membrane (NADH dehydrogenase, cytochrome *c* reductase and cytochrome *c* oxidase or COX) use the release of energy from high-energy electrons to pump protons (H^+) into the intermembrane space.[15] As the proton concentration increases in the intermembrane space, a strong electrochemical gradient is established across the inner membrane. The protons return to the matrix through the ATP synthase complex and their potential energy is used to synthesize ATP from ADP and inorganic phosphate.[15] This process, called chemiosmosis, was described by Peter Mitchell, winner of 1978 Nobel Prize in Chemistry.[18]

Deficiency in the two key enzymes of rate limiting step of the TCA cycle, PDHC and α-ketoglutarate dehydrogenase complex (KGDHC), can explain the observed defects in glucose metabolism in the AD brains.[19,20] Other alterations observed in AD brains include decreased activity of isocitrate dehydrogenase and increased activity of succinate dehydrogenase and malate dehydrogenase.[20] All the changes observed in TCA cycle activities correlated with clinical state, suggesting that they may lead to diminished brain metabolism resulting in the decline in brain function.[20]

A recent transcriptomic study compared the expression of 80 metabolically relevant nuclear genes in brains of AD cases and normal controls.[21] AD cases had significantly lower expression of the nuclear genes encoding subunits of the mitochondrial ETC in posterior cingulate cortex, in the middle temporal gyrus, in hippocampal CA1 cells, in entorhinal cortex and in other brain areas.[21] Western blots confirmed underexpression of the ETC subunits.[21] Thus, metabolic abnormalities in PET studies of AD are associated with reduced neuronal expression of nuclear genes encoding subunits of the mitochondrial ETC.

In aerobic organisms the ATP needed for biological functions is mainly produced in the mitochondria via the ETC. During this process, a small percentage of electrons may prematurely reduce O_2, forming reactive oxygen species (ROS) (Fig. 1). ROS include a variety of free radicals such as superoxide anion ($O_2^{-\cdot}$) and hydroxyl radical ($^{\cdot}OH$) and also derivatives of oxygen that do not contain unpaired electrons, such as hydrogen peroxide (H_2O_2).[12] Because of the presence of unpaired valence shell electrons, ROS are highly reactive.[22] Oxidative stress is caused by an imbalance between the production of ROS and a biological system's ability to readily detoxify the reactive intermediates or easily repair the resulting damage (see Fig. 1). All forms of life maintain a reducing environment within their cells. This reducing environment is preserved by enzymes that maintain the reduced state through a constant input of metabolic energy. Disturbances in this normal redox state can cause toxic effects through the production of peroxides and free radicals that damage all components of the cell, including proteins, lipids and DNA.[23] The central nervous system is especially vulnerable to ROS because of its high O_2 consumption rate, its abundant lipid content and the relative paucity of antioxidant enzymes compared with other tissues.

Oxidative Stress and Mitochondrial Dysfunction in AD

The free radical hypothesis of aging, which was proposed many years ago, posits that the age-related accumulation of ROS results in damage to major components of cells, including nDNA and mtDNA, membranes and cytoplasmic proteins. The generation of free radicals may be involved in the pathogenesis of neurodegenerative disorders, including AD.[12] The fact that age is a key risk factor in AD provides support for the free radical hypothesis because effects of the attacks by ROS can accumulate over the years. Moreover, the hypothesis of oxidative stress in AD is strengthened by: (i) increased brain iron, aluminium and mercury in AD brains, capable of stimulating ROS generation; (ii) increased lipid peroxidation and increased protein and DNA oxidation in AD brains; (iii) diminished energy metabolism and decreased COX activity in tissues from AD patients.[12]

Many studies have reported a direct toxic effect of Aβ on cultures of neurons. If this reflects the situation in vivo, the phenomenon may offer an explanation for the role of amyloidogenesis in the pathogenesis of AD. Behl et al[24] showed that Aβ toxicity on cell lines is prevented by antioxidants, such as vitamin E and that H_2O_2 mediates this toxicity. Aβ toxicity–resistant cell lines contain high concentrations of the antioxidant enzymes catalase and glutathione peroxidase.[25] Conversely, oxidative stress may transform non aggregated Aβ into the aggregated form.[26]

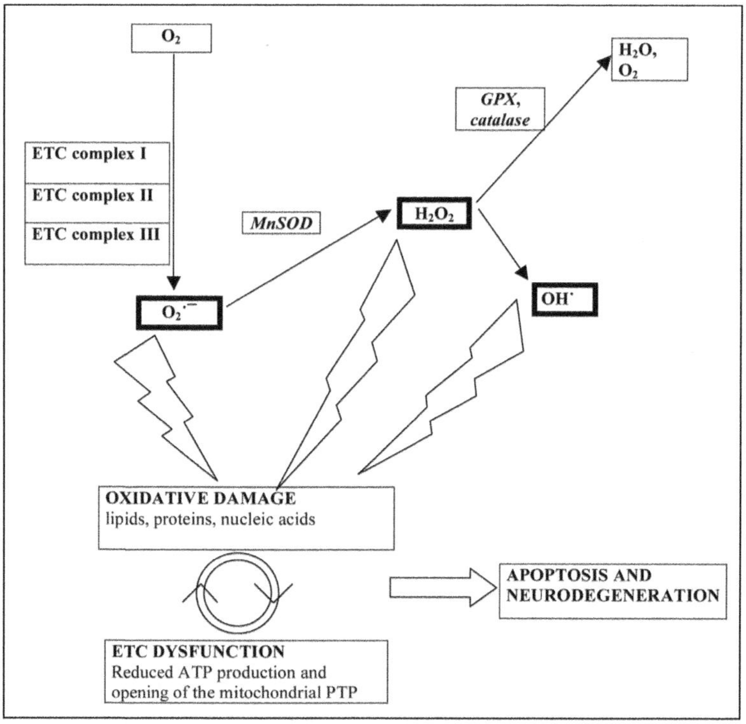

Figure 1. The vicious circle mitochondrial dysfunction—oxidative stress, leading to bioenergetic dysfunction, oxidative damage and apoptosis. In case of electron transport chain (ETC) dysfunction, complexes I-III leak electrons to oxygen (O_2) producing primarily superoxide anion ($O_2^{\cdot-}$). $O_2^{\cdot-}$ can be converted to hydrogen peroxide (H_2O_2) by manganese superoxide dismutase (MnSOD). Increased concentrations of $O_2^{\cdot-}$ can reduce transition metals (i.e., Fe^{2+}), which in turn react with H_2O_2 producing hydroxyl radicals ($^{\cdot}OH$). $O_2^{\cdot-}$ may also react with nitric oxide to form peroxynitrite, which is severely damaging for DNA and RNA. Both $OH^{\cdot}$ and peroxynitrite are strong oxidants and react with nucleic acids, lipids and proteins, finally leading to cell dysfunction and neurodegeneration. The mitochondrial DNA is particularly sensitive to oxidative damage. Because several its genes encode for subunits of the ETC, mitochondrial DNA damage results in mutations and deletions disrupting the function of genes involved in the production of ATP, ultimately leading to mitochondrial dysfunction, increased production of oxidants and cell death. For further details, see text. In Italics antioxidant enzymes, in thick boxes reactive species. GPX, glutathione peroxidase; H_2O, water; PTP, permeability transition pore.

Aβ could cause cellular dysfunction through its action on the mitochondrion. In transgenic mouse model for AD it has been observed that APP, the source of toxic Aβ, is targeted to neuronal mitochondria.[27] Before plaques are observed, intracellular aggregates of Aβ form early in mice overexpressing APP and strongly correlate with the cognitive impairment.[28] Chemical cross-linking together with immunoelectron microscopy showed that the mitochondrial APP exists in NH_2-terminal inside transmembrane orientation and in contact with mitochondrial translocase proteins.[27] Accumulation of full-length APP in the mitochondrial compartment in a transmembrane-arrested form caused mitochondrial dysfunction and impaired energy metabolism.[27]

Several recent studies on MCI patients suggest that oxidative damage could be one of the earliest events in the neurodegenerative process leading to AD.[29-31] Our group performed a study by a comet assay analysis to evaluate the level of primary and oxidative DNA damage in leukocytes of AD, MCI and healthy control individuals. We observed that the amounts of both primary DNA damage and oxidized bases were significantly higher in AD and MCI patients, compared to controls.[31] This gives a further indication that oxidative damage, at least at the DNA level, is an earlier event in AD pathogenesis. It is still unclear whether the increased oxidative damage in MCI and AD represents an acceleration of the "normal" age-related oxidative stress or is caused by alternative pathways.[12]

Increased levels of lipid peroxidation markers, as well as protein, DNA and RNA oxidation markers, have been reported in AD brains, as well as higher nitrotyrosination levels.[32,33] A recent study showed that mitochondrial DNA (mtDNA) had approximately 10-fold higher levels of oxidized bases than nuclear DNA (nDNA), that guanine was the most vulnerable base to DNA damage and that multiple oxidized bases were significantly higher in AD brain specimens in comparison to controls.[34] The use of proteomics identified oxidatively modified proteins, whose altered function is consistent with the disease.[35]

Because RNA is mostly single-stranded and probably less protected than DNA, it may be more susceptible to oxidative insults. Immunocytochemical studies revealed that the regional distribution of RNA oxidation in the brain was consistent with the selective neuronal vulnerability.[36] There were increased levels of 8-hydroxyguanosine (8-OHG) in the hippocampus and cerebral neocortex in AD, whereas no alteration in the 8-OHG level was found in the cerebellum.[36] Oxidized RNAs were localized predominantly in neuronal and endothelial cells compared with glial cells.[37] The early involvement of RNA oxidation in disease pathophysiology suggests that oxidized RNA in nervous and endothelial cells may have a key role in neurodegeneration. Moreover, in AD brains the cytoplasm of hippocampal neurons showed significantly higher redox activity and iron(II) staining than age-matched controls.[38] Both were susceptible to RNase.[38] This suggests a physical association of iron(II) with RNA. rRNA probably provides the greatest number of iron binding sites among cytoplasmic RNA species. Reflecting the difference of iron binding capacity, oxidation of rRNA by the Fenton reaction formed 13 times more 8-OHG than tRNA.[38] Thus, iron bound to RNA (and especially rRNA) seems to play a pivotal role for RNA oxidation in vulnerable neurons in AD brain.[38]

Aliyev et al[39] observed that ultrastructural features of vascular lesions and mitochondria in vascular wall cells from human AD brain biopsies were also suggestive of oxidative damage. As expected, there was a higher degree of Aβ deposition in the vascular walls in AD compared to aged-matched controls.[39] Moreover, increased levels of 8-OHG and significantly more mitochondrial abnormalities (including mtDNA deletions) were seen in the vascular endothelium and in the perivascular cells of microvessels where atherosclerotic lesions occurred.[39] These features were absent in undamaged regions of human AD tissues or in age-matched control subjects.[39] Therefore, mitochondria may be a central target for oxidative damage before the development of AD pathology.

Morphological alterations in neuronal mitochondria in AD have been reported. In 22 AD brains, the majority of the mitochondria in cerebellar and cerebral cortex and in basal nuclei presented disruption of the cristae or osmiophilic inclusions.[40] Further, ultrastructural examination showed that AD cybrid cells contained a significantly increased percentage of enlarged or swollen mitochondria that had a pale matrix and few remaining cristae.[41] The mean velocity of mitochondrial and lysosomal movement, as well as the percentage of moving mitochondria, was significantly reduced in AD cybrids.[41] Thus, delivery of mitochondria to the synapses might be impaired in AD. Other pathological features such as crystal-like intra-mitochondrial inclusions and cytoplasmic inclusion bodies were also found.[41]

Mitochondrial dysfunction seems to be an early event in AD animal models, which accelerates substantially with aging.[42] The most consistent defect in mitochondrial enzymes activity reported in AD regards the complex IV of the ETC, or COX.[43] Defects in ETC inhibit production of ATP and increase production of ROS. Deficient COX activity has been reported in different brain regions,[44-46] as well as in fibroblasts and platelets of AD patients.[45,47] Our group reported that COX, but not F1F0-ATPase (complex V) activity, was decreased in hippocampus and platelets of sporadic AD cases[45,48] and that the impaired COX activity could have functional consequence on energy metabolism.[48]

In a study on fresh AD and control brains, it was observed that nonglycosylated full-length and C-terminal truncated APP accumulated exclusively in the protein import channels of mitochondria of human AD brains, but not in age-matched controls.[49] In AD brains APP formed stable complexes with the translocase of the outer mitochondrial membrane-40 (TOM40) import channel and a super complex with both TOM40 and the translocase of the inner mitochondrial membrane-23 (TIM23).[49] Accumulation of APP across mitochondrial import channels inhibited the entry of nuclear-encoded COX subunits IV and Vb.[49] This could explain the observed decrease in COX activity and the increased levels of H_2O_2. Mitochondrial APP had higher levels in AD-vulnerable brain regions, such as the frontal cortex, hippocampus and amygdale.[49] The levels of translocationally arrested mitochondrial APP directly correlated with mitochondrial dysfunction and with the severity of the disease. Moreover, AD subjects with the *ApoE3/ApoE4* alleles had the highest content of mitochondrial APP.[49] Therefore, abnormal accumulation of APP across mitochondrial import channels causes mitochondrial oxidative stress, mitochondrial dysfunction and seems to be a hallmark of human AD.

Aβ binding alcohol dehydrogenase (ABAD), which is localized in the mitochondrial matrix, may be a direct molecular link from Aβ and mitochondrial toxicity.[50] Interaction between these two proteins promotes leakage of ROS, mitochondrial dysfunction, cytochrome *c* release, DNA fragmentation and apoptosis.[50] Further, γ-secretase, which is essential to cleave APP and create Aß, is also present in mitochondria.[51] In mitochondria Aβ can be removed by insulin degrading enzyme (IDE) and by presequence peptidase (PreP).[52,53] Complex IV inhibition by Aβ might also be mediated by sequestration of heme by Aβ.[54] This seems to promote also ROS production. At the same time, mitochondrial dysfunction and oxidative stress may alter APP processing leading to intracellular accumulation of Aβ.[55] Further, oxidative stress increases the activity of the β-secretase, the enzyme responsible for N-terminal cleavage of Aβ from the APP.[56]

To assess whether a genetically induced complex IV (COX) deficiency affects oxidative stress and accumulation of Aβ, Fukui et al[57] generated neuron-specific conditional COX10 knockout mice and then crossed them with a mouse model of AD. Thus they obtained a complex IV-deficient AD model (COXd/AD mice).[57] COX10 is an nDNA-encoded complex IV subunit (whereas the mtDNA encodes the subunits COXI, COXII, COXIII). Contrary to expectation, brains of COXd/AD mice exhibited less amyloid plaque, reduced Aβ deposition and reduced oxidative damage, as compared to AD mice.[57] Thus, a primary complex IV defect does not seem to contribute to oxidative stress and AD pathology.[58] In our opinion, this observation does not exclude a role for a secondary COX deficiency (i.e., caused by mtDNA oxidative damage) in the vicious circle leading to neurodegeneration.

Involvement of other mitochondrial ETC complexes is less documented and more controversial. For example mitochondria isolated from AD patients platelets have decrease ATP levels, but seem not to differ versus controls in NADH-ubiquinone oxidoreductase (complex I) or succinate dehydrogenase—cytochrome *c* reductase (complex II/III) activities.[59]

The Role of Mitochondrial DNA

In the cytoplasmic hybrid ("cybrid") technique, culturable cells depleted of endogenous mtDNA are repopulated with mitochondria (with their own mtDNA) from patients. Phenotypic differences among cybrid lines are caused by the donor mtDNA and not by nDNA or environmental factors. Cybrid models, obtained transfecting mtDNA from platelets of AD patients into neuronal cultured cells deprived of their own mtDNA, revealed decreased COX activity, increased ROS generation, increased Aβ production and morphological abnormalities (for a complete discussion, see our recent review).[12] In long-term culture AD cybrids develop populations of abnormal and damaged mitochondria, worsen the bioenergetic impairment and increase apoptosis and mtDNA replication.[60] The expression of AD mtDNA produces a more complex bioenergetic impairment beyond simple reduction in COX catalytic activity and may result in impaired ATP synthesis and mitochondrial calcium extrusion.[60]

Therefore, cybrid studies suggested a mitochondrial genomic contribution to mitochondria dysfunction in AD. Studies attempting to identify mtDNA mutations in brains of AD patients had limited success. Elson et al[61] sequenced the complete coding regions of 145 autoptic AD brain samples and 128 normal controls and observed that for both synonymous and nonsilent changes the overall numbers of nucleotide substitutions were the same for the AD and control sequences.

Different groups have analyzed the frequencies of polymorphisms and/or mutations in mtDNA in AD patients, with conflicting results. Polymorphisms in mtDNA may cause subtle differences in the encoded proteins and, subsequently, minimum changes in mitochondrial respiratory chain activity and free radical overproduction. This could predispose an individual, or a population sharing the same mtDNA genotype, to an earlier onset of apoptotic processes, such as accumulation of somatic mtDNA mutations and mitochondrial impairment. The opposite could be true for different polymorphism(s), which could be beneficial increasing mitochondrial respiration and/or reducing ROS production. Common mtDNA polymorphisms determine classes of continent-specific genotypes, called haplogroups. In Europe, nine Caucasian haplogroups (H,I,J,K,T,U,V,W,X) account for more than 90% of all mitochondrial genome of the entire population. Chagnon et al[62] reported that haplogroup T was under-represented in AD patients, while haplogroup J seemed to be over-represented. By studying an Italian sample of subjects, Carrieri et al[63] found that haplogroups K and U were present at a lower frequency in *ApoE4* carriers than in noncarriers AD patients (while in controls there were independence between *E4* allele and mtDNA haplogroups). Therefore, K and U may act by neutralizing the effect of the major AD risk factor *E4* allele. Another report showed that males classified as haplogroup U had a significant increase in risk of AD, while females demonstrated a significant decrease in risk with the same U haplogroup.[64] Therefore, the inheritance of haplogroup U may have negative effect on aging in Caucasian males.[64] Two studies including only neuropathologically proven cases of AD of European descent indicated that mtDNA haplogroups were not associated with AD.[61,65] Very recently, Maruszak et al[66] evaluated the involvement of mitochondrial haplogroups, haplogroup clusters (HV, UK, TJ, IWX) and of two functional mtDNA single nucleotide polymorphism in the pathogenesis of AD in Polish population. These authors observed that HV cluster seemed to be significantly associated with the risk of AD.[66]

In our laboratory we did not find any evidence for an etiological role of haplogroup-associated polymorphisms.[67] We studied the frequency of the European mtDNA haplogroups in a clinically well defined group of 209 unrelated patients and 191 controls, both with clear Tuscan origin (in order to minimize the risk of false associations between gene markers and disease). The frequency of haplogroups was not significantly different between the patients and control groups. Further, there was no significant difference between genders as far as mtDNA haplogroups distribution in both AD patients and control groups.[67] Therefore, it has been suggested that inherited haplogroups K and U may influence AD risk in Caucasians, but this is still an unresolved question. To date, mtDNA haplogroups do not seem to play any major role in AD and the question whether the mtDNA play a primary role in AD pathogenesis, or is damaged by somatic factors (such as oxidative damage) remains to be answered.[12]

Conclusion

The aetiology of AD is complex and only a minority of cases appears to be primarily genetic. Multiple interactions among genetic and environmental factors appear to be causative of the remaining forms. The "mitochondrial cascade hypothesis" could explain many of the biochemical features of sporadic AD. Accumulation of somatic mtDNA mutations accelerates normal aging, leads to oxidative damage, causes energy failure, increased production of ROS (see Fig. 1) and accumulation of Aβ, which in a vicious manner reinforces the mtDNA damage, the impairment of the mitochondrial respiration and the oxidative stress. These events open the permeability transition pore, inducing the release of cytochrome *c* and the induction of caspase-mediated apoptosis. Mitochondrial ROS generation seems to induce apoptosis also in a PARP-1 and AIF mediated manner. The presented data strengthen the idea that APP, Aβ-induced mitochondrial toxicity, oxidative stress, somatic mtDNA damage, mitochondrial dysfunction and apoptosis might be interconnected in the cascade leading to neurodegeneration and dementia. The APP "stocked" in the TOM transporters seems pivotal, able to cause mitochondrial impairment, respiratory deficiency and oxidative stress. Most likely, the mtDNA does not play a primary role and, therefore, it could be involved subsequently. Moreover, mtDNA deletions themselves seem to contribute to aging and pathology, but the exact mechanism of that is still unclear.

It will be important to develop a better understanding of the role of oxidative stress and mitochondrial energy metabolism and its link with the amyloid hypothesis in aging and AD, since it may lead to the development of more effective treatment strategies for this devastanting disorder.

References

1. Blenno K, de Leon MJ, Zetterberg H. Alzheimer's disease. Lancet 2006; 368:387-403.
2. Hardy J, Selkoe DJ. The amyloid hypothesis of Alzheimer's disease: progress and problems on the road to therapeutics. Science 2002; 297:353-356.
3. Riedl V, Honey CJ. Alzheimer's disease: a search for broken links. J Neurosci 2008; 28:8148-8149.
4. Jellinger KA, Janetzky B, Attems J et al. Biomarkers for early diagnosis of Alzheimer disease: 'ALZheimer ASsociated gene'—a new blood biomarker? J Cell Mol Med 2008; 2:1094-1117.
5. Gauthier S, Reisberg B, Zaudig M et al. Mild cognitive impairment. Lancet 2006; 367:1262-1270.
6. Ray S, Britschgi M, Herbert C et al. Classification and prediction of clinical Alzheimer's diagnosis based on plasma signaling proteins. Nat Med 2007; 13:1359-1362.
7. Goedert M, Spillantini MG. A Century of Alzheimer's Disease. Science 2006; 314:777-781.
8. Blass JP, Sheu RK, Gibson GE. Inherent abnormalities in energy metabolism in Alzheimer disease. Interaction with cerebrovascular compromise. Ann NY Acad Sci 2000; 903:204-221.
9. Mosconi L, Pupi A, De Leon MJ. Brain glucose hypometabolism and oxidative stress in preclinical Alzheimer's disease. Ann N Y Acad Sci 2008; 1147:180-195.
10. Farlow MR, Miller ML, Pejovic V. Treatment options in Alzheimer's disease: maximizing benefit, managing expectations. Dement Geriatr Cogn Disord 2008; 25:408-422.
11. Migliore L, Fontana I, Colognato R et al. Searching for the role and the most suitable biomarkers of oxidative stress in Alzheimer's disease and in other neurodegenerative diseases. Neurobiol Aging 2005; 26:587-595.
12. Mancuso M, Orsucci D, Siciliano G et al. Mitochondria, Mitochondrial DNA and Alzheimer's Disease. What Comes First? Curr Alzh Res 2008; 5:457-468.
13. Henze K, Martin W. Evolutionary biology: essence of mitochondria. Nature 2003; 426:127-128.
14. Chan DC. Mitochondria: Dynamic Organelles in Disease, Aging and Development. Cell 2006; 125:1241-1252.
15. Noji H, Yoshida M. The rotary machine of the cell ATP synthase, J Biol Chem 2001; 276:1665-1668.
16. DiMauro S, Schon EA. Mitochondrial respiratory-chain diseases N Engl J Med 2003; 348:2656-2668.
17. Lackner LL, Nunnari JM. The molecular mechanism and cellular functions of mitochondrial division. Biochim Biophys Acta, In press.
18. Mitchell P, Moyle J. Chemiosmotic hypothesis of oxidative phosphorylation. Nature 1967; 213:137-139.
19. Gibson GE, Haroutunian V, Zhang H et al. Mitochondrial damage in Alzheimer's disease varies with apolipoprotein E genotype. Ann Neurol 2000; 48:297-303.
20. Bubber P, Haroutunian V, Fisch G et al. Mitochondrial abnormalities in Alzheimer brain: mechanistic implications. Ann Neurol 2005; 57:695-703.
21. Liang WS, Reiman EM, Valla J et al. Alzheimer's disease is associated with reduced expression of energy metabolism genes in posterior cingulate neurons. Proc Natl Acad Sci USA 2008; 105:4441-4446.
22. Valko M, Leibfritz D, Moncol J et al. Free radicals and antioxidants in normal physiological functions and human disease. Int J Biochem Cell Biol 2006; 39:44-84.
23. Siciliano G, Piazza S, Carlesi C et al. Antioxidant capacity and protein oxidation in cerebrospinal fluid of amyotrophic lateral sclerosis. J Neurol 2007; 254:575-80.
24. Behl C, Davis JB, Lesley R et al. Hydrogen peroxide mediates amyloid b protein toxicity Cell 1994; 77:817-827.
25. Sagara Y, Dargusch R, Klier FG et al. Increased antioxidant enzyme activity in amyloid b protein-resistant cells. J Neurosci 1996; 16:497-505.
26. Dyrks T, Dyrks E, Harmann R et al. Amyloidogenicity of bA4 and bA4-bearing amyloid protein precursor fragments by metal-catalyzed oxidation. J Biol Chem 1992; 267:18210-18217.
27. Anandatheerthavarada HK, Biswas G, Robin MA et al. Mitochondrial targeting and a novel transmembrane arrest of Alzheimer's amyloid precursor protein impairs mitochondrial function in neuronal cells. J Cell Biol 2003; 161:41-54 .

28. Oddo S, Caccamo A, Shepherd JD et al. Triple-transgenic model of Alzheimer's disease with plaques and tangles: intracellular Abeta and synaptic dysfunction. Neuron 2003; 39:409-421.
29. Keller JN, Schmitt FA, Scheff SW et al. Evidence of increased oxidative damage in subjects with mild cognitive impairment. Neurology 2005; 64:1152-1156.
30. Wang J, Markesbery WR, Lovell MA. Increased oxidative damage in nuclear and mitochondrial DNA in mild cognitive impairment. J Neurochem 2006; 96: 825-832.
31. Migliore L, Fontana I, Trippi F et al. Oxidative DNA damage in peripheral leukocytes of mild cognitive impairment and AD patients. Neurobiol Aging 2005; 26:567-573.
32. Mecocci P, MacGarvey U, Beal MF. Oxidative damage to mitochondrial DNA is increased in Alzheimer's disease. Ann Neurol 1994; 36:747-751.
33. Guix FX, Uribesalgo I, Coma M et al. The physiology and pathophysiology of nitric oxide in the brain. Prog Neurobiol 2005; 76:126-152.
34. Wang J, Xiong S, Xie C et al. Increased oxidative damage in nuclear and mitochondrial DNA in Alzheimer's disease. J Neurochem 2005; 93:953-962.
35. Butterfield DA. Proteomics: a new approach to investigate oxidative stress in Alzheimer's disease brain. Brain Res 2004; 1000:1-7.
36. Nunomura A, Perry G, Pappolla MA et al. RNA oxidation is a prominent feature of vulnerable neurons in Alzheimer's disease. J Neurosci 1999; 19:1959-1964.
37. Nunomura A, Chiba S, Lippa CF et al. Neuronal RNA oxidation is a prominent feature of familial Alzheimer's disease. Neurobiol Dis 2004; 17:108-113.
38. Honda K, Smith MA, Zhu X et al. Ribosomal RNA in Alzheimer disease is oxidized by bound redox-active iron. J Biol Chem 2005; 280:20978-20986.
39. Aliyev A, Chen SG, Seyidova D et al. Mitochondria DNA deletions in atherosclerotic hypoperfused brain microvessels as a primary target for the development of Alzheimer's disease. J Neurol Sci 2005; 229-230:285-292 .
40. Baloyannis SJ. Mitochondrial alterations in Alzheimer's disease. J Alzheimers Dis 2006; 9:119-126.
41. Trimmer PA, Borland MK. Differentiated Alzheimer's disease transmitochondrial cybrid cell lines exhibit reduced organelle movement. Antioxid Redox Signal 2005; 7:1101-1109.
42. Hauptmann S, Scherping I, Dröse S et al. Mitochondrial dysfunction: an early event in Alzheimer pathology accumulates with age in AD transgenic mice. Neurobiol Aging, in press.
43. Castellani R, Hirai K, Aliev G et al. Role of mitochondrial dysfunction in Alzheimer's disease. J Neurosci Res 2002; 70:357-360.
44. Mutisya EM, Bowling AC, Beal MF. Cortical cytochrome oxidase activity is reduced in Alzheimer's disease. J Neurochem 1994; 63:2179-2184.
45. Bosetti F, Brizzi F, Barogi S et al. Cytochrome c oxidase and mitochondrial F1F0-ATPase (ATP synthase) activities in platelets and brain from patients with Alzheimer's disease. Neurobiol Aging 2002; 23:371-376.
46. Wong-Riley M, Antuono P, Ho KC et al. Cytochrome oxidase in Alzheimer's disease: biochemical, histochemical and immunohistochemical analyses of the visual and other systems. Vision Res 1997; 37:3593-3608.
47. Valla J, Schneider L, Niedzielko T et al. Impaired platelet mitochondrial activity in Alzheimer's disease and mild cognitive impairment. Mitochondrion 2006; 6:323-330.
48. Mancuso M, Filosto M, Bosetti F et al. Decreased platelet cytochrome c oxidase activity is accompanied by increased blood lactate concentration during exercise in patients with Alzheimer disease. Exp Neurol 2003; 182:421-426.
49. Devi L, Prabhu BM, Galati DF et al. Accumulation of amyloid precursor protein in the mitochondrial import channels of human Alzheimer's disease brain is associated with mitochondrial dysfunction. J Neurosci 2006; 26:9057-9068.
50. Lustbader JW, Cirilli M, Lin C et al. ABAD directly links Abeta to mitochondrial toxicity in Alzheimer's disease. Science 2004; 304:448-452.
51. Hansson CA, Frykman S, Farmery MR et al. Nicastrin, presenilin, APH-1 and PEN-2 form active gamma-secretase complexes in mitochondria. J Biol Chem 2004; 279:51654-51660.
52. Leissring MA, Farris W, Wu X et al. Alternative translation initiation generates a novel isoform of insulin-degrading enzyme targeted to mitochondria. Biochem J 2004; 383:439-446.
53. Falkevall A, Alikhani N, Bhushan S et al. Degradation of the amyloid beta-protein by the novel mitochondrial peptidasome, PreP. J Biol Chem 2006; 281:29096-29104.
54. Atamna H, Frey WH II. A role for heme in Alzheimer's disease: heme binds amyloid b and has altered metabolism. Proc Natl Acad Sci USA 2004; 101:11153-11158.
55. Ohyagi Y, Yamada T, Nishioka K et al. Selective increase in cellular A beta 42 is related to apoptosis but not necrosis. Neuroreport 2000; 11:167-171.

56. Drake J, Link CD, Butterfield DA. Oxidative stress precedes fibrillar deposition of Alzheimer's disease amyloid betapeptide (1-42) in a transgenic Caenorhabditis elegans model. Neurobiol Aging 2003; 24:415-420.
57. Fukui H, Diaz F, Garcia S et al. Cytochrome c oxidase deficiency in neurons decreases both oxidative stress and amyloid formation in a mouse model of Alzheimer's disease. Proc Natl Acad Sci USA 2007; 104:14163-14168.
58. Fukui H, Moraes CT. The mitochondrial impairment, oxidative stress and neurodegeneration connection: reality or just an attractive hypothesis? Trends Neurosci, 2008; 31:251-256.
59. Cardoso SM, Proenca MT, Santos S et al. Cytochrome c oxidase is decreased in Alzheimer's disease platelets. Neurobiol Aging 2004; 25:105-110.
60. Trimmer PA, Keeney PM, Borland MK et al. Mitochondrial abnormalities in cybrid cell models of sporadic Alzheimer's disease worsen with passage in culture. Neurobiol Dis 2004; 15:29-39.
61. Elson JL, Herrnstadt C, Preston G et al. Does the mitochondrial genome play a role in the etiology of Alzheimer's disease? Hum Genet 2006; 119:241-254.
62. Chagnon P, Gee M, Filion M et al. Phylogenetic analysis of the mitochondrial genome indicates significant differences between patients with Alzheimer disease and controls in a French-Canadian founder population. Am J Med Genet 1999; 85:20-30.
63. Carrieri G, Bonafè M, De Luca M et al. Mitochondrial DNA haplogroups and ApoE4 allele are non-independent variables in sporadic Alzheimer's disease. Hum Genet 2001; 108:194-198.
64. van der Walt JM, Dementieva YA, Martin ER et al. Analysis of European mitochondrial haplogroups with Alzheimer disease risk. Neurosci Lett 2004; 365:28-32.
65. Chinnery PF, Taylor GA, Howell N et al. Mitochondrial DNA haplogroups and susceptibility to AD and dementia with Lewy bodies. Neurology 2000; 55:302-304.
66. Maruszak A, Canter JA, Styczyńska M et al. Mitochondrial haplogroup H and Alzheimer's disease-Is there a connection? Neurobiol Aging, In press.
67. Mancuso M, Nardini M, Micheli D et al. Lack of association between mtDNA haplogroups and Alzheimer's disease in Tuscany. Neurol Sci 2007; 28:142-147.

Chapter 5

Huntington's Disease

Emmanuel Roze,* Cecilia Bonnet, Sandrine Betuing and Jocelyne Caboche

Abstract

Huntington's disease (HD) is one of the most frequently found neurodegenerative disorders. Its main clinical manifestations are chorea, cognitive impairment and psychiatric disorders. It is an autosomal-dominant disorder with almost complete penetrance. The mutation responsible for HD, unstable expansion of a CAG repeat, is located in the 5′ terminal section of the gene (*IT15*) that encodes huntingtin protein (Htt). The pathophysiology of HD is not entirely clear. One intriguing characteristic of HD is the special vulnerability of the striatum tomutated Htt, despite similar expression of the mutated protein in other brain regions. Aggregation of mutated Htt, transcriptional dysregulation, altered energy metabolism, excitotoxicity, impaired axonal transport and altered synaptic transmission culminate in neuronal dysfunction and death. There is currently no way of preventing or slowing down the disease progression and death usually occurs at about 20 years after diagnosis.

History

In 1872 a 22-year-old American physician, named George Huntington, in the first of only two research papers he published, reported on a disease afflicting several generations of a family in East Hampton on Long Island. This first detailed description of HD included chorea, cognitive and psychiatric impairment and gradual deterioration finally leading to death.[1] Huntington also noted the autosomal-dominant hereditary nature of the disease. In 1952, Americo Negrette, a Venezuelan physician, observed a "dancing epidemic" in a small town on the banks of Lake Maracaibo (Venezuela) and suspected that HD was responsible for this epidemic.[2] Nancy Wexler, whose mother died of HD, subsequently investigated these families as part of a collaborative project. The gene responsible for HD was finally identified in 1993.[3,4]

Clinical Aspects

The prevalence of HD is usually 4-8 per 100,000 population but can reach up to 700 in some parts of the world.[5,6] HD is an unremitting, devastating and invariably fatal disease. Age at onset is generally between 35 and 50 years but can range from 2 to 85 years or even more. Onset occurs before 30 years of age in about 15% of the cases, in which case the disease is referred to as juvenile HD.

The natural course of HD is variable. Life expectancy after HD onset is about 15-20 years; it is however shorter in juvenile HD.[7,8] The most common HD-related causes of death are pneumonia (due to swallowing disorders), falls, suicide and violence.

Before symptom onset, subtle alterations of eye movements, gait kinetics and finger-tapping speed can be detected with specific methods.[9-11] Sensitive rating scales can also detect subclinical behavioral and affective disorders such as apathy, irritability, obsessive-compulsive disorders and depression.[12-15]

*Corresponding Author: Emmanuel Roze—UMRS 952, INSERM, UMR 7224-CNRS Université Pierre et Marie Curie-Paris-6, Paris, France. Email: emmanuel.roze@psl.aphp.fr

Diseases of DNA Repair, edited by Shamim I. Ahmad. ©2010 Landes Bioscience and Springer Science+Business Media.

The clinical features of HD can be divided into three groups: movement disorders, cognitive impairment and psychiatric disorders.

Movement Disorders

Chorea, a Greek word meaning "dance", is the most characteristic movement disorder in classical HD. Chorea is characterized by brief, involuntary, abnormal and unpredictable movements that can affect any and all parts of the body (the four extremities, the trunk and face and respiratory and pharyngolaryngeal muscles).[16] The movements are typically distal and resemble finger-drumming. They occur against a background of generalized hypotonia with typical pendular reflexes. Initially, patients tend to incorporate their involuntary movements into apparently purposeful activities. Balism, which can be considered as a severe proximal type of chorea, can also occur. The consequences of chorea tend to be more cosmetic and social than functional,[16] but these movement disorders can interfere with walking, sitting, eating and even sleep in severe forms. As the disease progresses, choreic movements tend to be replaced by akineto-rigid parkinsonism that can be associated with dystonic postures of the trunk.[17,18] Many patients with advanced-stage HD suffer from falls[19] and lose their mobility, sometimes precipitating their admission to a nursing home.[20] Other movement disorders, occasionally observed in the course of the disease, include tics[21,22] and myoclonia.[23-25] Contrary to patients with typical HD, those with the juvenile form or the Westphal variant do not have chorea. These forms are characterized by a combination of progressive akineto-rigid parkinsonism, dystonia, ataxia, dementia and psychiatric disorders.[26,27] Seizures can also occur. Patients with childhood-onset HD can develop non specific encephalopathy resulting in seizures, myoclonus and rapid cognitive deterioration.[26-28]

Cognitive Impairment

Cognitive impairment is a major factor in HD patients' functional decline and loss of autonomy. It can precede or follow the onset of motor symptoms and usually leads to dementia. However, some patients with late-onset HD never develop dementia.[29,30] Cognitive changes in this setting have a subcortical profile, with predominant impairment of executive/attentional functions.[31-34] Instrumental functions (language, praxia and gnosis) and memory are generally better-preserved in HD than in other types of dementia such as Alzheimer's disease, despite elective defects in the retrieval processes.[35-38] One classical manifestation of the dysexecutive syndrome is motor impersistence—the inability to maintain an initiated voluntary movement that account for classical clinical signs of the disease, namely milkmaid's grip and darting tongue.

Psychiatric Disorders

Neurobehavioral disorders are very frequent in HD. They can occur at any time during the disease course and may be the initial manifestation.[39,40] Irritability, agitation, apathy, anxiety, social withdrawal, impulsiveness, alcohol abuse, obsessive-compulsive disorder,[41,42] hostility and sexual disorders are all common.[43-45] Mood disorders are also very frequent during the course of HD. Most HD patients, with mood disorders, have classical symptoms of depression, namely feelings of sadness, guilt, uselessness and anhedonia.[46-50] Other psychiatric symptoms include maniac episodes[51] and poorly systematized paranoia, delusions and psychotic states resembling various types of schizophrenia.[49,52] Because of all these psychiatric disturbances, the risk of suicide is 10 times higher among HD patients than in the general population. Suicidal ideation predominates at two different phases of the disease: first during the interval between symptom onset and diagnosis; and second towards the midpoint of the disease course when the patient starts to become dependent on others for help with daily activities.[53-55]

Other Clinical Manifestations

Sleep disorders seen in HD patients include insomnia, advanced sleep phase, REM sleep abnormalities and increased motor activity during sleep. These sleep disorders can be present before HD diagnosis.[56-59] Another consistent feature of HD is weight loss that may result from a hypermetabolic state.[60] It is directly linked to the length of the CAG repeat and appears to correlate with the severity of movement disorders.[61]

Genetic Aspects

The mutation responsible for HD is situated in the 5' terminal part of the *IT15* gene,[4] located on chromosome arm 4p16.3.[3] It contains 67 exons and encodes huntingtin (Htt), a 348-kDa protein composed of between 3127 and 3156 amino acids. The mutation consists of an unstable expansion of the CAG repeat sequence located in exon 1, that encodes polyglutamines in the NH2-terminal part of the protein. The mutated protein causes neuronal dysfunction and death, particularly in the striatum and cortex, even though it is ubiquitously expressed.

Penetrance is virtually complete. The *IT15* gene is considered normal when it contains fewer than 27 CAG repeats. CAG repeats numbering between 27 and 35 do not cause HD but may expand in successive generations. The 36 → 39 repeats (intermediate alleles) are usually associated with late disease onset and may have variable penetrance; even some patients dying before the disease has a chance to develop. Individuals with 39 CAG repeats or more, invariably develop HD. A controversial case of possible HD in an individual with 29 CAG repeats has been reported.[62-64]

About 10% of patients have no family history of HD and some may have de novo mutations.[65,66] More often the mutant alleles are inherited from an asymptomatic father with an "intermediate allele". Such alleles do not cause HD but are unstable during replication, like "pathogenic alleles" and tend to expand in successive generations.[67,68] This instability is greater during spermatogenesis than during oogenesis. These intermediate alleles can have as few as 27 CAG repeats.[69] Instability increases with advancing paternal age. This increase in the number of CAG repeats, over successive generations, leads to a phenomenon of "genetic anticipation",[65,70,71] defined by a trend towards earlier disease onset in successive generations.

Age at onset cannot be accurately predicted from the CAG repeat length. The number of repeats correlates negatively with age of onset;[72-76] this correlation is mostly due to studies on the small fraction of HD patients who have very large number of CAG repeats and very early disease onset. The correlation between CAG repeat length and the disease course is controversial; in classical HD cases the mean number of repeats is similar in patients with psychiatric disorders and in patients with motor disorders.[77-84]

Mutant huntingtin gene is expressed to a similar extent in many CNS and peripheral tissues, although the length of the CAG repeat is particularly unstable in the brain.[85-88] The mutation length profiles in different regions of the brain correlate with the chronological order in which the brain regions are affected. In particular, marked expansion-based changes in mutation length occur during aging in the striatum and, to a lesser extent, in the cortex. Striatal cells with the largest HD mutation expansions may be preferentially lost during the disease process. Neurons tend to show longer mutation length gains than glial cells. Mutation length gains occur early in the disease and continue to accumulate as the disease progresses. This is consistent with the view that mutation length variability in somatic tissues may contribute to both the progressive and the cell-selective nature of HD.[89]

Individuals with identical repeat lengths may have major differences in HD manifestations, progression and outcome, possibly owing to modifier genes and environmental factors[72] that affect excitotoxicity, dopamine toxicity, metabolism, gene transcription, protein folding or oxidative stress. Suspected genetic modifiers include the *HD* gene itself,[90-92] and the genes encoding the GluR6 kainate glutamate receptor (*GRIK2*),[92-95] apolipoprotein E (*APOE*),[96,97] transcriptional coactivator CA150 (*TCERG1*),[98] the ubiquitin carboxy-terminal hydrolase L1 (*UCHL1*),[95] p53 (*TP53*),[99] caspase-activated DNase (*DFFB*),[99] subunits of the NR2A and NR2B glutamate receptors (*GRIN2A*, *GRIN2B*)[100,101] and M441-HAP1.[102] Sequence variations in the *GRIN2A* gene, which encodes the NR2A subunit of the NMDA-type glutamate receptor, seem to have the strongest association with age of HD onset.[103] Interestingly, this subunit is expressed in medium spiny striatal neurons, which seem to be particularly sensitive to glutamate excitotoxicity.

Neuropathology

Brain weight may be reduced by as much as 25-30% in advanced HD cases. Gross pathological changes associated with HD are limited to the brain, where atrophy predominates in the caudate-putamen and, to a lesser extent, the cerebral cortex. The neuropathological signature of HD is prominent striatal neuron loss and intranuclear inclusion bodies, which mainly consist of accumulated polyglutamines. Vonsattel's grading system of striatal neuropathology is based on macroscopic and microscopic criteria[104] and defines five grades (0 to 4) of increasing severity. The grade correlates closely with the degree of clinical disability.

The most vulnerable neuronal population consists of medium spiny neurons in the striatum, which account for 90% of striatal neurons and can be separated into striato-nigral and striato-pallidal populations. In Vonsattel's system the striato-pallidal cell population, which expresses enkephalin and dopaminergic D2 receptors, degenerate first (Grade 2), followed by striato-nigral neurons, which expresses substance P and dopaminergic D1 receptors (Grade 3). Degeneration of medium spiny striatal neurons occurs along a stereotyped dorso-ventral and medio-lateral gradient and is associated with reduced expression of substance P, leuenkephalin, calcineurin, calbinbidin, histamine H2 receptors, dopamine receptors, cannabinoid receptors and adenosine A2 receptors.[105-108] Striatal interneurons, aspiny striatal cholinergic and somatostine-containing neurons are relatively spared.[109-113] Another characteristic neuropathological change affects the dentritic arborisation of spiny neurons, dystrophic neuritis preceding cell death.[114,115] These degenerative changes are associated with a reduction in the size of the striatum.

The cerebral atrophy is otherwise diffuse, mainly affecting the prefrontal dorso-lateral cortex, the parahippocampal gyrus and subcortical white matter.[116,117] Cortical neuron degeneration occurs mainly in deep cortical layers III, V and VI.[118,119] Neuron loss can also be found in the lateral pallidum (GPe) and medial pallidum (GPi),[104,113] the substantia nigra pars compacta, substantia nigra pars reticulata,[120] thalamic mediodorsal nucleus,[121,122] subthalamic nucleus, hypothalamic lateral tuberal nucleus,[113,123] and cerebellum.[124,125]

Intracellular deposition of aggregated and ubiquitynated proteins is a prominent cytopathological feature. The aggregates are ubiquitinated, non soluble and present in all brain regions but particularly in the cortex; they occur almost exclusively in neurons and can be seen in all cell compartments, including the nucleus, cytoplasm, axons and synaptic terminals.[114,126,127] They mainly consist of expanded huntingtin (exp-Htt) containing polyglutamine repeats. They have a modified three-dimensional structure, mostly stabilized by hydrogen bonds. Alteration of the Htt structure is directly related to the size of the polyglutamine expansion and occurs beyond a threshold length of 36 glutamines.[128,129] These aggregates modify cell functions in several ways.

Molecular Mechanisms

It is unclear whether the degeneration of neurons and more specifically striatal neurons, in HD is due to a loss of normal huntingtin properties or to a gain of toxic functions due to the poly CAG repeat. In addition, "normal" age-related changes in cell function may accelerate the pathogenesis of HD.[130] Despite being identified more than a decade ago, the function of huntingtin is largely unknown. Wild-type Htt is expressed in most cells and in virtually all cell compartments.[114,126,127,131-135] Huntingtin also has multiple interacting partners, some of which bind more efficiently to the mutant protein than to wild-type huntingtin.[136,137] These binding partners include a dozen transcription factors, which appear to affect the transcriptional profile in HD brain tissues and cells.[138-144] Huntingtin is thought to be a scaffold protein that orchestrates intracellular trafficking, signaling pathways and transcriptional activity required for normal cellular functioning. Importantly, Htt is required for normal embryonic development, as Htt knock-out mouse embryos die about 8.5 days after conception[145,146] and selective Htt knockdown in neurons and testis produces apoptosis.[147] In its normal and mutated versions, Htt is cleaved by various intracellular proteases, including caspases. This proteolysis plays a key role in the pathophysiology of HD, as the cleaved N-terminal fragment is much more toxic than the full-length mutant protein. These cleaved versions of the mutant protein can also form aggregates,[128,129,148] which are thought to hinder normal cellular functioning.

Aggregates

Aggregates of insoluble proteins are found in the brain of HD patients, as well as in various HD models and in other trinucleotide repeat disorders. Aggregates appear early[149] and are present in all cellular compartments of neurons.[114,126] These aggregates affect normal cellular functioning by sequestering wild-type Htt[150] or other proteins involved in transcription[151,152,153] or transport.[154,155] Not only can these aggregates not be cleared by the ubiquitin-proteasome system,[156-159] they also alter normal protein clearance by this system.[160,161] The low basal activity of the ubiquitin-proteasome system in neurons, as compared with glial cells, may account for the preferential accumulation of aggregates in neurons.[162] The toxicity of these aggregates is highly controversial,[163-167] some authors arguing that smaller oligomeric aggregates may be more toxic than larger ones.[168] In addition, their toxicity may depend on the cellular compartment in which they are located; for example, they could be more toxic in neurites than in the nucleus.[169-171] Aggregate toxicity has been attributed to the induction of defects in RNA synthesis, cell survival activity, microtubule-dependent trafficking and the ubiquitin-proteasome system.[136,137,160,172] Inhibition of poly (Q) aggregation can alleviate disease manifestations in various models of HD.[173-181]

Transcriptional Dysregulation

Downregulation of protein expression in HD patients' brain tissue is compatible with transcriptional dysregulation observed early in the neuropathological process. Altered mRNA levels of dopaminergic receptors and neuropeptides observed in HD patients[182,183] have also been found in presymptomatic transgenic HD model mice, suggesting that changes in transcription are responsible for the observed neurodegeneration rather than simply reflecting non specific degradation of all RNAs in affected neurons.[184,185] Subsequently, multiple genes encoding neurotransmitter receptors, enzymes and proteins, involved in neuron structure, stress responses and axonal transport were found to be dysregulated,[138,141,142,186,187] with several overlaps between mouse models and HD patients.[144] Coordinated transcriptional dysregulation can be found in large genomic regions and is associated with disease progression. Attempts have been made to use transcriptional dysregulation as a biomarker in HD. Significant differences in gene transcription were found in the blood of symptomatic patients in one study,[188] but not in moderate-stage patients in another study.[187] Thus, these biomarkers need to be validated before being adopted for use in clinical trials.

Regarding the molecular mechanisms underlying the transcriptional dysregulation, nuclear mutant huntingtin, in its soluble and aggregated forms, has been shown to interact with and inhibit proteins involved in the normal transcriptional machinery; these include TATA binding protein (TBP), transcription factor II F (TFIIF) and tyrosine-aminotransferase II (TATII) 130.[189-192] Mutant huntingtin also sequesters transcription factors involved in cell viability, including cyclic AMP-response element binding (CREB) binding protein (CBP), p53, Sp1 (specificity protein 1), nuclear factor-kappa B (NF-kB), nuclear receptor corepressor (NCoR) and CA150.[152,153,189,190,193-195] Mutant huntingtin interferes with transcription of brain-derived neurotrophic factor (BDNF), which is deficient in HD brains and plays a key role in neuronal functions and survival.[196-199] One function of huntingtin is to sequester R element-1 silencing transcription factor (REST), a transcriptional repressor of neuron survival factors, including BDNF, within the cytoplasm. Huntingtin mutation leads to REST release within the nucleus, where it exerts a potent inhibitory effect on the transcription of *BDNF* and other neuronal genes.[197,200,201] Importantly, restoral of BDNF levels by a reduction in REST activity or by AMPA glutamate receptor modulation (with ampakine) prevents neurodegeneration in HD models.[199,202,203]

Chromatin remodeling, a mechanism underlying DNA decompaction, also a prime transcriptional event in postmitotic mature neurons was first described in dividing cells.[204] This "above the genome" (epigenetic) molecular mechanism, is crucial for DNA access and hence transcription. It is critically controlled by post-translational modifications of histones (H2A and H2B, H3 and

H4), a group of highly basic proteins that are tightly linked to DNA. Chromatin remodelling, particularly in the striatum, is likely to play a key role in the transcriptional deregulation observed in HD. In particular, histone methylation and acetylation status is closely linked to transcriptional activity in this setting, by regulating transcription factor access to DNA promoter regions. Mutant huntingtin interacts with CBP and blocks its histone acetyltransferase activity.[205] Administration of histone deacetylase (HDAC) inhibitors, including SAHA, sodium butyrate and phenylbutyrate, is beneficial in several HD models,[205-208] improving behavior and survival. Interestingly, HDACi 4b, a new benzamide-type HDAC inhibitor with less toxicity than other HDAC inhibitors, also restores the transcription of critical striatal genes and improves the motor and neuropathological phenotype of R6/2 HD mice.[209] All these inhibitors broadly act on various classes of HDAC and inhibitors targeting specific HDAC classes might have better risk-benefit ratios.[210] Finally, it must be emphasized that acetylated histones are not depleted globally in HD mouse models but rather selectively in the promoters of genes that are specifically down-regulated in this setting.[211]

Histone methylation has an inhibitory effect on transcription. One of the proteins involved in methyltransferase activity at histone H3 (K9) is ESET (ERG-associated protein with SET domain) whose expression is increased in HD patients and in transgenic R6/2 HD mice.[212] Sp1 acts as a transcriptional activator of the ESET promoter at guanosine-cytosine (GC)-rich DNA binding sites.[213] Inhibition of Sp1 binding to these sites by using mitramycin (a clinically approved antitumor drug) suppresses basal ESET promoter activity in a dose-dependent manner. Combined treatment with mithramycin and cystamine downregulates ESET gene expression and decreases hypertrimethylation of histone H3. This treatment significantly improves the behavioral and neuropathological phenotype of R6/2 HD mice and also improves their survival. In general, anthracyclins (DNA/RNA binding agents) are thought to act by correcting pathological nucleosome changes and by realigning transcription. Two such agents, chromomycin and mithramycin, were found to improve nucleosome homeostasis, normalizing the balance between methylation and acetylation in HD mice, regulating a subset of down-regulated genes and inducing significant behavioral and neuropathological improvements.[214]

Although less extensively studied, histone H3 phosphorylation and histone H2 ubiquitylation are also critical for the nucleosomal response and gene transcription at some promoter sites. Mitogen and stress-activated protein kinase 1 (MSK1), a nuclear protein kinase involved in histone H3 phosphorylation, is deficient specifically in the striatum of R6/2 mice and HD patients.[215] Restoration of MSK1 expression and subsequent striatal H3 phosphorylation in an in vitro HD model system protects against neuronal changes induced by exp-Htt including neuritic retraction, aggregate formation and neuronal death.[215] Histone H2A ubiquitination is increased in R6/2 HD mice and association between ubiquitinated H2A and promoters of downregulated genes is increased in an in vitro model of HD.[216] This transcriptional repression is rescued by restoring the normal level of ubiquitinated H2A. In addition, Histone H2B ubiquitination is decreased in R6/2 HD mice and association of ubiquitinated H2B, with promoters, correlates positively with transcriptional activity in R6/2 mice. This transcriptional modulation by H2 ubiquitination is thought to be due to downstream interference with histone H3 methylation.[216]

Excitotoxicity

Excitotoxicity has long been implicated in neuronal death. Excitotoxicity mainly involves stimulation of glutamatergic N-methyl-D-aspartic acid (NMDA) subtype receptors.[217-219] In rodents, NMDA receptor agonist administration mimics certain clinical and pathological features of HD.[220-222] Several studies have provided insights into the excitotoxicity in HD by showing that neuronal death can result from a combination of increased glutamate or glutamate agonist release from cortical afferents,[223,224] reduced glutamate uptake by glial cells,[225-227] NMDA receptor hypersensitivity,[228,229] and altered intracellular calcium homeostasis associated with mitochondrial dysfunction.[133,230-232] In particular, modified interaction between Exp-Htt and postsynaptic density 95 (PSD-95), a scaffold protein necessary for NMDA receptor stability and activation, results in decreased cytoplasmic retention of PSD95 and in hypersensitivity of NMDA receptors expressed at the membrane.[233,234] Altered NMDA receptor trafficking could also be involved in the potentiation of the NMDA receptor-induced current and toxicity, as accelerated NMDA receptor trafficking, with increased cell-surface expression, has been found in striatal neurons of YAC transgenic mice.[235] Htt-interacting protein 1 (HIP1), which normally interacts with Htt, is involved in the intracellular trafficking of AMPA glutamate receptors. Exp-Htt, which interacts less efficiently than its normal counterpart with HIP1, participates in the increased membrane expression of AMPA receptors and, hence, in excitotoxic neuronal death in HD.[236] Altered calcium homeostasis in HD could be due to abnormal interaction between Exp-Htt and the Type 1 inositol 1,4,5-trisphosphate receptor (InsP3R1), which regulates cytoplasmic calcium clearance by the endoplasmic reticulum.[237,238] Disruption of the pathogenic interaction between InsP3R1 and Exp-Htt normalizes calcium signaling, protects striatal neurons from glutamate-induced apoptosis in vitro and reduces neuronal and motor disorders in a HD mouse model.[239] Although studies of animal models suggest that a reduction in glutamatergic tone might be an interesting approach, clinical trials of glutamate inhibitors (remacemide, amantadine, lamotrigine, riluzole and memantine) have given disappointing results,[240,241] despite possible symptomatic effects on movement disorders.[242,243] In addition to glutamatergic stimulation, there is evidence that dopamine (DA) stimulation may play a key role in the excitotoxicity associated with HD.[224,244-249] DA released from nigrostriatal inputs is present in high concentrations within the striatum and enhances sensitivity to glutamatergic inputs. DA may also trigger the production of reactive oxygen species (ROS) and thus augment oxidative stress, which normally increases with aging.[250] In particular, cultured striatal neurons from R6/2 HD brain are sensitized to DA-induced oxidative stress, leading to autophagy.[251] ROS produced by low concentrations of DA potentiate Exp-Htt-induced activation of the pro-apoptotic cJun-N-terminal kinase (JNK) pathway.[247,252] Conversely, pharmacological inhibition of the JNK pathway is neuroprotective in R6/2 HD mice.[253] Dopamine may also make striatal neurons more vulnerable to Exp-Htt via dopamine receptor-mediated mechanisms.[247,249] Depending on the HD cell model (expressing either full-length or truncated mutant Htt), D1 or D2 receptor stimulation seems to be critical for HD induced toxicity. Full-length exp-Htt is required to induce alteration of calcium signaling.[254] In striatal neurons from YAC128 transgenic and Q111 knock-in mice, both of which express full-length Htt with expanded polyglutamine repeats, D1 receptor stimulation potentiates calcium influx via NMDA receptors and hence excitotoxic processes, including mitochondrial depolarization and caspase activation.[228,229,248,255,256] Calcium-dependent calpain activation was recently shown to be involved in the degenerescence of striatal neuron after activation of NMDA and D1 receptors in HD cell models.[257] Increase in calcium influx leads to cleavage of the Cdk5 co-activator p35 into P25, enabling aberrant toxic activation of protein kinase Cdk5.[257] By contrast, when associated with p35 as co-activator, Cdk5 has a protective effect via Exp-Htt phosphorylation and blockade of caspase-induced cleavage, resulting in less aggregate formation and toxicity.[258,259] Cdk5/p35 suppresses aggregate formation induced by a short fragment of Exp-Htt (exon 1), pointing to an unexpected role of Cdk5/p35 in microtubule stability and, hence, in inclusion formation.[260] Together, these data highlight the complexity of Cdk5 activity and the need to target p25 selectively in order to block the formation of Exp-Htt inclusions and neuronal death. By contrast, D1 receptor stimulation has no effect on the inherent toxicity of Exp-Htt exon 1 overexpression in cell models, whereas D2 receptor stimulation seems to be critical.[247] Of interest, D2 receptor agonist treatment, without concomitant glutamate treatment, potentiates striatal death induced by Exp-Htt in this cell model, as well as aggregate formation. In a HD rat model, based on lentivirus-mediated expression of Exp-Htt exon 1 in the striatum,[261] early long-term treatment with the D2 antagonist haloperidol decanoate protects striatal neurons from expHtt-induced dysfunction and also from aggregate formation.[249] Thus, cleaved Exp-Htt is more sensitive to D2 receptor stimulation than to D1 receptor stimulation. One possible explanation is that the cleaved version is not sensitive to calcium overload (or probably to calcium-dependent proteolytic processes) produced by D1 receptor stimulation. Together, these data indicate that anti-exicitotoxic therapy in HD should use a combination of anti-glutamatergic and anti-dopaminergic agents.

Changes in Axonal Transport and Synaptic Function

Neuronal vesicular transport is altered in HD, leading to synaptic dysfunction.[134,154,155,262-264] Trafficking defects are mainly due to impaired interaction between huntingtin and the motor proteins. In addition, neuritic aggregates can act as physical roadblocks[154,155,264] and sequester proteins that govern axonal transport.[171,264] Normal Htt contributes to the control of intracellular dynamics, such as secretion and trafficking from the Golgi apparatus to vesicles, vesicular transfer to microtubules, axonal transport along microtubules and synaptic endo/exocytosis.[265,266] Normal Htt protein positively regulates vesicular transport by interacting with the molecular motor complex that transports organelles along microtubules within axons.[134,262] Post-translational modification of huntingtin and specifically phosphorylation of a residue (Ser421), critical for the kinase Akt, is crucial for vesicle trafficking in neurons. When phosphorylated, normal huntingtin recruits kinesin-1 to dynactin complexes on vesicles and microtubules, thereby participating in anterograde transport. When Htt is not phosphorylated, kinesin-1 detaches and vesicles are more likely to undergo retrograde transport.[267] In HD, the polyglutamine expansion modifies the interaction of Htt with components of the motor complex.[262] This is thought to contribute to the altered axonal transport of BDNF from cortical neurons to their terminals within the striatum, hence decreasing the survival of striatal neurons. Interestingly, increasing tubulin acetylation with an HDAC6 inhibitor restores the recruitment of motor proteins to microtubules, including kinesin-1 and dynein and enhances axonal BDNF transport in cortical neurons.[268] Another way of compensating for defective BDNF transport is to phosphorylate the Ser 421 of exp-Htt, thus restoring the interaction between exp-Htt and dynactin and their association with microtubules. Huntingtin is also involved in the trafficking and secretion of vesicles from the Golgi apparatus.[135] In particular, acting in concert with transglutaminase 2 (TGase 2) and HJSJ1b, it promotes the budding of BDNF-containing vesicles from the Golgi into the cytoplasm.[269,270] Modulation of TGase 2 and HJSJ1b expression by cystamine or cysteamine increases BDNF release in HD mouse and monkey models.[269] Htt-interacting protein 14 (HIP14) normally interacts with Htt to regulate the trafficking of neuronal proteins and their synaptic release, through palmitoylation of cysteine string protein (CSP).[271,272] Increased synaptic transmission could also participate in the early stages of HD through defective Ca^{2+} homeostasis.[273]

Mitochondrial Dysfunction and Altered Energy Metabolism

Defective energy metabolism in brain and muscle has long been suspected in HD, based on clinical,[274,275] biochemical,[276,277] and neuroimaging studies.[278] Plasma biomarkers of altered energy metabolism have been identified.[279] Most clinical and neuropathological characteristics of HD, including selective striatal degeneration, can be reproduced by administering 3-nitropropionic acid, a mitochondrial toxin that inhibits mitochondrial complex II and more specifically the activity of succinate dehydrogenase, thereby reducing metabolic energy production.[280] Conversely, restoration of complex II activity is neuroprotective in a cellular model of HD.[281] A link between dopamine signaling and regulation of mitochondrial complex II activity may account for the vulnerability of striatal neurons in HD.[245] Proposed mechanisms include direct interaction between Exp-Htt and mitochondrial membranes, that could disrupt calcium homeostasis during metabolic stress and lead to indirect excitotoxicity.[133,230-232,282] Mitochondrial dysfunction might also be related to changes in nuclear-encoded mitochondrial proteins. Exp-Htt represses the transcription of PGC1-alpha transcriptional coactivator, which regulates mitochondrial biogenesis and oxidative phosphorylation.[283,284] Similarly, abnormal interaction between exp-Htt and the transcription factor p53 leads to transcriptional changes and mitochondrial dysfunction.[193] Another possible mechanism of exp-Htt-induced mitochondrial toxicity consists of altered mitochondrial dynamics[285,286] and dysfunction of mitochondrial fission and fusion machinery, leading to increased mitochondrial fragmentation.[287] Exp-Htt increases transglutaminase activity in brain necropsy samples of HD patients and in animal models of HD.[288-290] This may result in altered modulation of the mitochondrial respiratory chain in response to apoptotic stimuli. Genetic suppression of transglutaminase activity improves the motor phenotype of HD mouse models.[291-293] Finally, extra-mitochondrial pathways may be involved in the impaired energy metabolism in HD.[294] These proposed mechanisms of mitochondrial dysfunction in HD are not all mutually exclusive. It is not clear whether mitochondrial defects are critical exp-Htt-induced disease triggers, but they are major components of HD neuronal dysfunction and are likely to amplify changes in neuronal functions that rely on energy production. Several agents targeting mitochondrial functions have shown some benefits in mouse models of HD,[295-299] but they failed to produce significant improvements in clinical trials.[300-302]

Conclusion

Huntington's disease is a relentless neurodegenerative disease resulting in severe disability through a triad of motor, cognitive and psychiatric symptoms. Drugs used to the disease act on the symptoms but do not slow the disease process itself. Basic research is providing novel insights into the complex molecular pathways that mediate neuronal dysfunction and death in Huntington's disease. Since several mechanisms have been identified, successful neuroprotective therapy for Huntington's disease patients is likely to involve a combined approach targeting both cellular and molecular mediators that account for the toxicity of mutated huntingtin.

References

1. Huntington G. On chorea. Med Surg Rep 1872; 317-321.
2. Okun MS, Thommi N. Americo Negrette (1924 to 2003): diagnosing Huntington disease in Venezuela. Neurology 2004; 63:340-343.
3. Gusella JF, Wexler NS, Conneally PM et al. A polymorphic DNA marker genetically linked to Huntington's disease. Nature 1983; 306:234-238.
4. A novel gene containing a trinucleotide repeat that is expanded and unstable on Huntington's disease chromosomes. The Huntington's Disease Collaborative Research Group. Cell 1993; 72:971-983.
5. Paradisi I, Hernandez A, Arias S. Huntington disease mutation in Venezuela: age of onset, haplotype analyses and geographic aggregation. J Hum Genet 2008; 53:127-135.
6. Harper PS. The epidemiology of Huntington's disease. Hum Genet 1992; 89:365-376.
7. Foroud T, Gray J, Ivashina J et al. Differences in duration of Huntington's disease based on age at onset. Journal of neurology, neurosurgery and psychiatry 1999; 66:52-56.
8. Mahant N, McCusker EA, Byth K et al. Huntington's disease: clinical correlates of disability and progression. Neurology 2003; 61:1085-1092.
9. Rao AK, Muratori L, Louis ED et al. Spectrum of gait impairments in presymptomatic and symptomatic Huntington's disease. Mov Disord 2008; 23:1100-1107.
10. Hicks SL, Robert MP, Golding CV et al. Oculomotor deficits indicate the progression of Huntington's disease. Prog Brain Res 2008; 171:555-558.
11. Paulsen JS, Langbehn DR, Stout JC et al. Detection of Huntington's disease decades before diagnosis: the Predict-HD study. Journal of neurology, neurosurgery and psychiatry 2008; 79:874-880.
12. Julien CL, Thompson JC, Wild S et al. Psychiatric disorders in preclinical Huntington's disease. Journal of neurology, neurosurgery and psychiatry 2007; 78:939-943.
13. Kingma EM, van Duijn E, Timman R et al. Behavioural problems in Huntington's disease using the Problem Behaviours Assessment. Gen Hosp Psychiatry 2008; 30:155-161.
14. Beglinger LJ, Langbehn DR, Duff K et al. Probability of obsessive and compulsive symptoms in Huntington's disease. Biol Psychiatry 2007; 61:415-418.
15. Gargiulo M, Lejeune S, Tanguy ML et al. Long-term outcome of presymptomatic testing in Huntington disease. Eur J Hum Genet 2009; 17:165-171.
16. Quinn N, Schrag A. Huntington's disease and other choreas. J Neurol 1998; 245:709-716.
17. Penney JB Jr, Young AB, Shoulson I et al. Huntington's disease in Venezuela: 7 ears of follow-up on symptomatic and asymptomatic individuals. Mov Disord 1990; 5:93-99.
18. Kremer B, Weber B, Hayden MR. New insights into the clinical features, pathogenesis and molecular genetics of Huntington disease. Brain Pathol 1992; 2:321-335.
19. Busse ME, Wiles CM, Rosser AE. Mobility and falls in people with Huntington's disease. Journal of neurology, neurosurgery and psychiatry 2009; 80:88-90.
20. Wheelock VL, Tempkin T, Marder K et al. Predictors of nursing home placement in Huntington disease. Neurology 2003; 60:998-1001.
21. Kerbeshian J, Burd L, Leech C et al. Huntington disease and childhood-onset Tourette syndrome. Am J Med Genet 1991; 39:1-3.
22. Jankovic J, Beach J, Ashizawa T. Emotional and functional impact of DNA testing on patients with symptoms of Huntington's disease. J Med Genet 1995; 32:516-518.

23. Vogel CM, Drury I, Terry LC et al. Myoclonus in adult Huntington's disease. Ann Neurol 1991; 29:213-215.
24. Carella F, Scaioli V, Ciano C et al. Adult onset myoclonic Huntington's disease. Mov Disord 1993; 8:201-205.
25. Thompson PD, Bhatia KP, Brown P et al. Cortical myoclonus in Huntington's disease. Mov Disord 1994; 9:633-641.
26. Siesling S, Vegter-van der Vlis M, Roos RA. Juvenile Huntington disease in the Netherlands. Pediatr Neurol 1997; 17:37-43.
27. Gonzalez-Alegre P, Afifi AK. Clinical characteristics of childhood-onset (juvenile) Huntington disease: report of 12 patients and review of the literature. J Child Neurol 2006; 21:223-229.
28. Gambardella A, Muglia M, Labate A et al. Juvenile Huntington's disease presenting as progressive myoclonic epilepsy. Neurology 2001; 57:708-711.
29. MacMillan JC, Morrison PJ, Nevin NC et al. Identification of an expanded CAG repeat in the Huntington's disease gene (IT15) in a family reported to have benign hereditary chorea. J Med Genet 1993; 30:1012-1013.
30. Britton JW, Uitti RJ, Ahlskog JE et al. Hereditary late-onset chorea without significant dementia: genetic evidence for substantial phenotypic variation in Huntington's disease. Neurology 1995; 45:443-447.
31. Ho AK, Sahakian BJ, Brown RG et al. Profile of cognitive progression in early Huntington's disease. Neurology 2003; 61:1702-1706.
32. Caine ED, Hunt RD, Weingartner H et al. Huntington's dementia. Clinical and neuropsychological features. Arch Gen Psychiatry 1978; 35:377-384.
33. Bamford KA, Caine ED, Kido DK et al. A prospective evaluation of cognitive decline in early Huntington's disease: functional and radiographic correlates. Neurology 1995; 45:1867-1873.
34. Peinemann A, Schuller S, Pohl C et al. Executive dysfunction in early stages of Huntington's disease is associated with striatal and insular atrophy: a neuropsychological and voxel-based morphometric study. J Neurol Sci 2005; 239:11-19.
35. Caine ED, Ebert MH, Weingartner H. An outline for the analysis of dementia. The memory disorder of Huntingtons disease. Neurology 1977; 27:1087-1092.
36. Hodges JR, Salmon DP, Butters N. Differential impairment of semantic and episodic memory in Alzheimer's and Huntington's diseases: a controlled prospective study. Journal of neurology, neurosurgery and psychiatry 1990; 53:1089-1095.
37. Pillon B, Deweer B, Agid Y et al. Explicit memory in Alzheimer's, Huntington's and Parkinson's diseases. Arch Neurol 1993; 50:374-379.
38. Pillon B, Deweer B, Michon A et al. Are explicit memory disorders of progressive supranuclear palsy related to damage to striatofrontal circuits? Comparison with Alzheimer's, Parkinson's and Huntington's diseases. Neurology 1994; 44:1264-1270.
39. Shiwach RS, Norbury CG. A controlled psychiatric study of individuals at risk for Huntington's disease. Br J Psychiatry 1994; 165:500-505.
40. Kirkwood SC, Siemers E, Viken R et al. Longitudinal personality changes among presymptomatic Huntington disease gene carriers. Neuropsychiatry Neuropsychol Behav Neurol 2002; 15:192-197.
41. Cummings JL, Cunningham K. Obsessive-compulsive disorder in Huntington's disease. Biol Psychiatry 1992; 31:263-270.
42. Patzold T, Brune M. Obsessive compulsive disorder in Huntington disease: a case of isolated obsessions successfully treated with sertraline. Neuropsychiatry Neuropsychol Behav Neurol 2002; 15:216-219.
43. Fedoroff JP, Peyser C, Franz ML et al. Sexual disorders in Huntington's disease. J Neuropsychiatry Clin Neurosci 1994; 6:147-153.
44. Pflanz S, Besson JA, Ebmeier KP et al. The clinical manifestation of mental disorder in Huntington's disease: a retrospective case record study of disease progression. Acta Psychiatr Scand 1991; 83:53-60.
45. Paulsen JS, Ready RE, Hamilton JM et al. Neuropsychiatric aspects of Huntington's disease. Journal of neurology, neurosurgery and psychiatry 2001; 71:310-314.
46. Caine ED, Shoulson I. Psychiatric syndromes in Huntington's disease. Am J Psychiatry 1983; 140:728-733.
47. Folstein SE, Franz ML, Jensen BA et al. Conduct disorder and affective disorder among the offspring of patients with Huntington's disease. Psychol Med 1983; 13:45-52.
48. Di Maio L, Squitieri F, Napolitano G et al. Onset symptoms in 510 patients with Huntington's disease. J Med Genet 1993; 30:289-292.
49. Cummings JL. Behavioral and psychiatric symptoms associated with Huntington's disease. Adv Neurol 1995; 65:179-186.
50. Jensen P, Fenger K, Bolwig TG et al. Crime in Huntington's disease: a study of registered offences among patients, relatives and controls. Journal of neurology, neurosurgery and psychiatry 1998; 65:467-471.
51. Mendez MF. Mania in neurologic disorders. Curr Psychiatry Rep 2000; 2:440-445.

52. Rosenblatt A, Leroi I. Neuropsychiatry of Huntington's disease and other basal ganglia disorders. Psychosomatics 2000; 41:24-30.
53. Farrer LA. Suicide and attempted suicide in Huntington disease: implications for preclinical testing of persons at risk. Am J Med Genet 1986; 24:305-311.
54. Paulsen JS, Hoth KF, Nehl C et al. Critical periods of suicide risk in Huntington's disease. Am J Psychiatry 2005; 162:725-731.
55. Lipe H, Schultz A, Bird TD. Risk factors for suicide in Huntingtons disease: a retrospective case controlled study. Am J Med Genet 1993; 48:231-233.
56. Taylor N, Bramble D. Sleep disturbance and Huntingdon's disease. Br J Psychiatry 1997; 171:393.
57. Morton AJ, Wood NI, Hastings MH et al. Disintegration of the sleep-wake cycle and circadian timing in Huntington's disease. J Neurosci 2005; 25:157-163.
58. Arnulf I, Nielsen J, Lohmann E et al. Rapid eye movement sleep disturbances in Huntington disease. Arch Neurol 2008; 65:482-488.
59. Videnovic A, Leurgans S, Fan W et al. Daytime somnolence and nocturnal sleep disturbances in Huntington disease. Parkinsonism Relat Disord 2008.
60. Aziz NA, van der Burg JM, Landwehrmeyer GB et al. Weight loss in Huntington disease increases with higher CAG repeat number. Neurology 2008; 71:1506-1513.
61. Pratley RE, Salbe AD, Ravussin E et al. Higher sedentary energy expenditure in patients with Huntington's disease. Ann Neurol 2000; 47:64-70.
62. Kenney C, Powell S, Jankovic J. Autopsy-proven Huntington's disease with 29 trinucleotide repeats. Mov Disord 2007; 22:127-130.
63. Semaka A, Warby S, Leavitt BR et al. Re: Autopsy-proven Huntington's disease with 29 trinucleotide repeats. Mov Disord 2008; 23:1794-1795; author reply 1793.
64. Reynolds N. Re: Autopsy-proven Huntington's disease with 29 trinucleotide repeats. Mov Disord 2008; 23:1795-1796; author reply 1793.
65. Goldberg YP, Rommens JM, Andrew SE et al. Identification of an Alu retrotransposition event in close proximity to a strong candidate gene for Huntington's disease. Nature 1993; 362:370-373.
66. Davis MB, Bateman D, Quinn NP et al. Mutation analysis in patients with possible but apparently sporadic Huntington's disease. Lancet 1994; 344:714-717.
67. Zuhlke C, Riess O, Bockel B et al. Mitotic stability and meiotic variability of the (CAG)n repeat in the Huntington disease gene. Hum Mol Genet 1993; 2:2063-2067.
68. Ranen NG, Stine OC, Abbott MH et al. Anticipation and instability of IT-15 (CAG)n repeats in parent-offspring pairs with Huntington disease. Am J Hum Genet 1995; 57:593-602.
69. Nance MA, Westphal B, Nugent S. Diagnosis of patients presenting to a Huntington disease (HD) clinic without a family history of HD. Neurology 1996; 47:1578-1580.
70. Myers RH, MacDonald ME, Koroshetz WJ et al. De novo expansion of a (CAG)n repeat in sporadic Huntington's disease. Nat Genet 1993; 5:168-173.
71. Alford RL, Ashizawa T, Jankovic J et al. Molecular detection of new mutations, resolution of ambiguous results and complex genetic counseling issues in Huntington disease. Am J Med Genet 1996; 66:281-286.
72. Wexler NS, Lorimer J, Porter J et al. Venezuelan kindreds reveal that genetic and environmental factors modulate Huntington's disease age of onset. Proc Natl Acad Sci USA 2004; 101:3498-3503.
73. Andresen JM, Gayan J, Djousse L et al. The relationship between CAG repeat length and age of onset differs for Huntington's disease patients with Juvenile onset or Adult onset. Ann Hum Genet 2007; 71:295-301.
74. Snell RG, MacMillan JC, Cheadle JP et al. Relationship between trinucleotide repeat expansion and phenotypic variation in Huntington's disease. Nat Genet 1993; 4:393-397.
75. Andrew SE, Goldberg YP, Kremer B et al. The relationship between trinucleotide (CAG) repeat length and clinical features of Huntington's disease. Nat Genet 1993; 4:398-403.
76. Duyao M, Ambrose C, Myers R et al. Trinucleotide repeat length instability and age of onset in Huntington's disease. Nat Genet 1993; 4:387-392.
77. Marder K, Sandler S, Lechich A et al. Relationship between CAG repeat length and late-stage outcomes in Huntington's disease. Neurology 2002; 59:1622-1624.
78. Ashizawa T, Wong LJ, Richards CS et al. CAG repeat size and clinical presentation in Huntington's disease. Neurology 1994; 44:1137-1143.
79. Illarioshkin SN, Igarashi S, Onodera O et al. Trinucleotide repeat length and rate of progression of Huntington's disease. Ann Neurol 1994; 36:630-635.
80. Kieburtz K, MacDonald M, Shih C et al. Trinucleotide repeat length and progression of illness in Huntington's disease. J Med Genet 1994; 31:872-874.
81. Claes S, Van Zand K, Legius E et al. Correlations between triplet repeat expansion and clinical features in Huntington's disease. Arch Neurol 1995; 52:749-753.

82. Brandt J, Bylsma FW, Gross R et al. Trinucleotide repeat length and clinical progression in Huntington's disease. Neurology 1996; 46:527-531.
83. Penney JB Jr, Vonsattel JP, MacDonald ME et al. CAG repeat number governs the development rate of pathology in Huntington's disease. Ann Neurol 1997; 41:689-692.
84. Squitieri F, Cannella M, Simonelli M. CAG mutation effect on rate of progression in Huntington's disease. Neurol Sci 2002; 23(Suppl 2):S107-108.
85. Telenius H, Kremer B, Goldberg YP et al. Somatic and gonadal mosaicism of the Huntington disease gene CAG repeat in brain and sperm. Nat Genet 1994; 6:409-414.
86. Aronin N, Chase K, Young C et al. CAG expansion affects the expression of mutant Huntingtin in the Huntington's disease brain. Neuron 1995; 15:1193-1201.
87. Kahlem P, Djian P. The expanded CAG repeat associated with juvenile Huntington disease shows a common origin of most or all neurons and glia in human cerebrum. Neurosci Lett 2000; 286:203-207.
88. Ishiguro H, Yamada K, Sawada H et al. Age-dependent and tissue-specific CAG repeat instability occurs in mouse knock-in for a mutant Huntington's disease gene. J Neurosci Res 2001; 65:289-297.
89. Shelbourne PF, Keller-McGandy C, Bi WL et al. Triplet repeat mutation length gains correlate with cell-type specific vulnerability in Huntington disease brain. Hum Mol Genet 2007; 16:1133-1142.
90. Farrer LA, Cupples LA, Wiater P et al. The normal Huntington disease (HD) allele, or a closely linked gene, influences age at onset of HD. Am J Hum Genet 1993; 53:125-130.
91. Vuillaume I, Vermersch P, Destee A et al. Genetic polymorphisms adjacent to the CAG repeat influence clinical features at onset in Huntington's disease. Journal of neurology, neurosurgery and psychiatry 1998; 64:758-762.
92. Chattopadhyay B, Ghosh S, Gangopadhyay PK et al. Modulation of age at onset in Huntington's disease and spinocerebellar ataxia type 2 patients originated from eastern India. Neurosci Lett 2003; 345:93-96.
93. Rubinsztein DC, Leggo J, Chiano M et al. Genotypes at the GluR6 kainate receptor locus are associated with variation in the age of onset of Huntington disease. Proc Natl Acad Sci USA 1997; 94:3872-3876.
94. MacDonald ME, Vonsattel JP, Shrinidhi J et al. Evidence for the GluR6 gene associated with younger onset age of Huntington's disease. Neurology 1999; 53:1330-1332.
95. Naze P, Vuillaume I, Destee A et al. Mutation analysis and association studies of the ubiquitin carboxy-terminal hydrolase L1 gene in Huntington's disease. Neurosci Lett 2002; 328:1-4.
96. Panas M, Avramopoulos D, Karadima G et al. Apolipoprotein E and presenilin-1 genotypes in Huntington's disease. J Neurol 1999; 246:574-577.
97. Kehoe P, Krawczak M, Harper PS et al. Age of onset in Huntington disease: sex specific influence of apolipoprotein E genotype and normal CAG repeat length. J Med Genet 1999; 36:108-111.
98. Holbert S, Denghien I, Kiechle T et al. The Gln-Ala repeat transcriptional activator CA150 interacts with huntingtin: neuropathologic and genetic evidence for a role in Huntington's disease pathogenesis. Proc Natl Acad Sci USA 2001; 98:1811-1816.
99. Chattopadhyay B, Baksi K, Mukhopadhyay S et al. Modulation of age at onset of Huntington disease patients by variations in TP53 and human caspase activated DNase (hCAD) genes. Neurosci Lett 2005; 374:81-86.
100. Arning L, Kraus PH, Valentin S et al. NR2A and NR2B receptor gene variations modify age at onset in Huntington disease. Neurogenetics 2005; 6:25-28.
101. Arning L, Saft C, Wieczorek S et al. NR2A and NR2B receptor gene variations modify age at onset in Huntington disease in a sex-specific manner. Hum Genet 2007.
102. Metzger S, Rong J, Nguyen HP et al. Huntingtin-associated protein-1 is a modifier of the age-at-onset of Huntington's disease. Hum Mol Genet 2008; 17:1137-1146.
103. Andresen JM, Gayan J, Cherny SS et al. Replication of twelve association studies for Huntington's disease residual age of onset in large Venezuelan kindreds. J Med Genet 2007; 44:44-50.
104. Vonsattel JP, Myers RH, Stevens TJ et al. Neuropathological classification of Huntington's disease. J Neuropathol Exp Neurol 1985; 44:559-577.
105. Richfield EK, Herkenham M. Selective vulnerability in Huntington's disease: preferential loss of cannabinoid receptors in lateral globus pallidus. Ann Neurol 1994; 36:577-584.
106. Martinez-Mir MI, Probst A, Palacios JM. Adenosine A2 receptors: selective localization in the human basal ganglia and alterations with disease. Neuroscience 1991; 42:697-706.
107. Richfield EK, O'Brien CF, Eskin T et al. Heterogeneous dopamine receptor changes in early and late Huntington's disease. Neurosci Lett 1991; 132:121-126.
108. Goto S, Hirano A, Rojas-Corona RR. Immunohistochemical visualization of afferent nerve terminals in human globus pallidus and its alteration in neostriatal neurodegenerative disorders. Acta Neuropathol (Berl) 1989; 78:543-550.
109. Dawbarn D, De Quidt ME, Emson PC. Survival of basal ganglia neuropeptide Y-somatostatin neurones in Huntington's disease. Brain Res 1985; 340:251-260.

110. Ferrante RJ, Kowall NW, Beal MF et al. Selective sparing of a class of striatal neurons in Huntington's disease. Science 1985; 230:561-563.
111. Ferrante RJ, Kowall NW, Beal MF et al. Morphologic and histochemical characteristics of a spared subset of striatal neurons in Huntington's disease. J Neuropathol Exp Neurol 1987; 46:12-27.
112. Ferrante RJ, Kowall NW, Richardson EP Jr. Proliferative and degenerative changes in striatal spiny neurons in Huntington's disease: a combined study using the section-Golgi method and calbindin D28k immunocytochemistry. J Neurosci 1991; 11:3877-3887.
113. Lange H, Thorner G, Hopf A et al. Morphometric studies of the neuropathological changes in choreatic diseases. J Neurol Sci 1976; 28:401-425.
114. DiFiglia M, Sapp E, Chase KO et al. Aggregation of huntingtin in neuronal intranuclear inclusions and dystrophic neurites in brain. Science 1997; 277:1990-1993.
115. Graveland GA, Williams RS, DiFiglia M. Evidence for degenerative and regenerative changes in neostriatal spiny neurons in Huntington's disease. Science 1985; 227:770-773.
116. de la Monte SM, Vonsattel JP, Richardson EP Jr. Morphometric demonstration of atrophic changes in the cerebral cortex, white matter and neostriatum in Huntington's disease. J Neuropathol Exp Neurol 1988; 47:516-525.
117. Sotrel A, Paskevich PA, Kiely DK et al. Morphometric analysis of the prefrontal cortex in Huntington's disease. Neurology 1991; 41:1117-1123.
118. Heinsen H, Strik M, Bauer M et al. Cortical and striatal neurone number in Huntington's disease. Acta Neuropathol (Berl) 1994; 88:320-333.
119. Selemon LD, Rajkowska G, Goldman-Rakic PS. Evidence for progression in frontal cortical pathology in late-stage Huntington's disease. J Comp Neurol 2004; 468:190-204.
120. Oyanagi K, Takeda S, Takahashi H et al. A quantitative investigation of the substantia nigra in Huntington's disease. Ann Neurol 1989; 26:13-19.
121. Heinsen H, Rub U, Bauer M et al. Nerve cell loss in the thalamic mediodorsal nucleus in Huntington's disease. Acta Neuropathol (Berl) 1999; 97:613-622.
122. Heinsen H, Rub U, Gangnus D et al. Nerve cell loss in the thalamic centromedian-parafascicular complex in patients with Huntington's disease. Acta Neuropathol (Berl) 1996; 91:161-168.
123. Kremer HP, Roos RA, Dingjan G et al. Atrophy of the hypothalamic lateral tuberal nucleus in Huntington's disease. J Neuropathol Exp Neurol 1990; 49:371-382.
124. Jeste DV, Barban L, Parisi J. Reduced Purkinje cell density in Huntington's disease. Exp Neurol 1984; 85:78-86.
125. Rodda RA. Cerebellar atrophy in Huntington's disease. J Neurol Sci 1981; 50:147-157.
126. Hoffner G, Island ML, Djian P. Purification of neuronal inclusions of patients with Huntington's disease reveals a broad range of N-terminal fragments of expanded huntingtin and insoluble polymers. J Neurochem 2005; 95:125-136.
127. Gutekunst CA, Li SH, Yi H et al. Nuclear and neuropil aggregates in Huntington's disease: relationship to neuropathology. J Neurosci 1999; 19:2522-2534.
128. Scherzinger E, Lurz R, Turmaine M et al. Huntingtin-encoded polyglutamine expansions form amyloid-like protein aggregates in vitro and in vivo. Cell 1997; 90:549-558.
129. Huang CC, Faber PW, Persichetti F et al. Amyloid formation by mutant huntingtin: threshold, progressivity and recruitment of normal polyglutamine proteins. Somat Cell Mol Genet 1998; 24:217-233.
130. Diguet E, Petit F, Escartin C et al. Normal aging modulates the neurotoxicity of mutant huntingtin. PLoS ONE 2009; 4:e4637.
131. Kegel KB, Meloni AR, Yi Y et al. Huntingtin is present in the nucleus, interacts with the transcriptional corepressor C-terminal binding protein and represses transcription. J Biol Chem 2002; 277:7466-7476.
132. Kegel KB, Sapp E, Yoder J et al. Huntingtin associates with acidic phospholipids at the plasma membrane. J Biol Chem 2005; 280:36464-36473.
133. Rockabrand E, Slepko N, Pantalone A et al. The first 17 amino acids of Huntingtin modulate its sub-cellular localization, aggregation and effects on calcium homeostasis. Hum Mol Genet 2007; 16:61-77.
134. Caviston JP, Ross JL, Antony SM et al. Huntingtin facilitates dynein/dynactin-mediated vesicle transport. Proc Natl Acad Sci USA 2007; 104:10045-10050.
135. Strehlow AN, Li JZ, Myers RM. Wild-type huntingtin participates in protein trafficking between the Golgi and the extracellular space. Hum Mol Genet 2007; 16:391-409.
136. Harjes P, Wanker EE. The hunt for huntingtin function: interaction partners tell many different stories. Trends Biochem Sci 2003; 28:425-433.
137. Li SH, Li XJ. Huntingtin-protein interactions and the pathogenesis of Huntington's disease. Trends Genet 2004; 20:146-154.
138. Sugars KL, Rubinsztein DC. Transcriptional abnormalities in Huntington disease. Trends Genet 2003; 19:233-238.

139. Zhai W, Jeong H, Cui L et al. In vitro analysis of huntingtin-mediated transcriptional repression reveals multiple transcription factor targets. Cell 2005; 123:1241-1253.
140. Sipione S, Rigamonti D, Valenza M et al. Early transcriptional profiles in huntingtin-inducible striatal cells by microarray analyses. Hum Mol Genet 2002; 11:1953-1965.
141. Luthi-Carter R, Hanson SA, Strand AD et al. Dysregulation of gene expression in the R6/2 model of polyglutamine disease: parallel changes in muscle and brain. Hum Mol Genet 2002; 11:1911-1926.
142. Luthi-Carter R, Strand A, Peters NL et al. Decreased expression of striatal signaling genes in a mouse model of Huntington's disease. Hum Mol Genet 2000; 9:1259-1271.
143. Luthi-Carter R, Strand AD, Hanson SA et al. Polyglutamine and transcription: gene expression changes shared by DRPLA and Huntington's disease mouse models reveal context-independent effects. Hum Mol Genet 2002; 11:1927-1937.
144. Kuhn A, Goldstein DR, Hodges A et al. Mutant huntingtin's effects on striatal gene expression in mice recapitulate changes observed in human Huntington's disease brain and do not differ with mutant huntingtin length or wild-type huntingtin dosage. Hum Mol Genet 2007.
145. Duyao MP, Auerbach AB, Ryan A et al. Inactivation of the mouse Huntington's disease gene homolog Hdh. Science 1995; 269:407-410.
146. Zeitlin S, Liu JP, Chapman DL et al. Increased apoptosis and early embryonic lethality in mice nullizygous for the Huntington's disease gene homologue. Nat Genet 1995; 11:155-163.
147. Dragatsis I, Dietrich P, Zeitlin S. Expression of the Huntingtin-associated protein 1 gene in the developing and adult mouse. Neurosci Lett 2000; 282:37-40.
148. Perutz MF, Pope BJ, Owen D et al. Aggregation of proteins with expanded glutamine and alanine repeats of the glutamine-rich and asparagine-rich domains of Sup35 and of the amyloid beta-peptide of amyloid plaques. Proc Natl Acad Sci USA 2002; 99:5596-5600.
149. Weiss A, Klein C, Woodman B et al. Sensitive biochemical aggregate detection reveals aggregation onset before symptom development in cellular and murine models of Huntington's disease. J Neurochem 2007.
150. Martindale D, Hackam A, Wieczorek A et al. Length of huntingtin and its polyglutamine tract influences localization and frequency of intracellular aggregates. Nat Genet 1998; 18:150-154.
151. Kazantsev A, Preisinger E, Dranovsky A et al. Insoluble detergent-resistant aggregates form between pathological and nonpathological lengths of polyglutamine in mammalian cells. Proc Natl Acad Sci USA 1999; 96:11404-11409.
152. Nucifora FC, Jr, Sasaki M, Peters MF et al. Interference by huntingtin and atrophin-1 with cbp-mediated transcription leading to cellular toxicity. Science 2001; 291:2423-2428.
153. Steffan JS, Kazantsev A, Spasic-Boskovic O et al. The Huntington's disease protein interacts with p53 and CREB-binding protein and represses transcription. Proc Natl Acad Sci USA 2000; 97:6763-6768.
154. Gunawardena S, Her LS, Brusch RG et al. Disruption of axonal transport by loss of huntingtin or expression of pathogenic polyQ proteins in Drosophila. Neuron 2003; 40:25-40.
155. Trushina E, Dyer RB, Badger JD 2nd et al. Mutant huntingtin impairs axonal trafficking in mammalian neurons in vivo and in vitro. Mol Cell Biol 2004; 24:8195-8209.
156. Bence NF, Sampat RM, Kopito RR. Impairment of the ubiquitin-proteasome system by protein aggregation. Science 2001; 292:1552-1555.
157. Verhoef LG, Lindsten K, Masucci MG et al. Aggregate formation inhibits proteasomal degradation of polyglutamine proteins. Hum Mol Genet 2002; 11:2689-2700.
158. Venkatraman P, Wetzel R, Tanaka M et al. Eukaryotic proteasomes cannot digest polyglutamine sequences and release them during degradation of polyglutamine-containing proteins. Mol Cell 2004; 14:95-104.
159. Jana NR, Zemskov EA, Wang G et al. Altered proteasomal function due to the expression of polyglutamine-expanded truncated N-terminal huntingtin induces apoptosis by caspase activation through mitochondrial cytochrome c release. Hum Mol Genet 2001; 10:1049-1059.
160. Bennett EJ, Shaler TA, Woodman B et al. Global changes to the ubiquitin system in Huntington's disease. Nature 2007; 448:704-708.
161. Hunter JM, Lesort M, Johnson GV. Ubiquitin-proteasome system alterations in a striatal cell model of Huntington's disease. J Neurosci Res 2007.
162. Tydlacka S, Wang CE, Wang X et al. Differential activities of the ubiquitin-proteasome system in neurons versus glia may account for the preferential accumulation of misfolded proteins in neurons. J Neurosci 2008; 28:13285-13295.
163. Saudou F, Finkbeiner S, Devys D et al. Huntingtin acts in the nucleus to induce apoptosis but death does not correlate with the formation of intranuclear inclusions. Cell 1998; 95:55-66.
164. Slow EJ, Graham RK, Osmand AP et al. Absence of behavioral abnormalities and neurodegeneration in vivo despite widespread neuronal huntingtin inclusions. Proc Natl Acad Sci USA 2005; 102:11402-11407.
165. Arrasate M, Mitra S, Schweitzer ES et al. Inclusion body formation reduces levels of mutant huntingtin and the risk of neuronal death. Nature 2004; 431:805-810.

166. Sawada H, Ishiguro H, Nishii K et al. Characterization of neuron-specific huntingtin aggregates in human huntingtin knock-in mice. Neurosci Res 2007; 57:559-573.
167. Mitra S, Tsvetkov AS, Finkbeiner S. Single neuron ubiquitin-proteasome dynamics accompanying inclusion body formation in Huntington disease. J Biol Chem 2009; 284:4398-4403.
168. Gong B, Lim MC, Wanderer J et al. Time-lapse analysis of aggregate formation in an inducible PC12 cell model of Huntington's disease reveals time-dependent aggregate formation that transiently delays cell death. Brain Res Bull 2008; 75:146-157.
169. Li H, Li SH, Yu ZX et al. Huntingtin aggregate-associated axonal degeneration is an early pathological event in Huntington's disease mice. J Neurosci 2001; 21:8473-8481.
170. Li H, Li SH, Johnston H et al. Amino-terminal fragments of mutant huntingtin show selective accumulation in striatal neurons and synaptic toxicity. Nat Genet 2000; 25:385-389.
171. Lee WC, Yoshihara M, Littleton JT. Cytoplasmic aggregates trap polyglutamine-containing proteins and block axonal transport in a Drosophila model of Huntington's disease. Proc Natl Acad Sci USA 2004; 101:3224-3229.
172. DiFiglia M. Clinical Genetics, II. Huntington's disease: from the gene to pathophysiology. Am J Psychiatry 1997; 154:1046.
173. Herbst M, Wanker EE. Small molecule inducers of heat-shock response reduce polyQ-mediated huntingtin aggregation. A possible therapeutic strategy. Neuro-degenerative diseases 2007; 4:254-260.
174. Jana NR, Tanaka M, Wang G et al. Polyglutamine length-dependent interaction of Hsp40 and Hsp70 family chaperones with truncated N-terminal huntingtin: their role in suppression of aggregation and cellular toxicity. Hum Mol Genet 2000; 9:2009-2018.
175. Carmichael J, Chatellier J, Woolfson A et al. Bacterial and yeast chaperones reduce both aggregate formation and cell death in mammalian cell models of Huntington's disease. Proc Natl Acad Sci USA 2000; 97:9701-9705.
176. Vacher C, Garcia-Oroz L, Rubinsztein DC. Overexpression of yeast hsp104 reduces polyglutamine aggregation and prolongs survival of a transgenic mouse model of Huntington's disease. Hum Mol Genet 2005; 14:3425-3433.
177. Perrin V, Regulier E, Abbas-Terki T et al. Neuroprotection by Hsp104 and Hsp27 in Lentiviral-based rat models of Huntington's disease. Mol Ther 2007; 15:903-911.
178. Seo H, Sonntag KC, Kim W et al. Proteasome activator enhances survival of Huntington's disease neuronal model cells. PLoS ONE 2007; 2:e238.
179. Ravikumar B, Berger Z, Vacher C et al. Rapamycin pretreatment protects against apoptosis. Hum Mol Genet 2006; 15:1209-1216.
180. Sarkar S, Perlstein EO, Imarisio S et al. Small molecules enhance autophagy and reduce toxicity in Huntington's disease models. Nature Chemical Biology 2007.
181. Chopra V, Fox JH, Lieberman G et al. A small-molecule therapeutic lead for Huntington's disease: preclinical pharmacology and efficacy of C2-8 in the R6/2 transgenic mouse. Proc Natl Acad Sci USA 2007; 104:16685-16689.
182. Augood SJ, Faull RL, Emson PC. Dopamine D1 and D2 receptor gene expression in the striatum in Huntington's disease. Ann Neurol 1997; 42:215-221.
183. Augood SJ, Faull RL, Love DR et al. Reduction in enkephalin and substance P messenger RNA in the striatum of early grade Huntington's disease: a detailed cellular in situ hybridization study. Neuroscience 1996; 72:1023-1036.
184. Cha JH, Kosinski CM, Kerner JA et al. Altered brain neurotransmitter receptors in transgenic mice expressing a portion of an abnormal human Huntington disease gene. Proc Natl Acad Sci USA 1998; 95:6480-6485.
185. Cha JH, Frey AS, Alsdorf SA et al. Altered neurotransmitter receptor expression in transgenic mouse models of Huntington's disease. Philos Trans R Soc Lond B Biol Sci 1999; 354:981-989.
186. Cha JH. Transcriptional signatures in Huntington's disease. Prog Neurobiol 2007.
187. Runne H, Kuhn A, Wild EJ et al. Analysis of potential transcriptomic biomarkers for Huntington's disease in peripheral blood. Proc Natl Acad Sci USA 2007; 104:14424-14429.
188. Borovecki F, Lovrecic L, Zhou J et al. Genome-wide expression profiling of human blood reveals biomarkers for Huntington's disease. Proc Natl Acad Sci USA 2005; 102:11023-11028.
189. Dunah AW, Jeong H, Griffin A et al. Sp1 and TAFII130 transcriptional activity disrupted in early Huntington's disease. Science 2002; 296:2238-2243.
190. Li SH, Cheng AL, Zhou H et al. Interaction of Huntington disease protein with transcriptional activator Sp1. Mol Cell Biol 2002; 22:1277-1287.
191. Shimohata T, Nakajima T, Yamada M et al. Expanded polyglutamine stretches interact with TAFII130, interfering with CREB-dependent transcription. Nat Genet 2000; 26:29-36.
192. Suhr ST, Senut MC, Whitelegge JP et al. Identities of sequestered proteins in aggregates from cells with induced polyglutamine expression. J Cell Biol 2001; 153:283-294.

193. Bae BI, Xu H, Igarashi S et al. p53 mediates cellular dysfunction and behavioral abnormalities in Huntington's disease. Neuron 2005; 47:29-41.
194. Arango M, Holbert S, Zala D et al. CA150 expression delays striatal cell death in overexpression and knock-in conditions for mutant huntingtin neurotoxicity. J Neurosci 2006; 26:4649-4659.
195. Boutell JM, Thomas P, Neal JW et al. Aberrant interactions of transcriptional repressor proteins with the Huntington's disease gene product, huntingtin. Hum Mol Genet 1999; 8:1647-1655.
196. Ferrer I, Goutan E, Marin C et al. Brain-derived neurotrophic factor in Huntington disease. Brain Res 2000; 866:257-261.
197. Zuccato C, Ciammola A, Rigamonti D et al. Loss of huntingtin-mediated BDNF gene transcription in Huntington's disease. Science 2001; 293:493-498.
198. Strand AD, Baquet ZC, Aragaki AK et al. Expression profiling of Huntington's disease models suggests that brain-derived neurotrophic factor depletion plays a major role in striatal degeneration. J Neurosci 2007; 27:11758-11768.
199. Lynch G, Kramar EA, Rex CS et al. Brain-derived neurotrophic factor restores synaptic plasticity in a knock-in mouse model of Huntington's disease. J Neurosci 2007; 27:4424-4434.
200. Zuccato C, Tartari M, Crotti A et al. Huntingtin interacts with REST/NRSF to modulate the transcription of NRSE-controlled neuronal genes. Nat Genet 2003; 35:76-83.
201. Zuccato C, Belyaev N, Conforti P et al. Widespread disruption of repressor element-1 silencing transcription factor/neuron-restrictive silencer factor occupancy at its target genes in Huntington's disease. J Neurosci 2007; 27:6972-6983.
202. Rigamonti D, Bolognini D, Mutti C et al. Loss of huntingtin function complemented by small molecules acting as re1/nrse silencer modulators. J Biol Chem 2007.
203. Simmons DA, Rex CS, Palmer L et al. Up-regulating BDNF with an ampakine rescues synaptic plasticity and memory in Huntington's disease knockin mice. Proc Natl Acad Sci USA 2009; 106:4906-4911.
204. Taniura H, Sng JC, Yoneda Y. Histone modifications in the brain. Neurochemistry International 2007; 51:85-91.
205. Steffan JS, Bodai L, Pallos J et al. Histone deacetylase inhibitors arrest polyglutamine-dependent neurodegeneration in Drosophila. Nature 2001; 413:739-743.
206. Hockly E, Richon VM, Woodman B et al. Suberoylanilide hydroxamic acid, a histone deacetylase inhibitor, ameliorates motor deficits in a mouse model of Huntington's disease. Proc Natl Acad Sci USA 2003; 100:2041-2046.
207. Ferrante RJ, Kubilus JK, Lee J et al. Histone deacetylase inhibition by sodium butyrate chemotherapy ameliorates the neurodegenerative phenotype in Huntington's disease mice. J Neurosci 2003; 23:9418-9427.
208. Gardian G, Browne SE, Choi DK et al. Neuroprotective effects of phenylbutyrate in the N171-82Q transgenic mouse model of Huntington's disease. J Biol Chem 2005; 280:556-563.
209. Thomas EA, Coppola G, Desplats PA et al. The HDAC inhibitor 4b ameliorates the disease phenotype and transcriptional abnormalities in Huntington's disease transgenic mice. Proc Natl Acad Sci USA 2008; 105:15564-15569.
210. Pallos J, Bodai L, Lukacsovich T et al. Inhibition of specific HDACs and sirtuins suppresses pathogenesis in a Drosophila model of Huntington's disease. Hum Mol Genet 2008; 17:3767-3775.
211. Sadri-Vakili G, Bouzou B, Benn CL et al. Histones associated with downregulated genes are hypo-acetylated in Huntington's disease models. Hum Mol Genet 2007.
212. Ryu H, Lee J, Hagerty SW et al. ESET/SETDB1 gene expression and histone H3 (K9) trimethylation in Huntington's disease. Proc Natl Acad Sci USA 2006; 103:19176-19181.
213. Yang L, Mei Q, Zielinska-Kwiatkowska A et al. An ERG (ets-related gene)-associated histone methyltransferase interacts with histone deacetylases 1/2 and transcription corepressors mSin3A/B. Biochem J 2003; 369:651-657.
214. Stack EC, Del Signore SJ, Luthi-Carter R et al. Modulation of Nucleosome Dynamics in Huntington's Disease. Hum Mol Genet 2007.
215. Roze E, Betuing S, Deyts C et al. Mitogen- and stress-activated protein kinase-1 deficiency is involved in expanded-huntingtin-induced transcriptional dysregulation and striatal death. FASEB J 2008; 22:1083-1093.
216. Kim MO, Chawla P, Overland RP et al. Altered histone monoubiquitylation mediated by mutant huntingtin induces transcriptional dysregulation. J Neurosci 2008; 28:3947-3957.
217. Rothman SM, Olney JW. Excitotoxicity and the NMDA receptor—still lethal after eight years. Trends Neurosci 1995; 18:57-58.
218. Cull-Candy S, Brickley S, Farrant M. NMDA receptor subunits: diversity, development and disease. Curr Opin Neurobiol 2001; 11:327-335.
219. Dingledine R, Borges K, Bowie D et al. The glutamate receptor ion channels. Pharmacol Rev 1999; 51:7-61.

220. Beal MF, Kowall NW, Ellison DW et al. Replication of the neurochemical characteristics of Huntington's disease by quinolinic acid. Nature 1986; 321:168-171.
221. Sanberg PR, Calderon SF, Giordano M et al. The quinolinic acid model of Huntington's disease: locomotor abnormalities. Exp Neurol 1989; 105:45-53.
222. Ferrante RJ, Kowall NW, Cipolloni PB et al. Excitotoxin lesions in primates as a model for Huntington's disease: histopathologic and neurochemical characterization. Exp Neurol 1993; 119:46-71.
223. Cepeda C, Hurst RS, Calvert CR et al. Transient and progressive electrophysiological alterations in the corticostriatal pathway in a mouse model of Huntington's disease. J Neurosci 2003; 23:961-969.
224. Stack EC, Dedeoglu A, Smith KM et al. Neuroprotective effects of synaptic modulation in Huntington's disease R6/2 mice. J Neurosci 2007; 27:12908-12915.
225. Behrens PF, Franz P, Woodman B et al. Impaired glutamate transport and glutamate-glutamine cycling: downstream effects of the Huntington mutation. Brain 2002; 125:1908-1922.
226. Shin JY, Fang ZH, Yu ZX et al. Expression of mutant huntingtin in glial cells contributes to neuronal excitotoxicity. J Cell Biol 2005; 171:1001-1012.
227. Hassel B, Tessler S, Faull RL et al. Glutamate uptake is reduced in prefrontal cortex in Huntington's disease. Neurochem Res 2007.
228. Zeron MM, Hansson O, Chen N et al. Increased sensitivity to N-methyl-D-aspartate receptor-mediated excitotoxicity in a mouse model of Huntington's disease. Neuron 2002; 33:849-860.
229. Starling AJ, Andre VM, Cepeda C et al. Alterations in N-methyl-D-aspartate receptor sensitivity and magnesium blockade occur early in development in the R6/2 mouse model of Huntington's disease. J Neurosci Res 2005; 82:377-386.
230. Panov AV, Burke JR, Strittmatter WJ et al. In vitro effects of polyglutamine tracts on Ca2+-dependent depolarization of rat and human mitochondria: relevance to Huntington's disease. Archives of Biochemistry and Biophysics 2003; 410:1-6.
231. Panov AV, Gutekunst CA, Leavitt BR et al. Early mitochondrial calcium defects in Huntington's disease are a direct effect of polyglutamines. Nature Neuroscience 2002; 5:731-736.
232. Oliveira JM, Jekabsons MB, Chen S et al. Mitochondrial dysfunction in Huntington's disease: the bioenergetics of isolated and in situ mitochondria from transgenic mice. J Neurochem 2007; 101:241-249.
233. Sun Y, Savanenin A, Reddy PH et al. Polyglutamine-expanded huntingtin promotes sensitization of N-methyl-D-aspartate receptors via postsynaptic density 95. J Biol Chem 2001; 276:24713-24718.
234. Roche KW, Standley S, McCallum J et al. Molecular determinants of NMDA receptor internalization. Nature Neuroscience 2001; 4:794-802.
235. Fan MM, Fernandes HB, Zhang LY et al. Altered NMDA receptor trafficking in a yeast artificial chromosome transgenic mouse model of Huntington's disease. J Neurosci 2007; 27:3768-3779.
236. Metzler M, Gan L, Wong TP et al. NMDA receptor function and NMDA receptor-dependent phosphorylation of huntingtin is altered by the endocytic protein HIP1. J Neurosci 2007; 27:2298-2308.
237. Tang TS, Tu H, Chan EY et al. Huntingtin and huntingtin-associated protein 1 influence neuronal calcium signaling mediated by inositol-(1,4,5) triphosphate receptor type 1. Neuron 2003; 39:227-239.
238. Tang TS, Tu H, Orban PC et al. HAP1 facilitates effects of mutant huntingtin on inositol 1,4,5-trisphosphate-induced Ca release in primary culture of striatal medium spiny neurons. Eur J Neurosci 2004; 20:1779-1787.
239. Tang TS, Guo C, Wang H et al. Neuroprotective effects of inositol 1,4,5-trisphosphate receptor C-terminal fragment in a Huntington's disease mouse model. J Neurosci 2009; 29:1257-1266.
240. Landwehrmeyer GB, Dubois B, de Yebenes JG et al. Riluzole in Huntington's disease: a 3-year, randomized controlled study. Ann Neurol 2007; 62:262-272.
241. O'Suilleabhain P, Dewey RB Jr. A randomized trial of amantadine in Huntington disease. Arch Neurol 2003; 60:996-998.
242. Ondo WG, Mejia NI, Hunter CB. A pilot study of the clinical efficacy and safety of memantine for Huntington's disease. Parkinsonism Relat Disord 2007; 13:453-454.
243. Shen YC. Lamotrigine in motor and mood symptoms of Huntington's disease. World J Biol Psychiatry 2008; 9:147-149.
244. Reynolds DS, Carter RJ, Morton AJ. Dopamine modulates the susceptibility of striatal neurons to 3-nitropropionic acid in the rat model of Huntington's disease. J Neurosci 1998; 18:10116-10127.
245. Benchoua A, Trioulier Y, Diguet E et al. Dopamine determines the vulnerability of striatal neurons to the N-terminal fragment of mutant huntingtin through the regulation of mitochondrial complex II. Hum Mol Genet 2008; 17:1446-1456.
246. Cyr M, Sotnikova TD, Gainetdinov RR et al. Dopamine enhances motor and neuropathological consequences of polyglutamine expanded huntingtin. FASEB J 2006; 20:2541-2543.
247. Charvin D, Vanhoutte P, Pages C et al. Unraveling a role for dopamine in Huntington's disease: the dual role of reactive oxygen species and D2 receptor stimulation. Proc Natl Acad Sci USA 2005; 102:12218-12223.

248. Tang TS, Chen X, Liu J et al. Dopaminergic signaling and striatal neurodegeneration in Huntington's disease. J Neurosci 2007; 27:7899-7910.
249. Charvin D, Roze E, Perrin V et al. Haloperidol protects striatal neurons from dysfunction induced by mutated huntingtin in vivo. Neurobiol Dis 2008; 29:22-29.
250. Jakel RJ, Maragos WF. Neuronal cell death in Huntington's disease: a potential role for dopamine. Trends Neurosci 2000; 23:239-245.
251. Petersen AA, Larsen KE, Behr GG et al. Brain-derived neurotrophic factor inhibits apoptosis and dopamine-induced free radical production in striatal neurons but does not prevent cell death. Brain Res Bull 2001; 56:331-335.
252. Garcia M, Charvin D, Caboche J. Expanded huntingtin activates the c-Jun terminal kinase/c-Jun pathway prior to aggregate formation in striatal neurons in culture. Neuroscience 2004; 127:859-870.
253. Apostol BL, Simmons DA, Zuccato C et al. CEP-1347 reduces mutant huntingtin-associated neurotoxicity and restores BDNF levels in R6/2 mice. Mol Cell Neurosci 2008; 39:8-20.
254. Zhang H, Li Q, Graham RK et al. Full length mutant huntingtin is required for altered Ca2+ signaling and apoptosis of striatal neurons in the YAC mouse model of Huntington's disease. Neurobiol Dis 2008; 31:80-88.
255. Cepeda C, Ariano MA, Calvert CR et al. NMDA receptor function in mouse models of Huntington disease. J Neurosci Res 2001; 66:525-539.
256. Zeron MM, Fernandes HB, Krebs C et al. Potentiation of NMDA receptor-mediated excitotoxicity linked with intrinsic apoptotic pathway in YAC transgenic mouse model of Huntington's disease. Mol Cell Neurosci 2004; 25:469-479.
257. Paoletti P, Vila I, Rife M et al. Dopaminergic and glutamatergic signaling crosstalk in Huntington's disease neurodegeneration: the role of p25/cyclin-dependent kinase 5. J Neurosci 2008; 28:10090-10101.
258. Anne SL, Saudou F, Humbert S. Phosphorylation of huntingtin by cyclin-dependent kinase 5 is induced by DNA damage and regulates wild-type and mutant huntingtin toxicity in neurons. J Neurosci 2007; 27:7318-7328.
259. Luo S, Vacher C, Davies JE et al. Cdk5 phosphorylation of huntingtin reduces its cleavage by caspases: implications for mutant huntingtin toxicity. J Cell Biol 2005; 169:647-656.
260. Kaminosono S, Saito T, Oyama F et al. Suppression of mutant Huntingtin aggregate formation by Cdk5/p35 through the effect on microtubule stability. J Neurosci 2008; 28:8747-8755.
261. de Almeida LP, Ross CA, Zala D et al. Lentiviral-mediated delivery of mutant huntingtin in the striatum of rats induces a selective neuropathology modulated by polyglutamine repeat size, huntingtin expression levels and protein length. J Neurosci 2002; 22:3473-3483.
262. Gauthier LR, Charrin BC, Borrell-Pages M et al. Huntingtin controls neurotrophic support and survival of neurons by enhancing BDNF vesicular transport along microtubules. Cell 2004; 118:127-138.
263. Sinadinos C, Burbidge-King T, Soh D et al. Live axonal transport disruption by mutant huntingtin fragments in Drosophila motor neuron axons. Neurobiol Dis 2009.
264. Szebenyi G, Morfini GA, Babcock A et al. Neuropathogenic forms of huntingtin and androgen receptor inhibit fast axonal transport. Neuron 2003; 40:41-52.
265. Borrell-Pages M, Zala D, Humbert S et al. Huntington's disease: from huntingtin function and dysfunction to therapeutic strategies. Cell Mol Life Sci 2006; 63:2642-2660.
266. Smith R, Brundin P, Li JY. Synaptic dysfunction in Huntington's disease: a new perspective. Cell Mol Life Sci 2005; 62:1901-1912.
267. Colin E, Zala D, Liot G et al. Huntingtin phosphorylation acts as a molecular switch for anterograde/retrograde transport in neurons. EMBO J 2008; 27:2124-2134.
268. Dompierre JP, Godin JD, Charrin BC et al. Histone deacetylase 6 inhibition compensates for the transport deficit in Huntington's disease by increasing tubulin acetylation. J Neurosci 2007; 27:3571-3583.
269. Borrell-Pages M, Canals JM, Cordelieres FP et al. Cystamine and cysteamine increase brain levels of BDNF in Huntington disease via HSJ1b and transglutaminase. J Clin Invest 2006; 116:1410-1424.
270. del Toro D, Canals JM, Gines S et al. Mutant huntingtin impairs the postGolgi trafficking of brain-derived neurotrophic factor but not its Val66Met polymorphism. J Neurosci 2006; 26:12748-12757.
271. Yanai A, Huang K, Kang R et al. Palmitoylation of huntingtin by HIP14 is essential for its trafficking and function. Nature Neuroscience 2006; 9:824-831.
272. Ohyama T, Verstreken P, Ly CV et al. Huntingtin-interacting protein 14, a palmitoyl transferase required for exocytosis and targeting of CSP to synaptic vesicles. J Cell Biol 2007; 179:1481-1496.
273. Romero E, Cha GH, Verstreken P et al. Suppression of neurodegeneration and increased neurotransmission caused by expanded full-length huntingtin accumulating in the cytoplasm. Neuron 2008; 57:27-40.
274. Hamilton JM, Wolfson T, Peavy GM et al. Rate and correlates of weight change in Huntington's disease. Journal of Neurology, Neurosurgery and Psychiatry 2004; 75:209-212.
275. Djousse L, Knowlton B, Cupples LA et al. Weight loss in early stage of Huntington's disease. Neurology 2002; 59:1325-1330.

276. Arenas J, Campos Y, Ribacoba R et al. Complex I defect in muscle from patients with Huntington's disease. Ann Neurol 1998; 43:397-400.
277. Turner C, Cooper JM, Schapira AH. Clinical correlates of mitochondrial function in Huntington's disease muscle. Mov Disord 2007.
278. Jenkins BG, Rosas HD, Chen YC et al. 1H NMR spectroscopy studies of Huntington's disease: correlations with CAG repeat numbers. Neurology 1998; 50:1357-1365.
279. Mochel F, Charles P, Seguin F et al. Early energy deficit in Huntington disease: identification of a plasma biomarker traceable during disease progression. PLoS ONE 2007; 2:e647.
280. Beal MF, Brouillet E, Jenkins BG et al. Neurochemical and histologic characterization of striatal excitotoxic lesions produced by the mitochondrial toxin 3-nitropropionic acid. J Neurosci 1993; 13:4181-4192.
281. Benchoua A, Trioulier Y, Zala D et al. Involvement of mitochondrial complex II defects in neuronal death produced by N-terminus fragment of mutated huntingtin. Mol Biol Cell 2006; 17:1652-1663.
282. Fernandes HB, Baimbridge KG, Church J et al. Mitochondrial sensitivity and altered calcium handling underlie enhanced NMDA-induced apoptosis in YAC128 model of Huntington's disease. J Neurosci 2007; 27:13614-13623.
283. Cui L, Jeong H, Borovecki F et al. Transcriptional repression of PGC-1alpha by mutant huntingtin leads to mitochondrial dysfunction and neurodegeneration. Cell 2006; 127:59-69.
284. Weydt P, Pineda VV, Torrence AE et al. Thermoregulatory and metabolic defects in Huntington's disease transgenic mice implicate PGC-1alpha in Huntington's disease neurodegeneration. Cell Metab 2006; 4:349-362.
285. Chang DT, Rintoul GL, Pandipati S et al. Mutant huntingtin aggregates impair mitochondrial movement and trafficking in cortical neurons. Neurobiol Dis 2006; 22:388-400.
286. Orr AL, Li S, Wang CE et al. N-terminal mutant huntingtin associates with mitochondria and impairs mitochondrial trafficking. J Neurosci 2008; 28:2783-2792.
287. Wang H, Lim PJ, Karbowski M et al. Effects of overexpression of huntingtin proteins on mitochondrial integrity. Hum Mol Genet 2009; 18:737-752.
288. Lesort M, Chun W, Johnson GV et al. Tissue transglutaminase is increased in Huntington's disease brain. J Neurochem 1999; 73:2018-2027.
289. Karpuj MV, Garren H, Slunt H et al. Transglutaminase aggregates huntingtin into nonamyloidogenic polymers and its enzymatic activity increases in Huntington's disease brain nuclei. Proc Natl Acad Sci USA 1999; 96:7388-7393.
290. Dedeoglu A, Kubilus JK, Jeitner TM et al. Therapeutic effects of cystamine in a murine model of Huntington's disease. J Neurosci 2002; 22:8942-8950.
291. Battaglia G, Farrace MG, Mastroberardino PG et al. Transglutaminase 2 ablation leads to defective function of mitochondrial respiratory complex I affecting neuronal vulnerability in experimental models of extrapyramidal disorders. J Neurochem 2007; 100:36-49.
292. Mastroberardino PG, Iannicola C, Nardacci R et al 'Tissue' transglutaminase ablation reduces neuronal death and prolongs survival in a mouse model of Huntington's disease. Cell Death Differ 2002; 9:873-880.
293. Bailey CD, Johnson GV. Tissue transglutaminase contributes to disease progression in the R6/2 Huntington's disease mouse model via aggregate-independent mechanisms. J Neurochem 2005; 92:83-92.
294. Lee JM, Ivanova EV, Seong IS et al. Unbiased gene expression analysis implicates the huntingtin polyglutamine tract in extra-mitochondrial energy metabolism. PLoS Genet 2007; 3:e135.
295. Schilling G, Coonfield ML, Ross CA et al. Coenzyme Q10 and remacemide hydrochloride ameliorate motor deficits in a Huntington's disease transgenic mouse model. Neurosci Lett 2001; 315:149-153.
296. Smith KM, Matson S, Matson WR et al. Dose ranging and efficacy study of high-dose coenzyme Q10 formulations in Huntington's disease mice. Biochim Biophys Acta 2006; 1762:616-626.
297. Ferrante RJ, Andreassen OA, Dedeoglu A et al. Therapeutic effects of coenzyme Q10 and remacemide in transgenic mouse models of Huntington's disease. J Neurosci 2002; 22:1592-1599.
298. Dedeoglu A, Kubilus JK, Yang L et al. Creatine therapy provides neuroprotection after onset of clinical symptoms in Huntington's disease transgenic mice. J Neurochem 2003; 85:1359-1367.
299. Andreassen OA, Dedeoglu A, Ferrante RJ et al. Creatine increase survival and delays motor symptoms in a transgenic animal model of Huntington's disease. Neurobiol Dis 2001; 8:479-491.
300. A randomized, placebo-controlled trial of coenzyme Q10 and remacemide in Huntington's disease. Neurology 2001; 57:397-404.
301. Puri BK, Leavitt BR, Hayden MR et al. Ethyl-EPA in Huntington disease: a double-blind, randomized, placebo-controlled trial. Neurology 2005; 65:286-292.
302. Verbessem P, Lemiere J, Eijnde BO et al. Creatine supplementation in Huntington's disease: a placebo-controlled pilot trial. Neurology 2003; 61:925-930.

Chapter 6

Clinical Features and Molecular Mechanisms of Spinal and Bulbar Muscular Atrophy (SBMA)

Masahisa Katsuno,* Haruhiko Banno, Keisuke Suzuki, Hiroaki Adachi, Fumiaki Tanaka and Gen Sobue*

Abstract

Spinal and bulbar muscular atrophy (SBMA) is an adult-onset neurodegenerative disease characterized by slowly progressive muscle weakness and atrophy. The cause of this disease is the expansion of a trinucleotide CAG repeat, which encodes the polyglutamine tract, within the first exon of the *androgen receptor* (*AR*) gene. SBMA exclusively occurs in adult males, whereas both heterozygous and homozygous females are usually asymptomatic. Lower motor neurons in the anterior horn of the spinal cord and those in the brainstem motor nuclei are predominantly affected in SBMA, and other neuronal and nonneuronal tissues are also widely involved to some extent. Testosterone-dependent nuclear accumulation of the pathogenic AR protein has been considered to be a fundamental step of neurodegenerative process, which is followed by several molecular events such as transcriptional dysregulation, axonal transport disruption and mitochondrial dysfunction. Results of animal studies suggest that androgen deprivation and activation of protein quality control systems are potential therapies for SBMA.

Clinical Features

Spinal and bulbar muscular dystrophy (SBMA) chiefly affects adult males. The prevalence of this disease is estimated to be 1-2 per 100,000, although a considerable number of patients may be misdiagnosed as other neuromuscular diseases including amyotrophic lateral sclerosis (ALS).[1] Patients of various ethnic backgrounds have been reported throughout the world.

The major symptoms of SBMA are weakness, atrophy and fasciculations of bulbar, facial and limb muscles, which are attributable to degeneration of lower motor neurons in the spinal cord and brainstem.[2,3] Subclinical dysfunction of upper motor neurons has been suggested by clinical examinations such as magnetic resonance spectroscopy, although histopathological evidence is not sufficient.[4] In extremities, involvement is usually predominant in proximal musculature and is occasionally asymmetric. The onset of weakness is usually between 30 and 60 years of age, but is often preceded by nonspecific symptoms such as postural tremor and muscle cramps. Typically, affected individuals require a wheelchair 15-20 years after the onset of weakness.[5,6] Fasciculations are not apparent at rest, but become conspicuous upon voluntary muscle movement. These contraction fasciculations are especially noticeable in the face, neck and tongue. Neuromuscular symptoms are often worsened by coldness and by fatigue after exercise. Bilateral facial and masseter muscle

*Corresponding Authors: Masahisa Katsuno and Gen Sobue—Department of Neurology, Nagoya University Graduate School of Medicine, 65 Tsurumai-cho, Showa-ku, Nagoya 466-8550, Japan. Email: ka2no@med.nagoya-u.ac.jp, sobueg@med.nagoya-u.ac.jp

Diseases of DNA Repair, edited by Shamim I. Ahmad. ©2010 Landes Bioscience and Springer Science+Business Media.

weakness, poor uvula and soft palatal movements and atrophy of the tongue with fasciculations are often encountered. Speech has a nasal quality in most cases due to reduced velopharyngeal closure. Patients occasionally experience laryngospasm, a sudden sensation of dyspnea.[7] Advanced cases often develop dysphagia, eventually resulting in aspiration or choking. Muscle tone is usually hypotonic and no pyramidal signs are detected. The deep tendon reflex is diminished or absent with no pathological reflex. Sensory involvement is largely restricted to a sense of vibration, which is affected distally in the legs. Cerebellar symptoms, dysautonomia and cognitive impairment are absent. Patients occasionally demonstrate signs of androgen insensitivity such as gynecomastia, testicular atrophy, dyserection and decreased fertility, some of which are detected before the onset of motor symptoms. Abdominal obesity is common, whereas male pattern baldness is rare in patients with SBMA.

Electromyogram shows neurogenic abnormalities, and distal motor latencies are often prolonged in nerve conduction studies. Both sensory nerve action potentials and sensory evoked potentials are reduced or absent.[8] Reflecting muscle fatigability, decremental responses to repetitive nerve stimulation are often detected in SBMA patients.[9] Endocrinological examinations frequently reveal partial androgen resistance with elevated serum testosterone levels.[10] Serum creatine kinase levels are elevated in the majority of patients. Although less common, hyperlipidemia, liver dysfunction and glucose intolerance are also detected. Profound facial fasciculations, bulbar signs, gynecomastia and sensory disturbances are the main clinical features distinguishing SBMA from other motor neuron diseases, although genetic analysis is indispensable for diagnosis. Female cases are usually asymptomatic, but some express subclinical phenotypes including high amplitude motor unit potentials on electromyography.[11]

The progression of SBMA is usually slow, but life-threatening respiratory tract infection often occurs in the advanced stage of the disease, resulting in early death in some patients. The cardinal cause of death is aspiration pneumonia.[5] No specific therapy for SBMA has been established. Testosterone has been used in some patients, although it has no effects on the progression of SBMA.

Genetic Basis

The cause of SBMA is expansion of a trinucleotide CAG repeat, which encodes the polyglutamine tract, in the first exon of the *androgen receptor* (*AR*) gene.[12] The CAG repeat within the *AR* ranges in size from 9 to 36 in normal subjects and from 38 to 62 in SBMA patients.[13] Expanded polyglutamine tracts have been found to cause several neurodegenerative diseases including SBMA, Huntington's disease (HD), several forms of spinocerebellar ataxia and dentatorubral-pallidoluysian atrophy (DRPLA).[14] In these disorders, known as polyglutamine diseases, the CAG repeat has a strong tendency to further expand, accelerating the disease onset with in successive generations.[15] As documented in other polyglutamine diseases, the CAG repeat size correlates well with the age of onset in SBMA, but does not appear to dictate the rate of disease progression.[5,16] Moreover, the CAG repeat size has also been shown to correlate with motor- and sensory-dominant phenotypes in nerve conduction studies.[8]

The AR, the causative protein of SBMA, is a 110-kDa nuclear receptor which belongs to the steroid/thyroid hormone receptor family. The AR is essential for major androgen effects including normal male sexual differentiation and pubertal sexual development, although an AR-independent nongenomic function of androgen has been reported. AR proteins are expressed not only in primary and secondary sexual organs, but also in nonreproductive organs including the kidney, skeletal muscle, adrenal gland, skin and nervous system, suggesting its far-reaching influence on a variety of mammalian tissues. In the central nervous system, the AR expression level is relatively high in spinal and brainstem motor neurons, the same cells which are vulnerable in SBMA. The *AR* gene is located on chromosome Xq11-12. This 90 kb DNA contains eight exons coding for the functional domains specific to the nuclear receptor family. In addition to SBMA, there are various medical conditions that are likely associated with the CAG repeat length: prostate cancer diseases susceptibility, hirsutism, male infertility and cryptorchidism.[17]

Histopathology

Histopathological studies have provided important information on the pathogenesis of polyglutamine-mediated neurodegeneration. The fundamental histopathological finding in SBMA is loss of lower motor neurons in the anterior horn of the spinal cord as well as in the brainstem motor nuclei except for the third, fourth and sixth cranial nerves.[18] Sensory neurons in the dorsal root ganglia are less severely affected. Muscle histopathology includes both neurogenic and myogenic finding. A pathological hallmark of polyglutamine diseases is the presence of nuclear inclusions (NIs). In SBMA, NIs containing the pathogenic AR are found in the residual motor neurons in the brainstem and spinal cord as well as in nonneuronal tissues including the prostate, testes and skin.[19] Recent studies, however, indicate that NIs are likely formed as a result of cellular defense reactions coping with the pathogenic polyglutamine protein.[20] Instead, diffuse accumulation of misfolded proteins appears to play an important role in the initiation of neurodegeneration in polyglutamine disease.[21] In agreement with this view, an immunohistochemical study on autopsied SBMA patients, using an anti-polyglutamine antibody, 1C2, demonstrated that diffuse nuclear accumulation of the pathogenic AR is more frequently observed than NIs in the anterior horn of the spinal cord.[22] Intriguingly, the frequency of diffuse nuclear accumulation of the pathogenic AR in spinal motor neurons strongly correlates with the length of the CAG repeat in the AR gene. No such correlation has been found between NIs occurrence and the CAG repeat length. Similar findings have also been reported in other polyglutamine diseases. Taken together, it appears that the pathogenic AR containing an elongated polyglutamine tract principally accumulates within the nuclei of motor neurons in a diffusible form, leading to neuronal dysfunction and eventual cell death. In support of this hypothesis, neuronal dysfunction is halted by genetic modulation preventing nuclear import of the pathogenic polyglutamine-containing protein in cellular and animal models of polyglutamine diseases.[14]

Since the human AR is widely expressed in various organs, nuclear accumulation of the pathogenic AR protein is detected not only in the central nervous system, but also in nonneuronal tissues such as scrotal skin. The degree of pathogenic AR accumulation in scrotal skin epithelial cells tends to be correlated with that in the spinal motor neurons in autopsy specimens and it is well correlated with CAG repeat length and inversely correlated with the motor functional scale.[23] These findings indicate that scrotal skin biopsy with anti-polyglutamine immunostaining is a biomarker for monitoring SBMA pathogenic processes. Since SBMA is a slowly progressive disorder, appropriate biomarkers would help improve the power and cost-effectiveness of longitudinal clinical treatment trials.

Molecular Mechanisms

Aggregation of Mutant Androgen Receptor

The expanded polyglutamine tract in AR has been implicated in the pathogenesis of SBMA in two different ways, but not mutually exclusive: (1) loss of normal AR function inducing neuronal degeneration and (2) the pathogenic AR acquiring toxic properties damaging motor neurons. Since AR possesses trophic effects on neuronal cells, one can assume that loss of the AR function may play a role in the pathogenesis of SBMA. Expansion of the polyglutamine tract mildly suppresses the transcriptional activities of AR, probably because it disrupts interaction between the N-terminal transactivating domain of AR and transcriptional co-activators.[24] Although this loss of function of AR may contribute to the androgen insensitivity in SBMA, the pivotal cause of neurodegeneration in SBMA is believed to be a gain of toxic function of the pathogenic AR due to expansion of the polyglutamine tract. This hypothesis is supported by the observation that motor impairment has never been observed in severe testicular feminization (Tfm) patients lacking AR function or in AR knockout mice. Moreover, a transgenic mouse model carrying an elongated CAG repeat, driven by human AR promoter demonstrated motor impairment, suggesting that the expanded polyglutamine tract is sufficient to induce the pathogenic process of SBMA.[25]

Aggregation of abnormal protein has been considered to be central to the pathogenesis of neurodegenerative diseases such as Alzheimer's disease, Parkinson's disease, ALS and prion diseases. An expanded polyglutamine stretch alters the conformation of causative proteins, resulting in aggregation of the proteins. It is now widely accepted that aggregation of these abnormal proteins in neurons is the primary event in the pathogenesis of polyglutamine diseases. The rate-limiting step of aggregation has been proposed to be the formation of an oligomeric nucleus, which may form after a repeat length-dependent conformational change of the polyglutamine monomer from a random coil to a parallel, helical ß-sheet.[26] Several experimental observations indicate that the formation of toxic oligomers or intermediates of abnormal polyglutamine-containing protein instigates a series of cellular events which lead to neurodegeneration.[27] This hypothesis is also applicable in a mouse model of SBMA, in which soluble oligomers are detectable prior to the onset of neuromuscular symptoms.[28]

Testosterone-Dependent Neurodegeneration in SBMA

SBMA is unique among polyglutamine diseases in that the pathogenic protein, AR, has a specific ligand, testosterone, which alters the subcellular localization of the protein by favoring its nuclear uptake. The AR is normally confined to a multi-heteromeric inactive complex in the cell cytoplasm and translocates into the nucleus in a ligand-dependent manner. This ligand-dependent intracellular trafficking of AR appears to play an important role in the pathogenesis of SBMA. The phenotypic difference with gender, which is a specific feature of SBMA, has been recapitulated in animal models of SBMA.[29,30] As for a transgenic mouse model of SBMA, expressing the full-length human AR containing 97 CAGs (AR-97Q), neuromuscular symptoms are markedly pronounced and accelerated in the male mice, but either not observed or far less severe in the female counterparts. Male AR-97Q mice show markedly more abundant diffuse nuclear staining than females, in accordance with the symptomatic difference according to gender. Androgen deprivation through surgical castration substantially improved the symptoms, histopathological findings and nuclear accumulation of the pathogenic AR in the male AR-97Q mice. In contrast, subcutaneous injection of testosterone causes significant aggravation of symptoms, histopathological features and nuclear localization of the pathogenic AR in the female AR-97Q mice.[29] Since the nuclear translocation of AR is ligand-dependent, testosterone appears to show toxic effects by accelerating nuclear translocation of the pathogenic AR. On the other hand, the pathogenic AR, eliminated from the nucleus, appears to be degraded via autophagic pathway.[31] The ligand-dependent degeneration of motor neurons has also been reported in other animal models of SBMA.[32,33] It should be noted that testosterone deprivation by castration reverses motor dysfunction in transgenic mice of SBMA.[32,34] Lending support to the ligand-dependent hypothesis are the clinical observations that manifestation of symptoms is minimal even in the females homozygous for an expanded CAG repeat in the *AR* gene and that testosterone administration exacerbates neuromuscular symptoms in a patient with SBMA.[35,36]

Transcriptional Dysregulation

Previous reports demonstrated that transcriptional alteration is an early change in the pathogenesis in mouse models of polyglutamine diseases.[37,38] For example, the expression levels of various heat shock proteins were substantially decreased in animal models of SBMA and HD.[39,40] Transcriptional co-activators, such as CBP, are sequestrated into the polyglutamine-containing NIs through protein-protein interaction in mouse models and patients with polyglutamine diseases.[41] Alternatively, the interaction between transcriptional co-activators and soluble pathogenic protein has also been demonstrated in animal models of polyglutamine diseases as well as in postmortem tissues of patients.[42] CBP functions as a histone acetyltransferase (HAT), regulating gene transcription and chromatin structure. It has been indicated that the HAT activity of CBP is suppressed in cellular models of polyglutamine diseases. Taken together, transcriptional dysregulation due to decrease in histone acetylation is likely to underlie the pathogenesis of neurodegeneration in polyglutamine diseases. This hypothesis is exemplified by the experimental observation that acetylation of nuclear histone H3 is significantly diminished in the spinal cord of SBMA mice.[43]

Additionally, dysfunction of CBP results in a decreased expression of vascular endothelial growth factor (VEGF) in another mouse model of SBMA, indicating that transcriptional alteration is a trigger for neurodegeneration in this disease.[44] Testosterone-dependant transcriptional dysregulation has also been studied using a Drosophila model of SBMA, in which the pathogenic AR enhances an androgen-dependent association of AR with Retinoblastoma protein. This interaction appears to lead to an aberrant E2F transactivation through suppression of histone deacetylation.[45]

Disruption of Axonal Transport

Obstruction of axonal transport has attracted attention as a cause of neuronal dysfunction in a variety of neurodegenerative diseases including SBMA. Mutations in the genes for proteins regulating axonal trafficking, dynein and dynactin 1, have been shown to cause motor neuron degeneration in both humans and rodents.[46,47] The mRNA level of dynactin 1 is significantly reduced in the SBMA mice resulting from pathogenic AR-induced transcriptional dysregulation.[34] In addition, overexpression of dynactin 1 mitigates the neuronal toxicity of the pathogenic AR in a cell culture model of SBMA. These observations indicate that polyglutamine-dependent transcriptional dysregulation of dynactin 1 plays a crucial role in the reversible neuronal dysfunction in the early stage of SBMA. Pathogenic AR containing expanded polyglutamine has also been demonstrated to activate c-Jun N-terminal kinase (JNK), leading to inhibition of kinesin-1 microtubule-binding activity and eventual disruption of anterograde axonal transport.[48] It is noteworthy that JNK inhibitors reverse the suppression of neurite outgrowth by pathogenic AR in cultured cells.

Mitochondrial Dysfunction

Mitochondrial impairment and eventual caspase activation have been implicated in the pathogenesis of neurodegeneration in polyglutamine diseases. The N-terminal fragments of the pathogenic AR are shown to activate an apoptotic pathway in primary cortical neurons. This phenomenon is fully dependent on Bax, an apoptosis-mediating factor which stimulates cytochrome c release from mitochondria.[49] Expression of pathogenic AR in the presence of ligand is shown to result in mitochondrial membrane depolarization and an elevated level of reactive oxygen species, which is blocked by the treatment with the antioxidants co-enzyme Q10 and idebenone.[50] It is also demonstrated that pathogenic AR represses the transcription of subunits of peroxisome proliferator-activated receptor gamma coactivator-1 (PGC-1), a transcriptional co-activator that regulates the expression of various nuclear-encoded mitochondrial proteins, suggesting a link between polyglutamine-mediated transcriptional dysregulation and mitochondrial dysfunction.[50]

Noncell Autonomous Toxicity of Lower Motor Neurons

Although neurons are the primary target of polyglutamine-mediated toxicity, pathological changes in nonneuronal cells may play a role in the development of neurodegeneration in polyglutamine diseases.[51] In SBMA, myopathic changes in the muscle biopsy and elevated serum creatine kinase levels imply a direct involvement of the skeletal muscle. Although it is not clear how the muscle damage modifies pathologic processes in motor neurons of the patients, it has been demonstrated that skeletal muscle pathology precedes neurodegeneration in a knock-in mouse model of SBMA.[30] Intriguingly, an exclusive overexpression of wild-type AR in the skeletal muscle is shown to induce motor axon loss mimicking SBMA.[52] These observations suggest that AR-mediated myopathy might contribute to noncell autonomous degeneration of spinal motor neurons.

Therapeutic Strategies

Testosterone Deprivation Therapy

Leuprorelin is a potent luteinizing hormone-releasing hormone (LHRH) analogue that suppresses the releases of gonadotropins, luteinising hormone (LH) and follicle-stimulating hormone (FSH). This drug has been tried for a variety of sex hormone-dependent diseases including prostate cancer, endometriosis and prepuberty. Leuprorelin has been shown to inhibit nuclear accumulation of the pathogenic AR, resulting in a marked improvement of neuromuscular phenotypes seen in the AR-97Q transgenic mice.[53] Leuprorelin-treated AR-97Q mice show a longer lifespan, larger body size and better motor performance compared with vehicle-treated mice. Leuprorelin appears to improve neuronal dysfunction by preventing ligand-dependent nuclear translocation of the pathogenic AR in the same way as castration. In a Phase 2 clinical trial, 12-month treatment with leuprorelin significantly diminished the serum level of creatine kinase and suppressed nuclear accumulation of the pathogenic AR in the scrotal skin of the patients. Of note is the observation that the frequencies of 1C2-positive neurons in the anterior horn and brainstem of an autopsied patient, who received leuprorelin for 2 years, were less than those in nontreated SBMA patients (Fig. 1), suggesting that androgen deprivation intervenes in the pathogenic process of human SBMA, as demonstrated in animal studies.[54]

AR coregulators, such as ARA70, are alternative therapeutic targets, because they control the function and cellular distribution of AR. It has been shown that 5-hydroxy-1,7-bis (3,4-dimethoxyphenyl)-1,4,6-heptatrien-3-one (ASC-J9) dissociates AR and ARA70, resulting in suppression of pathogenic AR aggregation as well as amelioration of neuromuscular symptoms in a mouse model of SBMA.[55] On the other hand, Akt-induced phosphorylation of AR blocks ligand binding and thereby mitigates toxicity in cultured motor neurons. This phenomenon is enhanced by insulin-like growth factor-1 (IGF-1) stimulation, which activates several cell survival promoting pathway.[56]

Manipulation of Heat Shock Proteins and Ubiquitin Proteasome System

Many components of the ubiquitin proteasome system (UPS) and molecular chaperones are known to colocalize with polyglutamine-containing NIs, implying that failure of cellular defense mechanisms underlies neurodegeneration in polyglutamine diseases. Given that the activity of UPS is not dampened in SBMA transgenic mice even at their advanced stage, intensification of these cellular coping responses to the abnormal protein burden is another key to suppression of the pathogenesis of SBMA.[57] Inhibition of Hsp90 has been demonstrated to facilitate proteasomal degradation of its client proteins. The Hsp90-client protein complex is stabilized when it is associated with p23, a cochaperone interacting with Hsp90. Treatment with 17-allylamino geldanamycin (17-AAG), a potent Hsp90 inhibitor, dissociated p23 from the Hsp90-AR complex and thus facilitated proteasomal degradation of the pathogenic AR in cellular and mouse models of SBMA. 17-AAG thereby inhibits nuclear accumulation of this protein, leading to marked amelioration of motor phenotypes in the SBMA mouse model without detectable toxicity.[58] Oral administration of 17-DMAG, a derivative of 17-AAG, shows a similar effects.[57]

On the other hand, there is increasing evidence that heat shock proteins (HSPs), stress-inducible molecular chaperones, abrogate polyglutamine-mediated cytotoxicity by refolding and solubilizing the pathogenic proteins. Over-expression of the inducible form of human Hsp70 markedly ameliorated symptomatic and histopathological phenotypes in our transgenic mouse model of SBMA.[59] Hsp70 has also been shown to facilitate proteasomal degradation of abnormal AR protein in a cell culture model of SBMA.[60] Overexpression of CHIP, or C terminus of Hsc70 (heat shock cognate protein 70)-interacting protein has also been shown to prevent nuclear accumulation of pathogenic AR and thereby ameliorate motor symptoms in a transgenic mouse model of SBMA.[61] Favorable effects obtained by genetic modulation of HSPs suggest that pharmacological induction of molecular chaperones might be a promising approach to SBMA and other polyglutamine diseases.

Geranylgeranylacetone (GGA), an acyclic isoprenoid compound with a retinoid skeleton, has been shown to strongly induce HSP expression in various tissues. Oral administration of GGA upregulates the levels of Hsp70, Hsp90 and Hsp105 via activation of heat shock factor-1 (Hsf1) in the central nervous system and inhibits nuclear accumulation of the pathogenic AR protein, resulting in amelioration of polyglutamine-dependent neuromuscular phenotypes of SBMA transgenic mice.[40] Induction of autophagy, another cellular machinery coping with abnormal protein toxicity, is also shown to mitigate the toxicity of nuclear-residing pathogenic AR in cultured motor neurons.[31] Taken together, pharmacological activation of protein quality control systems appears to be a potential therapy for SBMA.

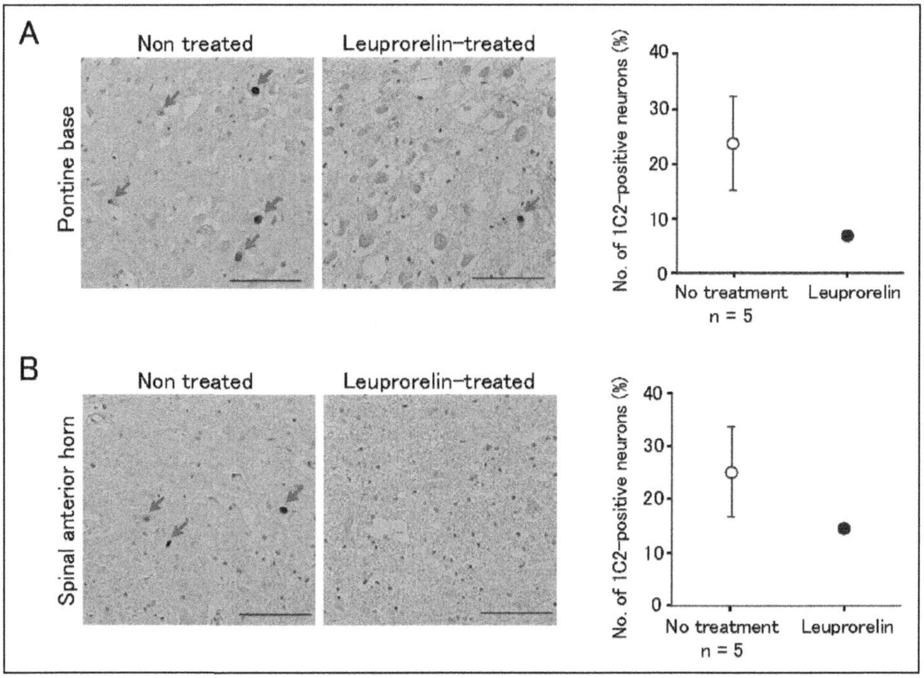

Figure 1. Effects of leuprorelin acetate on nuclear accumulation of mutant androgen receptor (AR) in motor neurons. The accumulation of pathogenic AR in neurons was remarkable, both in the pontine base (A) and in the spinal anterior horn (B) of all the control, nontreated autopsied patients, but the number of 1C2-positive neurons was relatively small in the leuprorelin-treated patient. Scale bars = 100 μm. Reproduced from Banno et al. Phase 2 trial of leuprorelin in patients with spinal and bulbar muscular atrophy. An Neurol 2009; 65(2):140-50,.

Restoration of Transcriptional Activity

The histone acetylation level is determined by an interplay between histone acetyltransferase and histone deacetylase (HDAC). Since suppression of HDAC activity results in augmentation of histone acetylation and subsequent restoration of gene transcription, HDAC inhibitors have been considered to be of therapeutic benefit in polyglutamine diseases. Butyrate, the first HDAC inhibitor to be discovered, has been shown to ameliorate symptomatic and histopathological phenotypes of the AR-97Q transgenic mouse model through upregulation of histone acetylation in nervous tissues.[43] This compound has also been shown to alleviate neurodegeneration in a mouse model of DRPLA.[62] Although sodium butyrate is likely to be a promising treatment of SBMA, this compound yielded beneficial effects only within a narrow therapeutic range of dose in the mouse model. Careful dose determination is therefore necessary when using HDAC inhibitors for the treatment of polyglutamine diseases.

Conclusion

The ligand-dependent accumulation of the pathogenic AR initiates the neurodegenerative process in SBMA (Fig. 2). This step is followed by several downstream molecular events such as transcriptional dysregulation, axonal transport disruption and mitochondrial insufficiency. Although the precise mechanism by which motor neurons die remains unclear androgen deprivation and activation of cellular defense reactions, UPS, HSPs and autophagy, are promising therapeutic approaches to SBMA.

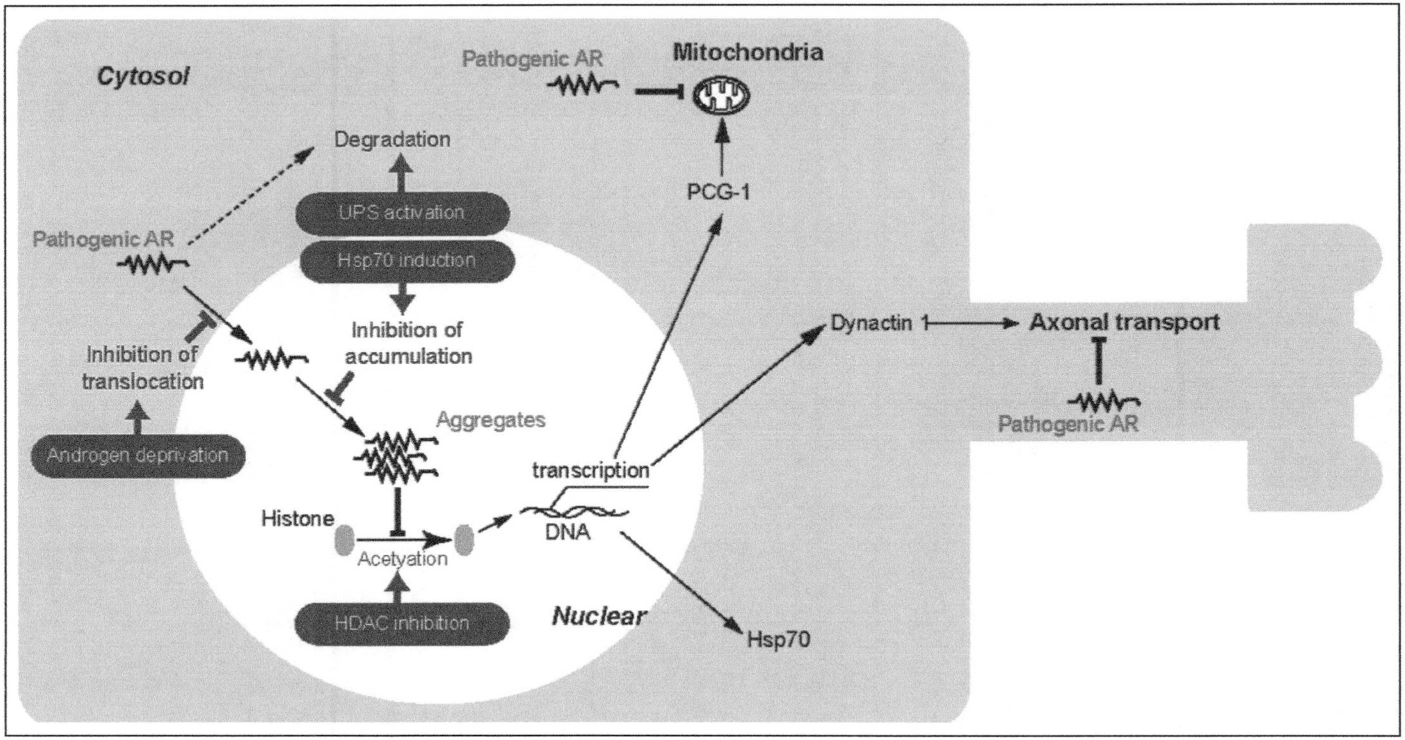

Figure 2. Molecular basis and therapeutic strategies in SBMA. Ligand-dependent nuclear accumulation of the pathogenic AR has been construed as the initial step in the neurodegeneration process of SBMA, which leads to transcriptional dysregulation of several genes including peroxisome proliferator-activated receptor gamma coactivator-1 (PGC-1), dynactin 1 and heat shock protein 70 (Hsp70). There is increasing evidence that androgen deprivation inhibits the pathogenic AR accumulation and thereby prevents symptomatic progression of SBMA. Sodium butyrate, a histone deacetylase (HDAC) inhibitor, is shown to restore transcriptional activity through histone acetylation in a mouse model of SBMA. Animal studies also suggest that activation of ubiquitin proteasome system (UPS) and induction of Hsp70 are potential therapeutic approaches to SBMA.

Acknowledgements

This work was supported by the Program for Improvement of Research Environment for Young Researchers from Special Coordination Funds for Promoting Science and Technology (SCF) commissioned by the Ministry of Education, Culture, Sports, Science and Technology of Japan, a Center-of-Excellence (COE) grant from the Ministry of Education, Culture, Sports, Science and Technology, Japan and grants from the Ministry of Health, Labour and Welfare, Japan.

References

1. Fischbeck KH. Kennedy disease. J Inherit Metab Dis 1997; 20:152-158.
2. Kennedy WR, Alter M, Sung JH. Progressive proximal spinal and bulbar muscular atrophy of late onset. A sex-linked recessive trait. Neurology 1968; 18:671-680.
3. Katsuno M, Adachi H, Waza M et al. Pathogenesis, animal models and therapeutics in spinal and bulbar muscular atrophy (SBMA). Exp Neurol 2006; 200:8-18.
4. Pachatz C, Terracciano C, Desiato MT et al. Upper motor neuron involvement in X-linked recessive bulbospinal muscular atrophy. Clin Neurophysiol 2007; 118:262-268.
5. Atsuta N, Watanabe H, Ito M et al. Natural history of spinal and bulbar muscular atrophy (SBMA): a study of 223 Japanese patients. Brain 2006; 129:1446-1455.
6. Takeuchi Y, Katsuno M, Banno H et al. Walking capacity evaluated by the 6-minute walk test in spinal and bulbar muscular atrophy. Muscle Nerve 2008; 38:964-971.
7. Sperfeld AD, Hanemann CO, Ludolph AC et al. Laryngospasm: an underdiagnosed symptom of X-linked spinobulbar muscular atrophy. Neurology 2005; 64:753-754.
8. Suzuki K, Katsuno M, Banno H et al. CAG repeat size correlates to electrophysiological motor and sensory phenotypes in SBMA. Brain 2008; 131:229-239.
9. Inoue K, Hemmi S, Miyaishi M et al. Muscular fatigue and decremental response to repetitive nerve stimulation in X-linked spinobulbar muscular atrophy. Eur J Neurol 2009; 16:76-80.
10. Dejager S, Bry-Gauillard H, Bruckert E et al. A comprehensive endocrine description of Kennedy's disease revealing androgen insensitivity linked to CAG repeat length. J Clin Endocr Metab 2002; 87:3893-3901.
11. Sobue G, Doyu M, Kachi T et al. Subclinical phenotypic expressions in heterozygous females of X-linked recessive bulbospinal neuronopathy. J Neurol Sci 1993; 117:74-78.
12. La Spada AR, Wilson EM, Lubahn DB et al. Androgen receptor gene mutations in X-linked spinal and bulbar muscular atrophy. Nature 1991; 352:77-79.
13. Andrew SE, Goldberg YP, Hayden MR. Rethinking genotype and phenotype correlations in polyglutamine expansion disorders. Hum Mol Genet 1997; 6:2005-2010.
14. Gatchel JR, Zoghbi HY. Diseases of unstable repeat expansion: mechanism and principles. Nat Rev Genet 2005; 6:743-755.
15. Tanaka F, Doyu M, Ito Y et al. Founder effect in spinal and bulbar muscular atrophy (SBMA). Hum Mol Genet 1996; 5:1253-1257.
16. Doyu M, Sobue G, Mukai E et al. Severity of X-linked recessive bulbospinal neuronopathy correlates with size of the tandem CAG repeat in androgen receptor gene. Ann Neurol 1992; 32:707-710.
17. Palazzolo I, Gliozzi A, Rusmini P et al. The role of the polyglutamine tract in androgen receptor. J Steroid Biochem Mol Biol 2008; 108:245-253.
18. Sobue G, Hashizume Y, Mukai E et al. X-linked recessive bulbospinal neuronopathy. A clinicopathological study. Brain 1989; 112:209-232.
19. Li M, Miwa S, Kobayashi Y et al. Nuclear inclusions of the androgen receptor protein in spinal and bulbar muscular atrophy. Ann Neurol 1998; 44:249-254.
20. Arrasate M, Mitra S, Schweitzer ES et al. Inclusion body formation reduces levels of mutant huntingtin and the risk of neuronal death. Nature 2004; 431:805-810.
21. Nagai Y, Inui T, Popiel HA et al. A toxic monomeric conformer of the polyglutamine protein. Nat Struct Mol Biol 2007; 14:332-340.
22. Adachi H, Katsuno M, Minamiyama M et al. Widespread nuclear and cytoplasmic accumulation of mutant androgen receptor in SBMA patients. Brain 2005; 128:659-670.
23. Banno H, Adachi H, Katsuno M et al. Mutant androgen receptor accumulation in SBMA scrotal skin: a pathogenic marker. Ann Neurol 2006; 59:520-526.
24. Poletti A. The polyglutamine tract of androgen receptor: from functions to dysfunctions in motor neurons. Front. Neuroendocrinol 2004; 25:1-26.
25. Adachi H, Kume A, Li M et al. Transgenic mice with an expanded CAG repeat controlled by the human AR promoter show polyglutamine nuclear inclusions and neuronal dysfunction without neuronal cell death. Hum Mol Genet 2001; 10:1039-1048.

26. Wyttenbach A. Role of heat shock proteins during polyglutamine neurodegeneration: mechanisms and hypothesis. J Mol Neurosci 2004; 23:69-96.
27. Muchowski PJ, Wacker JL. Modulation of neurodegeneration by molecular chaperones. Nat Rev Neurosci 2005; 6:11-22.
28. Li M, Chevalier-Larsen ES, Merry DE et al. Soluble androgen receptor oligomers underlie pathology in a mouse model of spinobulbar muscular atrophy. J Biol Chem 2007; 282:3157-3164.
29. Katsuno M, Adachi H, Kume A et al. Testosterone reduction prevents phenotypic expression in a transgenic mouse model of spinal and bulbar muscular atrophy. Neuron 2002; 35:843-854.
30. Yu Z, Dadgar N, Albertelli M et al. Abnormalities of germ cell maturation and sertoli cell cytoskeleton in androgen receptor 113 CAG knock-in mice reveal toxic effects of the mutant protein. Am J Pathol 2006; 168:195-204.
31. Montie HL, Cho MS, Holder L et al. Cytoplasmic Retention of Polyglutamine-Expanded Androgen Receptor Ameliorates Disease via Autophagy in a Mouse Model of Spinal and Bulbar Muscular Atrophy. Hum Mol Genet 2009 (Epub ahead of print).
32. Takeyama K, Ito S, Yamamoto A et al. Androgen-dependent neurodegeneration by polyglutamine-expanded human androgen receptor in Drosophila. Neuron 2002; 35:855-864.
33. Chevalier-Larsen ES, O'Brien CJ, Wang H et al. Castration restores function and neurofilament alterations of aged symptomatic males in a transgenic mouse model of spinal and bulbar muscular atrophy. J Neurosci 2004; 24:4778-4786.
34. Katsuno M, Adachi H, Minamiyama M et al. Reversible disruption of dynactin 1-mediated retrograde axonal transport in polyglutamine-induced motor neuron degeneration. J Neurosci 2006; 26:12106-12117.
35. Schmidt BJ, Greenberg CR, Allingham-Hawkins DJ et al. Expression of X-linked bulbospinal muscular atrophy (Kennedy disease) in two homozygous women. Neurology 2002; 59:770-772.
36. Kinirons P, Rouleau GA. Administration of testosterone results in reversible deterioration in Kennedy's disease. J Neurol Neurosurg Psychiatry 2008; 79:106-107.
37. Lin X, Antalffy B, Kang D et al. Polyglutamine expansion down-regulates specific neuronal genes before pathologic changes in SCA1. Nat Neurosci 2000; 3:157-163.
38. Luthi-Carter R, Strand A, Peters NL et al. Decreased expression of striatal signaling genes in a mouse model of Huntington's disease. Hum Mol Genet 2000; 9:1259-1271.
39. Hay DG, Sathasivam K, Tobaben S et al. Progressive decrease in chaperone protein levels in a mouse model of Huntington's disease and induction of stress proteins as a therapeutic approach. Hum Mol Genet 2004; 13:1389-1405.
40. Katsuno M, Sang C, Adachi H et al. Pharmacological induction of heat-shock proteins alleviates polyglutamine-mediated motor neuron disease. Proc Natl Acad Sci USA 2005; 102:16801-16806.
41. Nucifora FC Jr, Sasaki M, Peters MF et al. Interference by huntingtin and atrophin-1 with cbp-mediated transcription leading to cellular toxicity. Science 2001; 291:2423-2428.
42. Steffan JS, Bodai L, Pallos J et al. Histone deacetylase inhibitors arrest polyglutamine-dependent neurodegeneration in Drosophila. Nature 2001; 413:739-743.
43. Minamiyama M, Katsuno M, Adachi H et al. Sodium butyrate ameliorates phenotypic expression in a transgenic mouse model of spinal and bulbar muscular atrophy. Hum Mol Genet 2004; 13:1183-1192.
44. Sopher BL, Thomas PS Jr, LaFevre-Bernt MA et al. Androgen receptor YAC transgenic mice recapitulate SBMA motor neuronopathy and implicate VEGF164 in the motor neuron degeneration. Neuron 2004; 41:687-699.
45. Suzuki E, Zhao Y, Ito S et al. Aberrant E2F activation by polyglutamine expansion of androgen receptor in SBMA neurotoxicity. Proc Natl Acad Sci USA 2009; 106:3818-3822.
46. Puls I, Jonnakuty C, LaMonte BH et al. Mutant dynactin in motor neuron disease. Nat Genet 2003; 33:455-456.
47. Hafezparast M, Klocke R, Ruhrberg C et al. Mutations in dynein link motor neuron degeneration to defects in retrograde transport. Science 2003; 300:808-812.
48. Morfini G, Pigino G, Szebenyi G et al. JNK mediates pathogenic effects of polyglutamine-expanded androgen receptor on fast axonal transport. Nat Neurosci 2006; 9:907-916.
49. Young JE, Garden GA, Martinez RA et al. Polyglutamine-expanded androgen receptor truncation fragments activate a Bax-dependent apoptotic cascade mediated by DP5/Hrk. J Neurosci 2009; 29:1987-1997.
50. Ranganathan S, Harmison GG, Meyertholen K et al. Mitochondrial abnormalities in spinal and bulbar muscular atrophy. Hum Mol Genet 2009; 18:27-42.
51. Custer SK, Garden GA, Gill N et al. Bergmann glia expression of polyglutamine-expanded ataxin-7 produces neurodegeneration by impairing glutamate transport. Nat Neurosci 2006; 9:1302-1311.

52. Monks DA, Johansen JA, Mo K et al. Overexpression of wild-type androgen receptor in muscle recapitulates polyglutamine disease. Proc Natl Acad Sci USA 2007; 104:18259-18264.
53. Katsuno M, Adachi H, Doyu M et al. Leuprorelin rescues polyglutamine-dependent phenotypes in a transgenic mouse model of spinal and bulbar muscular atrophy. Nat Med 2003; 9:768-773.
54. Banno H, Katsuno M, Suzuki K et al. Phase 2 trial of leuprorelin in patients with spinal and bulbar muscular atrophy. Ann Neurol 2009; 65:140-150.
55. Yang Z, Chang YJ, Yu IC et al. ASC-J9 ameliorates spinal and bulbar muscular atrophy phenotype via degradation of androgen receptor. Nat Med 2007; 13:348-353.
56. Palazzolo I, Burnett BG, Young JE et al. Akt blocks ligand binding and protects against expanded polyglutamine androgen receptor toxicity. Hum Mol Genet 2007; 16:1593-1603.
57. Tokui K, Adachi H, Waza M et al. 17-DMAG ameliorates polyglutamine-mediated motor neuron degeneration through well-preserved proteasome function in an SBMA model mouse. Hum Mol Genet 2009; 18:898-910.
58. Waza M, Adachi H, Katsuno M et al. 17-AAG, an Hsp90 inhibitor, ameliorates polyglutamine-mediated motor neuron degeneration. Nat Med 2005; 11:1088-1095.
59. Adachi H, Katsuno M, Minamiyama M et al. Heat shock protein 70 chaperone overexpression ameliorates phenotypes of the spinal and bulbar muscular atrophy transgenic mouse model by reducing nuclear-localized mutant androgen receptor protein. J Neurosci 2003; 23:2203-2211.
60. Bailey CK, Andriola IF, Kampinga HH et al. Molecular chaperones enhance the degradation of expanded polyglutamine repeat androgen receptor in a cellular model of spinal and bulbar muscular atrophy. Hum Mol Genet 2002; 11:515-523.
61. Adachi H, Waza M, Tokui K et al. CHIP overexpression reduces mutant androgen receptor protein and ameliorates phenotypes of the spinal and bulbar muscular atrophy transgenic mouse model. J Neurosci 2007; 27:5115-5126.
62. Ying M, Xu R, Wu X et al. Sodium butyrate ameliorates histone hypoacetylation and neurodegenerative phenotypes in a mouse model for DRPLA. J Biol Chem 2006; 281:12580-12586.

Chapter 7

Spinocerebellar Ataxia with Axonal Neuropathy

Cheryl Walton, Heidrun Interthal, Ryuki Hirano, Mustafa A.M. Salih, Hiroshi Takashima and Cornelius F. Boerkoel*

Abstract

Spinocerebellar ataxia with axonal neuropathy (SCAN1) is an autosomal recessive disorder caused by a specific point mutation (c.1478A>G, p.H493R) in the tyrosyl-DNA phosphodiesterase (*TDP1*) gene. Functional and genetic studies suggest that this mutation, which disrupts the active site of the Tdp1 enzyme, causes disease by a combination of decreased catalytic activity and stabilization of the normally transient covalent Tdp1-DNA intermediate. This covalent reaction intermediate can form during the repair of stalled topoisomerase I-DNA adducts or oxidatively damaged bases at the 3′ end of the DNA at a strand break. However, our current understanding of the biology of Tdp1 function in humans is limited and does not allow us to fully elucidate the disease mechanism.

Introduction

Disorders of DNA repair can result in multiple pathological phenotypes, depending on the nature of the defect.[1,2] One of the most common features is neurological disease,[3] which can manifest as a developmental malformation or more commonly as a degenerative disorder during later life (Table 1).

Predisposing to these neurological manifestations are the poor renewal of neural tissues and the requirement that the tissue function for decades of life. As a consequence of their high oxygen requirement, neurons must cope with the DNA damage from oxidative and metabolic stress[5-7] and consequently require efficient DNA strand-break surveillance and repair mechanisms. Consistent with these observations, many studies link aging with a decline in DNA repair activity.[8-12] Also, individuals who incur genetic mutations inactivating these repair pathways show accelerated neuronal death.[4,13]

Spinocerebellar ataxia with axonal neuropathy (SCAN1) is an autosomal recessive disorder of DNA repair that clinically only affects the nervous system. Its neurodegenerative features include cerebellar atrophy with ataxia and axonal loss with peripheral neuropathy. The absence of affects on other tissues suggests that it is a good model for understanding the role of DNA repair in the nervous system.[14] SCAN1 is very rare and has only been reported for one extended family in Saudi Arabia.[14]

*Corresponding Author: Cornelius F. Boerkoel—Provincial Medical Genetics Program, Department of Medical Genetics, Children's and Women's Health Centre of BC, 4500 Oak St., Rm. C234, Vancouver, British Columbia, V6H 3N1 Canada. Email: boerkoel@interchange.ubc.ca

Diseases of DNA Repair, edited by Shamim I. Ahmad.

Table 1. ***DNA processing disorders with prominent neurological features***

Disorder	Disease Gene	Neurological Signs	Other Symptoms
Nucleotide excision repair			
Xeroderma pigmentosum	*XPA, ERCC3, ERCC2, ERCC4, ERCC5, POLH, XPC, DDB2*	Microcephaly Neurodegeneration Peripheral neuropathy	Photosensitivity; skin cancer; poikiloderma; hearing loss; cognitive impairment
Cockayne syndrome	*ERCC3, ERCC2, ERCC5, ERCC8, ERCC6*	Microcephaly Neurodegeneration Peripheral neuropathy	Photosensitivity; growth retardation; hearing loss; cognitive impairment; progeria
Trichothiodystrophy	*ERCC3, ERCC2, GTF2H5, TTDN1*	Microcephaly	Photosensitivity; short brittle hair; cognitive impairment; ichthyosis; decreased fertility; short stature
Cerebro-oculo-facio-skeletal syndrome	*ERCC6, ERCC2, ERCC5, ERCC1*	Microcephaly Microphthalmia	Spasticity; neurological impairment; growth retardation
DSB repair			
Ataxia telangiectasia	*ATM*	Ataxia Neurodegeneration Oculomotor apraxia Choreoathetosis Peripheral neuropathy	Infections; immune defects; malignancy
Ataxia telangiectasia-like disorder	*MRE11A*	Ataxia Neurodegeneration Oculomotor apraxia Choreoathetosis Peripheral neuropathy	

continued on next page

Table 1. Continued

Disorder	Disease Gene	Neurological Signs	Other Symptoms
Nijmegen breakage syndrome	*NBS1*	Neurodegeneration Microcephaly	Short stature; cognitive impairment; premature ovarian failure; infections; immune defects; malignancy; short stature
ATR-Seckel syndrome	*ATR*	Microcephaly	Dwarfism
Primary microcephaly	*MCPH1*	Microcephaly Neural migration defect	Short stature
LIG4 syndrome	*LIG4*	Microcephaly	Infections; immune defects; malignancy; cognitive impairment
Immunodeficiency with microcephaly	*NHEJ1*	Microcephaly	Immune defects; malignancy; growth retardation
SSB repair			
Spinocerebellar ataxia with axonal neuropathy Type 1 (scan 1)	*TDP1*	Ataxia Peripheral neuropathy	
Ataxia and oculomotor apraxia	*APTX*	Ataxia Neurodegeneration Oculomotor apraxia	Hypoalbuminemia Hypercholesterolemia

Symptoms of SCAN1

SCAN1 is a progressive neurodegenerative disorder that begins in late childhood with the gradual onset of ataxic gait and loss of touch, pain and vibration sensation in the extremities.[14] As the disease progresses patients develop areflexia, gaze nystagmus, cerebellar dysarthria and *pes cavus*. This leads to decreased foot dorsiflexion, a steppage gait and eventually loss of independent walking. Additional features identified by Takashima et al included mild hypercholesterolemia, borderline hypoalbuminemia and seizures.[14] Unlike many diseases caused by deficiencies in DNA repair, patients with SCAN1 do not develop neoplasia, immunodeficiency, or photosensitivity and have normal intellect, fertility and longevity.

Genetic Basis of SCAN1

SCAN1 was associated with a mutation in the *TDP1* gene by linkage analysis and positional cloning.[14] The *TDP1* mutation identified in SCAN1 patients is the missense change H493R that disrupts the active site of Tdp1.[14-16] As explained below, the unusual properties of this mutation likely account for the rarity of SCAN1 and the absence of detectable *TDP1* mutations among other diseases associated with ataxia and peripheral neuropathy.[14,17]

Tdp1 Function

TDP1 encodes the enzyme tyrosyl-DNA phosphodiesterase 1 (Tdp1) that participates in the resolution of DNA damage caused by stalling of topoisomerase I (Topo I).[16-18] It can also process protruding 3′-phosphoglycolate termini that form in response to oxidative stress, ionizing radiation and specific chemotherapeutic agents such as bleomycin.[19-23]

Topo I is an essential enzyme that cleaves supercoiled DNA in order to relieve the torsional stress generated by key nuclear processes such as replication and transcription.[24] The active site of Topo I contains a tyrosine residue which cleaves one strand of the DNA by a nucleophilic attack upon a phosphodiester bond in the DNA backbone. Normally the result is a transient DNA single strand break with the Topo I covalently bound to the 3′ phosphate terminus of the break via its nucleophilic tyrosine.[24-26] After DNA relaxation has occurred a nucleophilic attack by the 5′ hydroxyl group on the phosphotyrosyl linkage between Topo I and the 3′ end of the DNA at the nick usually religates the DNA and the topoisomerase dissociates. The anticancer drug camptothecin (CPT) or endogenous DNA damage, such as abasic sites, nicks and mismatched base pairs, can prevent removal of Topo I from the DNA, often by causing a misalignment of the 5′ hydroxyl end of the DNA and preventing it from acting as a nucleophile.[25,27,28]

Importantly, collision of the DNA replication machinery or RNA polymerase with the Topo I-DNA covalent intermediate can cause irreversible DNA breaks. In the former case, the collision results in replication fork arrest and formation of a double-strand DNA break,[29,30] whereas in the latter case, collision of RNA polymerase with a Topo I-DNA complex on the template strand results in transcription arrest at single-strand DNA breaks.[31]

Processing of stalled Topo I-DNA complexes likely requires proteolytic degradation of the stalled Topo I, removal of the peptide remnant from the DNA by Tdp1 and then repair of the break by the DNA single strand break repair complex.[32-35] Processing of dead-end Topo I-DNA complexes at double strand breaks is less well understood. Tdp1 has two modified conserved HxKx4Dx6G(G/S) motifs, known as the HKD motifs[15,36,37] and these two HKD motifs together form a single active site.[38] In the human enzyme, amino acids 263 and 493 are the conserved histidines of the HKD motifs.[15,39] In removal of the Topo I peptide from DNA, H263 acts as a nucleophile attacking the phosphotyrosyl bond between the topoisomerase and the 3′ end of the DNA. H493 acts as a general acid to protonate the leaving group tyrosine. In this reaction intermediate, Tdp1 is covalently bound to the DNA via H263; H493 then acts as a general base and activates a water molecule to hydrolyse the bond between H263 and the DNA 3′ phosphate releasing Tdp1 from the DNA.[15,35,39,40]

Molecular Basis of SCAN1

Since the SCAN1-associated mutation of Tdp1 (c.1478A>G) changed histidine 493 to arginine (H493R) disrupting the active site,[14] the hypothesis was that loss of functional Tdp1 gave rise to the neurodegenerative disease. To test this, three groups analyzed mouse models.[17,41,42] None of these Tdp1 null mice recapitulated the SCAN1 phenotype.[17,41,42] These results suggested that mice may have redundant pathways for stalled Topo I or other Tdp1 substrates similar to yeast,[43] or that SCAN1 does not arise simply from loss of Tdp1 enzymatic activity as suggested earlier by Interthal et al.[16] This hypothesis was based on biochemical analysis of the mutant Tdp1 ($Tdp1^{H493R}$) associated with SCAN1. Although $Tdp1^{H493R}$ showed an approximately 25-fold reduction in catalytic activity for its phosphotyrosyl substrate, it became trapped on the DNA as a covalent reaction intermediate and had an extended half-life of approximately 13 min.[16,17,41] Thus, $Tdp1^{H493R}$ essentially just replaced the stalled topoisomerase. Consistent with the autosomal recessive inheritance of SCAN1, the only identified enzyme capable of resolving this covalent $Tdp1^{H493R}$-DNA intermediate was wild type Tdp1.[16,35] This suggested that SCAN1 might arise, at least in part, from stabilization of the Tdp1-DNA reaction intermediate (Fig. 1).[16,17,41]

An alternative and less considered possibility is that Tdp1 has a function in humans that is not conserved in mice. In the comparative profiling of Tdp1 expression in mice and humans, Tdp1 always exhibited nuclear expression in the mice, whereas it had cytoplasmic expression in some human cell types.[17] Interestingly, the cell types with the most cytoplasmic expression are those most likely affected in SCAN1, namely spinal anterior horn motor neurons, cerebellar Purkinje cells and dentate nucleus neurons.[17] As precedent for a cytoplasmic function, *glaikit*, the *Drosophila* homolog of Tdp1, has only been detected in the cytoplasm. During *Drosophila* embryogenesis, *glaikit* was essential for epithelial polarity and for neuronal development; it localized proteins to the apical lateral membrane of epithelial cells and its deficiency led to a severe disruption of central nervous system architecture.[44] In contrast to *Drosophila*, however, Tdp1 does not have a major role in human neurodevelopment because SCAN1 patients have normal neurodevelopment.[14,17] Therefore, if human Tdp1 has a cytoplasmic function in human neurons analogous to that of *glaikit*, it is likely a maintenance function that leads to neurodegeneration when disrupted.

Current and Future Research

Current and future research on Tdp1 focuses on four areas: (1) the DNA repair processes interacting with and dependent on Tdp1, (2) Tdp1 as a cancer therapeutic target, (3) the role of Tdp1 in biological processes other than DNA repair and (4) further delineation of the mechanism underlying SCAN1. Each of these areas is rapidly advancing and should open new understanding of neurodegenerative diseases, aging, cancer and fundamental human biology.

Emerging data in mammalian systems and earlier studies in yeast suggest that Tdp1 interacts with many DNA repair processes.[43,45-47] Understanding of the distribution and function of these redundant pathways in the human brain may shed light on the peculiar sensitivity of the human nervous system to expression of $Tdp1^{H493R}$.[17] Additionally, if transgenic mice expressing human $Tdp1^{H493R}$ recapitulate the SCAN1 phenotype, this will suggest that it is the specific H493R mutation that is responsible for disease and not solely the loss of functional Tdp1. In this situation, SCAN1 is potentially treatable by directed inhibition of $Tdp1^{H493R}$.

Besides its role in neural maintenance, Tdp1 has been regarded as a promising therapeutic target for cancer. Tdp1 is a promising cotarget of Topo I in cancer therapy as it counteracts the effects of Topo I inhibitors, such as camptothecin (CPT) and its clinically used derivatives.[35] Also, resistance to CPT is frequently encountered in nonsmall cell lung cancer and has been attributed to overexpression of Tdp1.[48] It is hypothesized therefore that Tdp1 inhibitors can augment the anticancer activity of Topo I inhibitors by reducing the repair of Topo I-DNA lesions.[17,49,50] The DNA double strand breaks caused by replication forks that encounter CPT-trapped Topo I are considered to be the major cytotoxic lesion caused by CPT based cancer therapy.[29] Additionally, since Tdp1 deficiency increases sensitivity to radiation, bleomycin, oxidative DNA damage and radiation, Tdp1 is also a promising cotarget for several other cancer therapies.

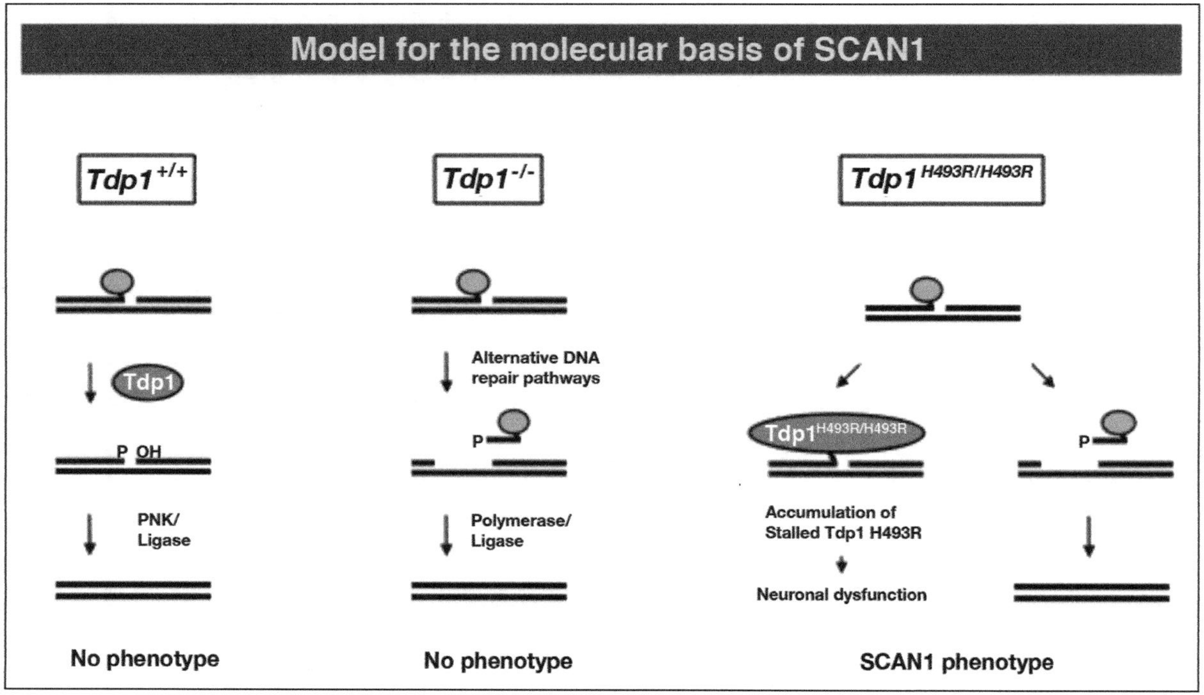

Figure 1. A model for the molecular basis of SCAN1. DNA breaks with blocked 3′ ends (e.g., Topo I or phosphoglycolate) undergo Tdp1-facilitated DNA repair via both DNA single-strand break repair and double-strand break repair mechanisms. With loss of functional Tdp1 (Tdp1$^{-/-}$), there is sufficient redundant activity for adequate DNA repair by alternative pathways (e.g., endonuclease-dependent pathways) unless the system is further stressed as by DNA damaging agents such as radiation, oxygen free radicals, or chemotherapy. In contrast, when Tdp1 carries the H493R mutation, it not only has a quantitative reduction in overall activity, but also a qualitative change resulting in accumulation of Tdp1-DNA complexes. These complexes are efficiently removed from the DNA by wild-type Tdp1 in all tissues of heterozygotes, whereas they are only removed in replicating cells of homozygotes by alternative DNA strand break repair mechanisms. According to this model, the transcriptional interference and/or apoptosis resulting from the Tdp1-DNA complexes in nondividing neurons causes SCAN1 via neurodegeneration.

Based on these precedents as well as the association of differences in Tdp1 expression with outcome in breast cancer,[51] identification of Tdp1 inhibitors is being actively pursued in order to treat cancers resistant to Topo I inhibitors and bleomycin as well as to predict outcome. A number of Tdp1 inhibitors have been described, including vanadate,[52,53] tungstate, the aminoglycoside neomycin,[54] NSC 305831, NSC 118695, NSC 88915[50] and furamidine.[55]

Conclusion

In summary, the discovery of the association of SCAN1 with a specific mutation of Tdp1 has spurred understanding of DNA repair in human biology and suggested a novel mechanism of human disease. SCAN1 may be the first example of a human genetic disease that results from a failure to repair DNA-protein covalent complexes. SCAN1 likely arises not only from a quantitative change in Tdp1 activity but also from a qualitative change that renders the enzyme different from wild type Tdp1 causing it to become covalently trapped on the DNA.[16,17,41] Additionally, the absence of detectable acute effects of Topo I inhibitors and bleomycin treatment on the nervous system of mice deficient for Tdp1 suggests that nonproliferating cells of the nervous system are sufficiently insensitive to Topo I-DNA complexes and 3′ phosphoglycolate-DNA damage that short-term administration of these chemotherapeutic agents is unlikely to induce neurological disease even in the absence of functional Tdp1.[17]

References

1. de Boer J, Hoeijmakers JH. Nucleotide excision repair and human syndromes. Carcinogenesis 2000; 21:453-60.
2. O'Driscoll M, Jeggo PA. The role of double-strand break repair—insights from human genetics. Nat Rev Genet 2006; 7:45-54.
3. Friedberg EC, Walker GC, Siede W et al. Disease states associated with defective biological responses to DNA damage. DNA Repair and Mutagenesis. Washington: ASM Press, 2006;863-1080.
4. Subba Rao K. Mechanisms of disease: DNA repair defects and neurological disease. Nat Clin Pract Neurol 2007; 3:162-72.
5. Barzilai A. The contribution of the DNA damage response to neuronal viability. Antioxid Redox Signal 2007; 9:211-8.
6. Chen L, Lee HM, Greeley GH Jr et al. Accumulation of oxidatively generated DNA damage in the brain: a mechanism of neurotoxicity. Free Radic Biol Med 2007; 42:385-93.
7. Lombard DB, Chua KF, Mostoslavsky R et al. DNA repair, genome stability and aging. Cell 2005; 120:497-512.
8. Gorbunova V, Seluanov A, Mao Z et al. Changes in DNA repair during aging. Nucleic Acids Res 2007; 35:7466-74.
9. Imam SZ, Karahalil B, Hogue BA et al. Mitochondrial and nuclear DNA-repair capacity of various brain regions in mouse is altered in an age-dependent manner. Neurobiol Aging 2006; 27:1129-36.
10. Intano GW, Cho EJ, McMahan CA et al. Age-related base excision repair activity in mouse brain and liver nuclear extracts. J Gerontol A Biol Sci Med Sci 2003; 58:205-11.
11. Lu T, Pan Y, Kao SY et al. Gene regulation and DNA damage in the ageing human brain. Nature 2004; 429:883-91.
12. Rutten BP, Schmitz C, Gerlach OH et al. The aging brain: accumulation of DNA damage or neuron loss? Neurobiol Aging 2007; 28:91-8.
13. Katyal S, McKinnon PJ. DNA strand breaks, neurodegeneration and aging in the brain. Mech Ageing Dev 2008; 129:483-91.
14. Takashima H, Boerkoel CF, John J et al. Mutation of TDP1, encoding a topoisomerase I-dependent DNA damage repair enzyme, in spinocerebellar ataxia with axonal neuropathy. Nat Genet 2002; 32:267-72.
15. Interthal H, Pouliot JJ, Champoux JJ. The tyrosyl-DNA phosphodiesterase Tdp1 is a member of the phospholipase D superfamily. Proc Natl Acad Sci USA 2001; 98:12009-14.
16. Interthal H, Chen HJ, Kehl-Fie TE et al. SCAN1 mutant Tdp1 accumulates the enzyme-DNA intermediate and causes camptothecin hypersensitivity. EMBO J 2005; 24:2224-33.
17. Hirano R, Interthal H, Huang C et al. Spinocerebellar ataxia with axonal neuropathy: consequence of a Tdp1 recessive neomorphic mutation? EMBO J 2007; 26:4732-43.
18. Pouliot JJ, Yao KC, Robertson CA et al. Yeast gene for a Tyr-DNA phosphodiesterase that repairs topoisomerase I complexes. Science 1999; 286:552-5.
19. Bertoncini CR, Meneghini R. DNA strand breaks produced by oxidative stress in mammalian cells exhibit 3′-phosphoglycolate termini. Nucleic Acids Res 1995; 23:2995-3002.

20. Henner WD, Rodriguez LO, Hecht SM et al. Gamma ray induced deoxyribonucleic acid strand breaks. 3′ Glycolate termini. J Biol Chem 1983; 258:711-3.
21. Povirk LF. DNA damage and mutagenesis by radiomimetic DNA-cleaving agents: bleomycin, neocarzinostatin and other enediynes. Mutat Res 1996; 355:71-89.
22. Inamdar KV, Pouliot JJ, Zhou T et al. Conversion of phosphoglycolate to phosphate termini on 3′ overhangs of DNA double strand breaks by the human tyrosyl-DNA phosphodiesterase hTdp1. J Biol Chem 2002; 277:27162-8.
23. Zhou T, Lee JW, Tatavarthi H et al. Deficiency in 3′-phosphoglycolate processing in human cells with a hereditary mutation in tyrosyl-DNA phosphodiesterase (TDP1). Nucleic Acids Res 2005; 33:289-97.
24. Champoux JJ. DNA topoisomerases: structure, function and mechanism. Annu Rev Biochem 2001; 70:369-413.
25. Pommier Y, Pourquier P, Fan Y et al. Mechanism of action of eukaryotic DNA topoisomerase I and drugs targeted to the enzyme. Biochim Biophys Acta 1998; 1400:83-105.
26. Stewart L, Redinbo MR, Qiu X et al. A model for the mechanism of human topoisomerase I. Science 1998; 279:1534-41.
27. Pommier Y, Redon C, Rao VA et al. Repair of and checkpoint response to topoisomerase I-mediated DNA damage. Mutat Res 2003; 532:173-203.
28. Pourquier P, Pommier Y. Topoisomerase I-mediated DNA damage. Adv Cancer Res 2001; 80:189-216.
29. Hsiang YH, Lihou MG, Liu LF. Arrest of replication forks by drug-stabilized topoisomerase I-DNA cleavable complexes as a mechanism of cell killing by camptothecin. Cancer Res 1989; 49:5077-82.
30. Tsao YP, Russo A, Nyamuswa G et al. Interaction between replication forks and topoisomerase I-DNA cleavable complexes: studies in a cell-free SV40 DNA replication system. Cancer Res 1993; 53:5908-14.
31. Wu J, Liu LF. Processing of topoisomerase I cleavable complexes into DNA damage by transcription. Nucleic Acids Res 1997; 25:4181-6.
32. Debethune L, Kohlhagen G, Grandas A et al. Processing of nucleopeptides mimicking the topoisomerase I-DNA covalent complex by tyrosyl-DNA phosphodiesterase. Nucleic Acids Res 2002; 30:1198-204.
33. El-Khamisy SF, Saifi GM, Weinfeld M et al. Defective DNA single-strand break repair in spinocerebellar ataxia with axonal neuropathy-1. Nature 2005; 434:108-13.
34. Yang SW, Burgin AB Jr, Huizenga BN et al. A eukaryotic enzyme that can disjoin dead-end covalent complexes between DNA and type I topoisomerases. Proc Natl Acad Sci USA 1996; 93:11534-9.
35. Interthal H, Chen HJ, Champoux JJ. Human Tdp1 cleaves a broad spectrum of substrates, including phosphoamide linkages. J Biol Chem 2005; 280:36518-28.
36. Ponting CP, Kerr ID. A novel family of phospholipase D homologues that includes phospholipid synthases and putative endonucleases: identification of duplicated repeats and potential active site residues. Protein Sci 1996; 5:914-22.
37. Stuckey JA, Dixon JE. Crystal structure of a phospholipase D family member. Nat Struct Biol 1999; 6:278-84.
38. Davies DR, Interthal H, Champoux JJ et al. The crystal structure of human tyrosyl-DNA phosphodiesterase, Tdp1. Structure (Camb) 2002; 10:237-48.
39. Raymond AC, Rideout MC, Staker B et al. Analysis of human tyrosyl-DNA phosphodiesterase I catalytic residues. J Mol Biol 2004; 338:895-906.
40. Davies DR, Interthal H, Champoux JJ et al. Insights into substrate binding and catalytic mechanism of human tyrosyl-DNA phosphodiesterase (Tdp1) from vanadate and tungstate- inhibited structures. J Mol Biol 2002; 324:917-32.
41. Hawkins AJ, Subler MA, Akopiants K et al. In vitro complementation of Tdp1 deficiency indicates a stabilized enzyme-DNA adduct from tyrosyl but not glycolate lesions as a consequence of the SCAN1 mutation. DNA Repair (Amst) 2009; 8:654-63.
42. Katyal S, el-Khamisy SF, Russell HR et al. TDP1 facilitates chromosomal single-strand break repair in neurons and is neuroprotective in vivo. EMBO J 2007; 26:4720-31.
43. Deng C, Brown JA, You D et al. Multiple endonucleases function to repair covalent topoisomerase I complexes in Saccharomyces cerevisiae. Genetics 2005; 170:591-600.
44. Dunlop J, Morin X, Corominas M et al. glaikit is essential for the formation of epithelial polarity and neuronal development. Curr Biol 2004; 14:2039-45.
45. Liu C, Pouliot JJ, Nash HA. Repair of topoisomerase I covalent complexes in the absence of the tyrosyl-DNA phosphodiesterase Tdp1. Proc Natl Acad Sci USA 2002; 99:14970-5.
46. Vance JR, Wilson TE. Yeast Tdp1 and Rad1-Rad10 function as redundant pathways for repairing Top1 replicative damage. Proc Natl Acad Sci USA 2002; 99:13669-74.

47. El-Khamisy SF, Katyal S, Patel P et al. Synergistic decrease of DNA single-strand break repair rates in mouse neural cells lacking both Tdp1 and aprataxin. DNA Repair (Amst) 2009; 8:760-6.
48. Liu C, Zhou S, Begum S et al. Increased expression and activity of repair genes TDP1 and XPF in nonsmall cell lung cancer. Lung Cancer 2007; 55:303-11.
49. McKinnon PJ, Caldecott KW. DNA strand break repair and human genetic disease. Annu Rev Genomics Hum Genet 2007; 8:37-55.
50. Dexheimer TS, Antony S, Marchand C et al. Tyrosyl-DNA phosphodiesterase as a target for anticancer therapy. Anticancer Agents Med Chem 2008; 8:381-9.
51. Turashvili G, Kalloger S, Gelmon K et al. Subcellular protein expression pattern of the DNA repair enzyme Tdp1 is prognostic in breast cancer 2009. In Press.
52. Davies DR, Interthal H, Champoux JJ et al. Crystal structure of a transition state mimic for Tdp1 assembled from vanadate, DNA and a topoisomerase I-derived peptide. Chem Biol 2003; 10:139-47.
53. Davies DR, Interthal H, Champoux JJ et al. Explorations of peptide and oligonucleotide binding sites of tyrosyl-DNA phosphodiesterase using vanadate complexes. J Med Chem 2004; 47:829-37.
54. Liao Z, Thibaut L, Jobson A et al. Inhibition of human tyrosyl-DNA phosphodiesterase by aminoglycoside antibiotics and ribosome inhibitors. Mol Pharmacol 2006; 70:366-72.
55. Antony S, Marchand C, Stephen AG et al. Novel high-throughput electrochemiluminescent assay for identification of human tyrosyl-DNA phosphodiesterase (Tdp1) inhibitors and characterization of furamidine (NSC 305831) as an inhibitor of Tdp1. Nucleic Acids Res 2007; 35:4474-84.

Chapter 8

Tuberous Sclerosis Complex and DNA Repair

Samy L. Habib*

Abstract

Tuberous sclerosis complex (TSC) is an autosomal dominant disorder in humans characterized by the development of hamartomas in several organs, including renal angiomyolipomas, cardiac rhabdomyomas and subependymal giant cell astrocytomas. TSC causes disabling neurologic disorders, including epilepsy, mental retardation and autism. Brain lesions, including subependymal and subcortical hamartomas, have also been reported in TSC patients. TSC is associated with hamartomas and renal cell carcinoma (RCC) as well as sporadic tumors in TSC patient. Renal angiomyolipomas associated with TSC tend to be larger, bilateral, multifocal and present at a younger age compared with sporadic forms. Tuberous sclerosis complex of 2 genes, TSC2 encodes a protein called tuberin that normally exists in an active state and forms a heterodimeric complex with hamartin, the protein encoded by the TSC1. Deficiency of TSC2 in Eker rat is associated with the development of tumors in several organs including kidney. The majority of renal cell tumors observed in the Eker rat originates from renal proximal tubules and are histologically similar to renal cell carcinoma in humans. On the other hand, mutations in DNA repair enzyme 8-oxoG-DNA glycosylase (OGG1) are associated with cancer. OGG1 gene is found somatically mutated in some cancer cells and is highly polymorphic among human cancers. Moreover, knockout mice in OGG1 developed spontaneously adenoma and carcinoma. We recently show that the constitutive expression of OGG1 in heterozygous ($TSC2^{+/-}$) Eker rat and in angiomyolipomas kidney tissue from human is 2-3fold less than in kidney from wild-type rats and control human subjects. In addition, we show that loss of TSC2 in kidney tumor of Eker rat is associated with loss of OGG1 and accumulation significant levels of oxidative DNA damage 8-oxo-deoxyguanine suggesting that TSC2 and OGG1 play a major role in renal tumorigenesis.

Introduction: Clinical Manifestations of TSC Disease

Tuberous sclerosis complex (TSC) is a group of genetic diseases caused by defects in one of two genes: TSC-1 and TSC2 described by Bourneville in 1880.[1] The TSC1 gene encodes a protein called hamartin and TSC2 encodes tuberin.[2-4] TSC1 and TSC2 genes were identified as the genetic loci mutated in the autosomal dominant tumour syndrome TSC.[2,3] Hereditary tumor syndrome caused by the TSC1 or TSC2 tumor suppressor genes located on chromosomes 9 and 16 respectively.[3,4] Cells undergo bi-allelic inactivation of either gene to give rise to tumors in a classic tumor suppressor "two-hit" paradigm.[5] About 60% of cases with TSC are sporadic, representing new mutations and TSC affects about 1 million individuals worldwide, with an estimated prevalence of up to 1 in

*Samy L. Habib—South Texas Veterans Healthcare System, Geriatric Research, Education and Clinical Center and Department of Medicine, The University of Texas Health Science Center, San Antonio, TX 78229, USA. Email: habib@uthscsa.edu

Diseases of DNA Repair, edited by Shamim I. Ahmad. ©2010 Landes Bioscience and Springer Science+Business Media.

6,000 newborns.[6] TSC causes disabling neurological disorders, including epilepsy, mental retardation and autism.[5-8] In additional, TSC is a multisystemic disorder characterized by the development of numerous benign tumours (e.g., hamartomas) most commonly affecting the brain, kidneys, skin, heart and lungs.[7] Malignant tumors progression is rarely detected in TSC patients. TSC mutations have also been found in sporadic (non-TSC-associated) cases of lymphangiomyomatosis, which primarily affects women and is characterized by smooth-muscle cell proliferation and cystic destruction of the lung.[5] These tumors grow during the second half of pregnancy and regress after birth. A cardiac rhabdomyoma can be discovered using echocardiography in approximately 50% of people with TSC.[7] However, the incidence in the newborn may be as high as 90% and in adults as low as 20%. The tumor size remains constant as the heart grows, which has much the same effect. The pathology of the lesions in TSC has features such as abnormal cellular proliferation, size, differentiation and migration.[7]

Renal Lesions in TSC-Deficient Mammals

Mouse

TSC2 homozygous mice ($TSC2^{-/-}$) have an embryonic lethal phenotype, failing to develop past embryonic days 9.5 to 12.5 due to hepatic hypoplasia.[9] All TSC2 homozygous died in utero confirming the essential role for tuberin function in development.[9] TSC2 heterozygotes display 100% incidence of multiple bilateral renal cystadenomas by 15 months of age.[9] The precise physiological or cellular role(s) of TSC2 and its role in renal carcinogenesis remain unclear. Loss of the second allele of this gene as a somatic event leads to the development of RCC in these animals.[9,10] Progression to renal carcinoma, fatal bleeding from the liver hemangiomas and extremity angiosarcomas all occur at a rate of less than 10%. The renal cystadenomas develop from intercalated cells of the cortical collecting duct and uniformly express gelsolin at high levels, enabling detection of early neoplastic lesions.[9] TSC2 heterozygous mutant ($TSC2^{+/-}$) mice developed renal carcinomas with a complete penetrance, as seen in the Eker rat, but not the angiomyolipomas characteristic of human TSC, confirming the existence of a species-specific mechanism of tumorigenesis caused by TSC1 deficiency.[11] On the other hand, TSC1 homozygous mice ($TSC1^{-/-}$) have an embryonic lethal phenotype, failing to develop past embryonic days 9.5 to 13.5.[9] TSC1 heterozygotes mice develop kidney cystadenomas and liver hemangiomas at high frequency, but the incidence of kidney tumors is somewhat lower than in TSC2 heterozygote mice.[9]

Rat

Eker homozygotes rats ($TSC2^{-/-}$) are embryonically lethal between 9 and 13 days of gestation.[12-14] Eker heterozygotes rats ($TSC2^{+/-}$) spontaneously develop multicentric, bilateral tumors scattered throughout the renal cortex and outer medulla (see Fig. 1B). Preneoplastic lesions in the renal tubules begin to appear in the Eker rats carrying mutation in TSC2 gene at around four months of age and by the age of one year the incidence of renal cell tumors approaches 100%.[15] Loss of the second allele of TSC2 gene as a somatic event leads to the development of RCC in these animals, fulfilling Kundson's "2 hit" hypothesis for the loss of tumor suppressor gene function in tumorgenesis. The majority of renal cell tumors observed in the Eker rat originates from renal proximal tubules and are histologically similar to renal tumors in humans.[15,16] Gross appearance of kidneys from the wild type ($TSC2^{+/+}$) (Fig. 1A) and multiple tumors in the kidney of Eker ($TSC2^{+/-}$) rats (Fig. 1B) can be seen at 12 months of age. In addition, H and E staining show normal histology of wild type rat (Fig. 1C) and renal cell carcinoma in the Eker rat kidneys (Fig. 1D). Magnetic resonance imaging (MRI) of Eker rat kidney shows several tumors that match the morphology and H and E staining (Fig. 1E,F,G).

Human

Renal angiomyolipomas, benign tumors composed of abnormal vessels, immature smooth-muscle cells and fat cells are bilateral, with multiple tumors in each kidney, in most patients with TSC (see Fig. 2). The estimated incidence of angiomyolipomas in TSC ranges from

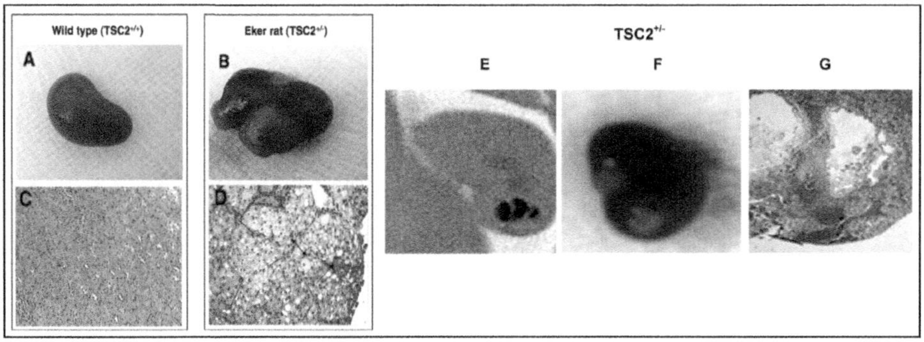

Figure 1. A) Normal appearing kidney from wild type, and B) kidney cancerous lesion in Eker rat. C) H and E staining of kidney section showing normal tubular architecture of kidney in wild type, and D) fibrovascular stroma with clear cytoplasm and excentric large dark nuclei (clear cell type) in Eker rat. E) MRI shows several tumors that matched (F) the gross appearance, and G) the H and E staining in kidney of Eker rat. A-D) From Habib et al. Mol Cancer 2008; 7:1-9.

55 to 75%.[6] The rate of growth of angiomyolipomas varies among patients and lesions. However, epithelial renal lesions that include epithelial cysts, polycystic kidney disease and renal-cell carcinomas may develop in patients with TSC.[17-19] Epithelial cysts are generally asymptomatic and are more often associated with hypertension and renal failure than are angiomyolipomas.[17,20] In addition, few patients with TSC carry a contiguous germline deletion that affects both the TSC2 gene and one of the genes that leads to autosomal dominant polycystic kidney disease on chromosome 16p13, resulting in a polycystic kidney phenotype that is detectable in infancy or early childhood and that generally leads to renal insufficiency.[2,19] The overall incidence of renal carcinoma in patients with TSC is similar to that in the general population, with a lifetime risk of 2 to 3%. However, the kidney cancer is diagnosed at a younger age in patients with TSC.[18] Clear-cell, papillary and chromophobe carcinoma subtypes, as well as oncocytomas, have all been reported in patients with TSC.[18,19] Kidney tumorangiomyolipoma of human with TSC disease is composed of three cell types: smooth muscle, fat and blood vessels cells (Fig. 2A). In addition,

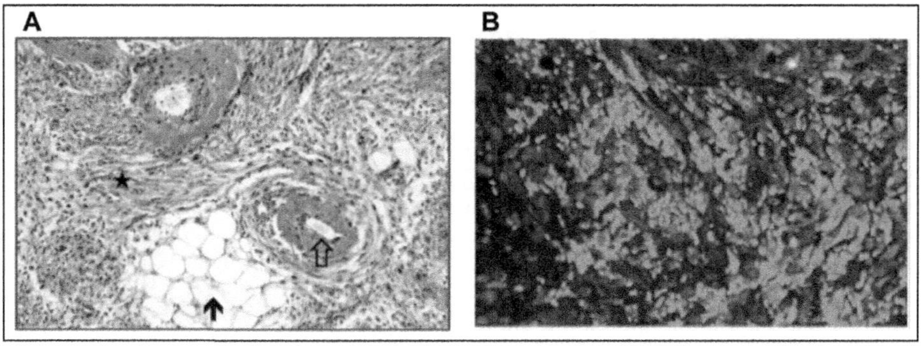

Figure 2. Angiomyolipomas kidney tissue from TSC patient. A) H and E staining of kidney section of TSC patient showing 3 cell types in angiomyolipomas tissue; fat cells (blocked arrowhead), smooth muscle cells (star) and vessel cells (open arrowhead). B) Immunostaining of phopho-S6K (Thr-389) in kidney tumor from patients with TSC. Phosphorylation of S6K was significantly higher (green signal) in tumor kidney tissue confirming the increase of mTOR activity. B) From Habib SL. Mol Cancer 2009; 8:1-10. A color version of this figure is available online at www.landesbioscience.com/curie.

phospho-p70S6K as a marker of mTOR activity was highly expressed in kidney angiomyolipomas tissue of TSC patients (Fig. 2B).

TSC1 and TSC2 Genes

Hamartin a protein product of TSC1 is an 1164 amino acid/130 KDa expressed in most human tissues, including brain, kidney, heart, liver, small and large intestine, prostate and testes.[22] The TSC2 gene encodes a protein called tuberin. Tuberin normally exists in an active state physically bound to hamartin. Tuberin (TSC2) is a 1784 amino acid/198 KDa protein showing significant homology to the Rap1-GAP protein. The TSC1-TSC2 (hamartin-tuberin) complex, through its GAP (GTPase-activating protein) activity towards the small G-protein Rheb (Ras homologue enriched in brain), is a critical negative regulator of mTORC1 (mammalian target of rapamycin complex 1).[23] The GAP domain spans residues 1517-1674, encoded by exons 34-38 and is shown to inhibit Ras-related family of small G proteins, such as Rap1, Rab5 and Rheb.[24-26] The effect of tuberin on cell proliferation and protein translation is usually associated with its GAP activity towards Rheb.[25,26] TSC1 is localized in the cytoplasm.[27,28] On the other hand, TSC2 is localized in the golgi apparatus and in the nucleus.[28] Tuberin has a single transmembrane domain at amino acids 127-144 and a predicted coiled-coil region at residues 719-998.[3] Coiled-coil regions are often responsible for protein-protein interactions; thus research was directed at identifying a binding partner for hamartin.[29]

TSC2 mouse embryo fibroblasts expressing several N-terminal, central and a C-terminal domains of tuberin, confirmed that S6 phosphorylation significantly decreased in the C-terminal domain (amino acids 1125-1784), where the GAP activity is located.[30,31] TSC2 stimulates the conversion of Rheb-GTP to Rheb-GDP, thereby inactivating Rheb (see Fig. 3). Loss of TSC2 function leads to enhanced Rheb-GTP signaling. TSC2 mutations in the GAP domain have been identified in unrelated patients with tuberous sclerosis suggesting that GAP is a target of mutations in tuberous sclerosis.[32] According to Knudson's hypothesis that two-hit tumor suppressor gene model, inactivation of both alleles of TSC1 or TSC2 are required for lesion formation in TSC patient.[33] Most second-hit mutations are large deletions involving the loss of surrounding loci. These mutations are referred to as loss of heterozygosity, since they affect neighboring heterozygous polymorphic markers. Loss of heterozygosity in TSC1 or TSC2 has been consistently observed in the majority of TSC-associated angiomyolipomas with fewer and less incidence of RCC.

TSC Genes and Cell Signals

TSC Genes and Akt

Tuberin-hamartin complex acts directly downstream of the serine/threonine kinase Akt within this pathway. Akt directly phosphorylates tuberin at Ser^{939}, Ser^{981}, Ser^{1130}, Ser^{1132} and Thr^{1462} residues in full-length human TSC2.[34] Phosphorylation of tuberin at these sites regulates tuberin-hamartin complexes.[35] Akt-mediated phosphorylation of tuberin at Ser^{939} and Thr^{1462} sites to activate the key downstream regulators of cell growth.[35] Increased tuberin phophorylation at Thr^{1462} site and the decreased total tuberin expression were detected in kidney angiomyolipoma tissue of TSC patients.[21] Phosphorylation of tuberin by Akt affects its function through at least two mechanisms: first, phosphorylation decreases the activity of tuberin; second, phosphorylation destabilizes tuberin by disrupting the complex formation between hamartin and tuberin resulting in ubiquitination of free tuberin and its degradation by the proteosome.[36] Another mechanism possibility involve in tuberin regulation that phosphorylation of tuberin at Ser^{939} and Ser^{981} sites may alter its subcellular localization and it can no longer act as a GAP for Rheb.[37] Loss of TSC2 function leads to enhanced Rheb–GTP signaling and activates the mammalian target of rapamycin (mTOR), which promotes activation of p70S6 kinase, inhibition of eukaryotic initiation factor 4E binding protein 1 (4E-BP1, an inhibitor of translation initiation) and eventual activation of translation.[38]

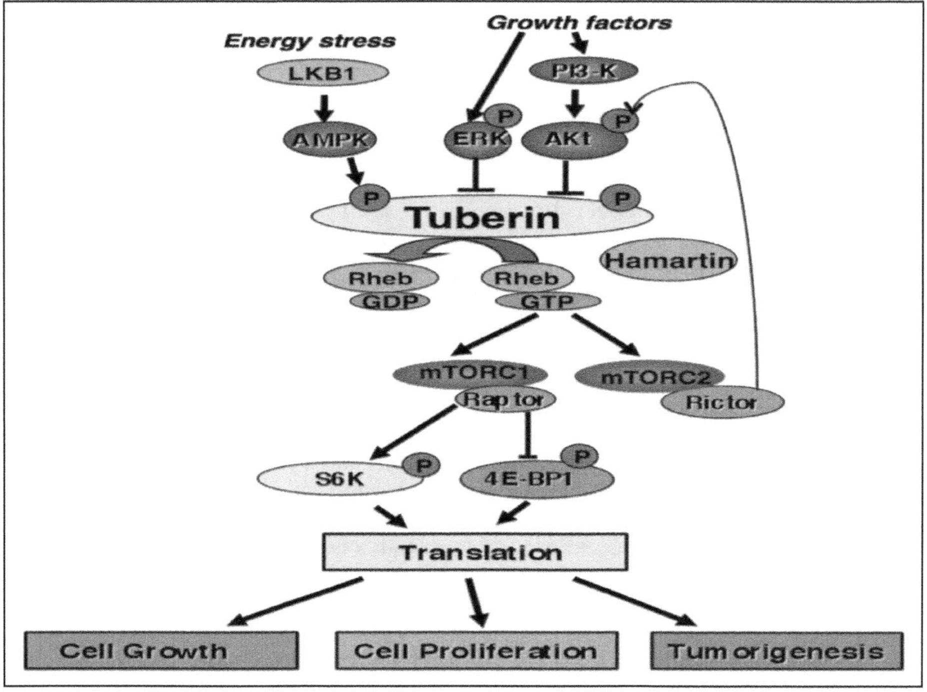

Figure 3. Role of tuberin and hamartin in cell signalling. Cells treated with growth factors show activation of PI3K and ERK and phosphorylation of tuberin that leads to disassociate tuberin/hamartin complex. On the other hand, cells under energy stress show increase in Lkb1 expression to activate AMPK and results in tuberin activation. Phosphorylation of tuberin by Akt and ERK results in activation of Rheb and mTORC1/2 pathway. Activation of mTORC2 through mTORC2/rictor results in activation of Akt through feedback loop signals. However, activation of mTORC1/raptor activates S6k and blocks 4E-BP1 to increase the protein translation and leads to cell growth, cell proliferation and cell tumorigenesis.

TSC genes and ERK/AMPK

Tuberin is regulated at least through two other major signaling pathways in the form of kinase-mediated phosphorylation events: the extracellular signal-regulated kinase (ERK)/mitogen-activated protein (MAP) kinase pathway and the AMP-activated protein kinase (AMPK).[39] Extracellular-signal-regulated kinase (Erk) plays a critical role in TSC progression through posttranslational inactivation of TSC2.[40] Erk-dependent phosphorylation leads to TSC1-TSC2 dissociation. Activation of ERK by growth factors leads to phosphorylates TSC2 at Ser^{540} and/or Ser^{664} sites and markedly impairs TSC2 function as a GTPase activating protein[40] (see Fig. 3). In addition, phosphoryaltion of TSC2 by ERK at Ser^{540} contributes to impair TSC2 ability to inhibit mTOR signaling, cell proliferation and oncogenic transformation.[41]

AMPK is the primary energy sensor in cells. Upon depletion of intracellular ATP, AMP levels rise and bind to an AMPK regulatory subunit. AMP binding places AMPK in a conformation that makes it accessible to its upstream activating kinase LKB1, which activates AMPK via phosphorylation of its catalytic subunit at Thr^{172}.[42] Once activated, AMPK phosphorylates tuberin at two sites (Ser^{1271}, Ser^{1387}) distinct from the Akt/PKB sites.[43] In contrast to the phosphorylation of tuberin by Akt/PKB, AMPK-mediated phosphorylation enhances the ability of the tuberin/hamartin complex to act as a Rheb-GAP and therefore blocks Rheb-dependent mTOR activation under conditions of energy stress (refs. 15,16, see Fig. 3). AMPK is inactivated in certain disease that leads to inactivation of tuberin and impair its ability to act as Rheb-GAP. AMPK functions as a serine/threonine kinase and directly phosphorylates TSC2.[44] Cells deficient for TSC1/TSC2, or producing a TSC2 variant that can't be phosphorylated by AMPK, fail to down-regulate mTOR in situations of energy deprivation.[44] Signaling by energy depletion also involves the Lkb1 tumor suppressor protein, which is inactivated in Peutz-Jeghers syndrome. Lkb1 is a serine/threonine kinase that phosphorylates a variety of substrates, including AMPK.[42,43,45] AMPK activation with consequent TSC1/TSC2-mediated inhibition of mTOR by energy depletion requires Lkb1.[46] Thus under conditions that are adverse for growth, such as in the presence of reduced energy stores, mTOR function is inhibited, thereby down-regulating protein synthesis and conserving energy (Fig. 3).

TSC Genes and mTOR

mTOR is a serine/threonine kinase, which increases protein translation and cell growth through phosphorylation of two effector molecules: p70 ribosomal protein S6 kinase (S6K) and eukaryotic-translation-initiation-factor-4E-binding protein 1 (4E-BP1).[47] S6K phosphorylates downstream substrates, such as ribosomal protein S6 and eIF4B, to promote mRNA translation.[47,48] In parallel, S6K also increases phosphorylation of eEF2 kinase, leading to its inactivation and resulting in the release of eEF2 inhibition, thus promoting the elongation phase of translation.[48] While mTOR signaling is best characterized for its role in regulation of translation. mTOR pathway also controls the transcription of several genes.[49] Majority of cellular mTOR is found in the cytoplasm but a cytoplasmic-nuclear shuttling behavior of mTOR has been identified,[50] indicating a nuclear function for mTOR. Several evidence has implicated the direct involvement of mTOR and its downstream S6K in the regulation of transcription. S6K phosphorylates cAMP-responsive element binding protein (CREB) and its coactivator cAMP-responsive element modulator (CREM) resulting in enhancement of their transcriptional activity.[51] Of interest, these effects are blocked by rapamycin, implicating mTOR is the upstream activator of S6K.

Rapamycin inhibits mTORC1 activity by blocking its interaction with Raptor.[48] Raptor is a conserved protein recruits S6K1 and 4E-BP1 that forms a rapamycin-sensitive complex with mTOR and the adaptor protein mLST8 named mTORC1.[34] While mTOR complex 2 (mTORC2) is constitutes of the adaptor protein called rictor.[35] mTORC2 complex is insensitive to rapamycin and is thought to regulate actin cytoskeleton and has been shown to control Akt Ser^{473} phosphorylation.[52] mTORC1 activation is a common molecular event in a number of hamartoma syndromes. mTORC1-dependent feedback mechanism, especially in cells lacking the TSC1-TSC2 complex, blocks growth-factor-stimulated phosphorylation of Akt (Fig. 3). The molecular mechanism by which the TSC1-TSC2 complex promotes mTORC2 activation and whether certain signals regulate the TSC complex to inhibit mTORC1 is not known.

mTOR serves a critical role in regulating the translational machinery that affects growth, proliferation and differentiation, all of which are abnormally manifested in TSC lesions. The importance of mTOR pathway in human pathology is reflected in the overexpression of p70S6K in a subset of cancer and its correlation with a poor prognosis.[35] Activation of S6 Kinase and its target S6 ribosomal protein (P70S6K) was demonstrated in cells lacking TSC2 expression.[35] Activation of mTOR in lymphangioleiomyomatosis-associated angiomyolipomas through phosphorylation of S6 protein at $Ser^{235/236}$.[53] Recent published data from our lab showed that phosphorylation of S6 Kinase at Thr^{389} was significantly higher in angiomyolipomas tissues compared to control samples indicating the increase of mTOR activity in tumor tissues.[21] Loss of the TSC1-TSC-2 appears to give rise to the highest levels of mTORC1 signalling. However, loss of a number of upstream tumour suppressors, which effect specific inputs into the pathway, are likely to lead to more moderate basal increases in mTORC1. Furthermore, increased phosphorylation of the TSC1-TSC2 complex on the Akt and ERK sites have been detected in human cancers with elevated mTORC1 signalling (ref. 53, Lewiz-Habib unpublished data). Therefore dysregulation of the TSC1-TSC2 complex and subsequent activation of mTORC1 is likely to be a common molecular event in tumorigenesis.

TSC and DNA Damage/Repair Pathway

DNA Damage/Repair Pathway and Cancer

DNA damage plays a fundamental role in neoplastic transformation. Oxidative DNA damage is one of the most common threats to genomic stability. Mutations that influence the repair of oxidized DNA modifications are expected to increase the steady-state (background) levels of these modifications and thus create a mutator phenotype that predisposes to malignant transformation.[54] Many of these mutations occur as a result of irreparable or incompletely repaired genomic DNA, which is constantly subject to assault from intrinsic and environmental insults. Oxidized forms of DNA in particular are produced in mammalian cells as a byproduct of normal oxidative metabolism or in response to exogenous sources of reactive oxygen species.[54]

8-Oxo-deoxyguanine (8-oxodG) is one of the major base lesions formed after oxidative damage to DNA. 8-oxodG is mutagenic since it pairs with adenine during DNA synthesis, increasing G:C to T:A transversions.[55] The steady-state levels of 8-oxoG, which reflect the balance between continuous generation and removal, are significantly higher in livers of $OGG1^{-/-}$ mice compared to wild-type animals.[56] Adenoma and carcinoma spontaneously developed five times higher in OGG1 knockout mice at age of 1.5 years compared to wild type mice.[56,57] Oxidative damage-induced mutations activate oncogenes or inactivate tumor suppressor genes such as TSC1/2, altering certain cell signals involved in tumor initiation. 8-OxodG in DNA is repaired primarily via the DNA base excision repair pathway. The enzyme that recognizes and excises 8-oxo-dG is 8-oxoG-DNA glycosylase (OGG1).[55] The OGG1 protein initiates the base excision repair process by recognizing and excising the modified base. Mice lacking a functional OGG1 protein accumulate abnormal levels of 8-oxoG in their genomes and display a moderately elevated spontaneous mutation rate in nonproliferative tissues.[58] Because inactivation of the OGG1 gene in mammalian cells causes a mutator phenotype, it can be expected that cells lacking OGG1 activity could have an enhanced probability of undergoing malignant transformation. The validation of this hypothesis requires the identification of human tumors where both alleles of the OGG1 gene are nonfunctional.

The human OGG1 gene is located on chromosome 3p25. Human chromosome 3p cytogenetic abnormalities and loss of heterozygosity (LOH) have been observed at high frequency in sporadic forms of RCC.[59] The 3p21.2-p21.3 locus is frequently deleted in renal cell carcinoma and also in other cancers, but to date, no candidate gene has been identified. Mutations in DNA repair genes are associated with cancer and the efficacy of DNA repair may determine the susceptibility to carcinogens. OGG1 gene has been found to be somatically mutated in some cancer cells and is highly polymorphic among human cancers. hOGG1 Ser326Cys polymorphism is significantly increased in lung,[61] kidney,[62,63] gastric,[64] gallbladder,[65] esophageal[66] and orolaryngeal cancer.[67] Loss of heterozygosity at the OGG1 allele, located on chromosome 3p25, was found in 85% of 99 human kidney clear cell carcinoma samples, identifying loss of OGG1 function as a possible contributor to tumorigenesis in the kidney.[62] Increased susceptibility to mutagens and impaired DNA repair can contribute to the genomic instability and, in consequence, to cancer. While extensive literature supports the evidence of accumulation of DNA damage and deficiency in DNA repair, the mechanism(s) by which the oxidized DNA accumulate or is efficiently removed by cellular DNA repair pathways has not been elucidated.

TSC2 Regulates DNA Damage/Repair Pathway

The generation of oxidative DNA damage is counteracted in all species by specific repair mechanisms. 8-OxodG induces mutation via misincorporation of DNA bases present in the unrepaired DNA adducts, or by slippage of DNA polymerase during replicative bypass. OGG1 is one of the major enzymes involved in the repair of 8-oxodG adducts in DNA and is highly expressed in kidney tissue. In mammalian cells, the base excision repair initiated by OGG1 represents the main defense mechanism against the mutagenic effects of 8-oxodG. Dysregulation of human DNA repair gene OGG1 is associated with an increased cancer risk. The constitutive expression of OGG1 in $TSC2^{+/-}$ heterozygous Eker rat kidneys is 3-fold less than in wild type rats suggesting that these proteins may be functionally linked.[68] In addition, the basal levels of oxidative DNA damage, 8-oxodG, in the kidney cortex of Eker rats ($TSC2^{+/-}$) were 2-fold higher than in kidney cortex of wild type rats ($TSC2^{+/+}$).[69] The decrease in OGG1 expression in $Tsc\text{-}2^{+/-}$ rats has important functional consequences, compromising the ability of these animals to respond to oxidative stress. Moreover, treatment of $TSC2^{+/-}$ rats with an oxidative DNA damaging agent greatly decreases renal OGG1 expression with concomitant increase of 47-folds of 8-oxo-dG compared to wild type rats. Indeed, the loss of OGG1 expression in kidney tumor tissue from Eker rat resulted in the accumulation of significant amounts of 8-oxodG (unrepaired DNA lesions), suggesting that loss of tuberin was biologically significant in affecting OGG1.[68]

Recently we showed that suppression of renal OGG1 in tuberin-deficient cells is sufficient to downregulate OGG1, at least in part through the transcription factor, NF-YA, the major transcription factor regulates OGG1 activity, with consequent decreased transcription of OGG1.[70] Increased tuberin phophorylation and decreased total tuberin levels is associated with decreased expression of mRNA and protein of OGG1 in kidney angiomyolipoma tissue of TSC patients.[21] Decrease OGG1 mRNA in kidney tumor samples suggests that decreased transcription is one potential mechanism responsible for downregulation of OGG1 protein.[21] In addition, deficiency of tuberin in null and heterozygous mouse embryonic fibroblast cells is associated with decreased in NF-YA expression.[70] NFYA expression is significantly decreased in most TSC tumor kidney samples compared to control kidney samples.[21] However, decrease OGG1 expression in angiomyolipomas tumor kidney resulted in 4-fold increase in 8-oxodG levels compared to the control kidney tissue.[21] In addition, phosphorylation of S6K on Thr^{389} is significantly increased in tumor tissues compared to normal samples suggesting that tuberin phosphorylation activates mTOR in tumor kidney tissue. In summary, loss of tuberin is associated with loss of OGG1 expression in kidney tumor tissue from Eker rat and resulted in the accumulation of significant amounts of 8-oxodG (unrepaired DNA lesions),[69] suggesting that loss of tuberin is biologically relevant in affecting OGG1. Impaired OGG1 function and loss of tuberin in tumor tissue predisposes to further genetic alterations as a result accumulation of mismatched DNA base lesions, a form of genomic instability that if left unrepaired promotes additional genetic alterations leading to the full blown tumor phenotype in kidney patients with TSC. These data suggest that tuberin and OGG1 are strong susceptibility candidate genes for tumorigenesis.

Conclusion

In conclusion, TSC disease is associated mainly with benign tumor in several organs and disabling neurologic disorders in human. Malignant tumors progress is rarely detected in TSC patients. On the other hand, deficiency of TSC in experimental animal model is associated with development of malignant tumors in several organs including kidney. Renal cell carcinoma in TSC-deficient rat is histologically similar to human suggesting that rat is a unique model to investigate mechanisms of carcinogenesis in the kidney. Loss of TSC2 in kidney tumor of Eker rat is associated with loss of OGG1 and accumulation significant levels of oxidative DNA damage 8-oxo-deoxyguanine suggesting that TSC2 and OGG1 play a major role in renal tumorigenesis.

Acknowledgements

The author would like to thank Dr. Daniel Riley and Dr. Anthony Valente for critical reading the chapter. This work was supported in part by grants from the American Diabetes Association and New Investigator Award from South Texas Veterans Healthcare System to SLH.

References

1. Bourneville DM. Sclerose tubereuse des circonvolutions cerebrales: idiotie et epilepsie hemiplegique. Arch Neurol (Paris) 1880; 1:81-91.
2. Kandt RS, Haines JL, Smith M et al. Linkage of an important gene locus for tuberous sclerosis to a chromosome 16 marker for polycystic kidney disease. Nat Genet 1992; 2:37-41.
3. Consortium. Identification and characterization of the tuberous sclerosis gene on chromosome 16. Cell 1993; 75:1305-1315.

4. Van Slegtenhorst M, de Hoogt R, Hermans C et al. Identification of the tuberous sclerosis gene TSC1 on chromosome 9q34. Science 1997; 277:805-808.
5. Carsillo T, Astrinidis A, Henske EP. Mutations in the tuberous sclerosis complex gene TSC2 are a cause of sporadic pulmonary lymphangioleiomyomatosis. Proc Natl Acad Sci USA 2000; 97:6085-6090.
6. Stillwell TJ, Gomez MR, Kelalis PP. Renal lesions in tuberous sclerosis. J Urol 1987; 138:477-481.
7. Al-Saleem T, Wessner LL, Scheithauer BW et al. Malignant tumors of the kidney, brain and soft tissues in children and young adults with the tuberous sclerosis complex. Cancer 1998; 83:2208-2216.
8. Shapiro RA, Skinner DG, Stanley P et al. Renal tumors associated with tuberous sclerosis. the case for aggressive surgical management. J Urol 1984; 132:1170-1174.
9. Onda H, Lueck A, Marks PW et al. $Tsc2^{(+/-)}$ mice develop tumors in multiple sites that express gelsolin and are influenced by genetic background. J Clin Invest 1999; 104:687-695.
10. Kobayashi T, Minowa O, Kuno J et al. Renal carcinogenesis, hepatic hemangiomatosis and embryonic lethality caused by a germ-line Tsc2 mutation in mice. Cancer Res 1999; 59:1206-1211.
11. Hino O, Mitani H, Katsuyama H et al. A novel cancer predisposition syndrome in the Eker rat model. Cancer Lett 1994; 83:117-121.
12. Hino O, Klein-Szanto AJ, Freed JJ et al. Spontaneous and radiation-induced renal tumors in the Eker rat model of dominantly inherited cancer. Proc Natl Acad Sci USA 1993; 90:327-331.
13. Rennebeck G, Kleymenova EV, Anderson R et al. Loss of function of the tuberous sclerosis 2 tumor suppressor gene results in embryoni c lethality characterized by disrupted neuroepithelial growth and development. Proc Natl Acad Sci USA 1998; 95:15629-15634.
14. Everitt JI, Goldsworthy TL, Wolf DC et al. Hereditary renal cell carcinoma in the Eker rat: A rodent familial cancer syndrome. J Urol 1992; 148:1932-1936.
15. Eker R, Mossige J, Johannessen JV et al. Hereditary renal adenomas and adenocarcinoma s in rats. Diagn Histopathol 1981; 4:99-110.
16. Everitt JI, Goldsworthy TL, Wolf DC et al. Hereditary renal cell carcinoma in the Eker rat: A unique animal model for the study of cancer susceptibility. Toxicol Lett 1995; 82-83:621-625.
17. Washecka R, Hanna M. Malignant renal tumors in tuberous sclerosis. Urology 1991; 37:340-343.
18. Robertson FM, Cendron M, Klauber GT et al. Renal cell carcinoma in association with tuberous sclerosis in children. J Pediatr Surg 1996; 31:729-730.
19. Bjornsson J, Short MP, Kwiatkowski DJ et al. Tuberous sclerosis-associated renal cell carcinoma: clinical, pathological and genetic features. Am J Pathol 1996; 149:1201-1208.
20. Plank TL, Logginidou H, Klein-Szanto A et al. The expression of hamartin, the product of the TSC1 gene, in normal human tissues and in TSC1- and TSC2-linked angiomyolipomas. Mod Pathol 1999; 12:539-545.
21. Habib SL. Insight into mechanism of oxidative DNA damage in angiomyolipomas from TSC patients. Mol Cancer 2009; 8:1-10.
22. Johnson MW, Kerfoot C, Bushnell T et al. Hamartin and tuberin expression in human tissues. Mod Pathol 2001; 14:202-210.
23. Lou D, Griffith N, Noonan DJ. The tuberous sclerosis 2 gene product can localize to nuclei in a phosphorylation-dependentmanner. Mol Cell Biol Res Commun 2001; 4:374-380.
24. Lamb RF, Roy C, Diefenbach TJ et al. The TSC1 tumour suppressor hamartin regulates cell adhesion through ERM proteins and the GTPase Rho. Nat Cell Biol 2000; 2:281-287.
25. Inoki K, Li Y, Xu T et al. Rheb GTPase is a direct target of TSC2 GAP activity and regulates mTOR signaling. Genes Dev 2003; 17:1829-1834.
26. Wienecke R, Konig A, DeClue JE. Identification of tuberin, the tuberous sclerosis-2 product. Tuberin possesses specific Rap1GAP activity. J Biol Chem 1995; 270:16409-16414.
27. Jansen FE, Notenboom RG, Nellist M et al. Differential localization of hamartin and tuberin and increased S6 phosphorylation in a tuber. Neurology 2004; 63:1293-1295.
28. Nellist M, van Slegtenhorst MA, Goedbloed M et al. Characterization of the cytosolic tuberin-hamartin complex. Tuberin is a cytosolic chaperone for hamartin. J Biol Chem 1999; 274:35647-35652.
29. van Slegtenhorst M, Nellist M, Nagelkerken B et al. Interaction between hamartin and tuberin, the TSC1 and TSC2 gene products. Hum Mol Genet 1998; 7:1053-7.
30. Maheshwar MM, Cheadle JP, Jones AC et al. The GAP-related domain of tuberin, the product of the TSC2 gene, is a target for missense mutations in tuberous sclerosis. Hum Mol Genet 1997; 6:1991-1996.
31. Nellist M, Verhaaf B, Goedbloed MA et al. TSC2 missense mutations inhibit tuberin phosphorylation and prevent formation of the tuberin-hamartin complex. Hum Mol Genet 2001; 10:2889-2898.
32. Li Y, Inoki K, Guan KL. Biochemical and functional characterizations of small GTPase Rheb and TSC2 GAP activity. Mol Cell Biol 2004; 24:7965-7975.
33. Knudson AG Jr. Mutation and cancer: statistical study of retinoblastoma. Proc Natl Acad Sci USA 1971; 68:820-823.

34. Manning BD, Tee AR, Logsdon MN et al. Identification of the tuberous sclerosis complex-2 tumor suppressor gene product tuberin as a target of the phosphoinositide 3-kinase/akt pathway. Mol Cell 2002; 10:151-162.
35. Huang J, Manning BD. The TSC1-TSC2 complex: a molecular switchboard controlling cell growth. Biochem J 2008; 412:179-190.
36. Inoki K, Li Y, Zhu T et al. TSC2 is phosphorylated and inhibited by Akt and suppresses mTOR signalling. Nat Cell Biol 2002; 4:648-657.
37. Cai SL, Tee AR, Short JD et al. Activity of TSC2 is inhibited by AKT-mediated phosphorylation and membrane partitioning. J Cell Biol 2006; 173:279-289.
38. Xiao GH, Shoarinejad F, Jin F et al. The tuberous sclerosis 2 gene product, tuberin, functions as a Rab5 GTPase activating protein (GAP) in modulating endocytosis. J Biol Chem 1997; 272:6097-6100.
39. Inoki K, Zhu T, Guan KL. TSC2 mediates cellular energy response to control cell growth and survival. Cell 2003; 115:577-590.
40. Ma L, Chen Z, Erdjument-Bromage H et al. Phosphorylation and functional inactivation of TSC2 by Erk implications for tuberous sclerosis and cancer pathogenesis. Cell 2005; 121:179-193.
41. Ma L, Teruya-Feldstein J, Bonner P et al. Identification of S664 TSC2 phosphorylation as a marker for extracellular signal-regulated kinase mediated mTOR activation in tuberous sclerosis and human cancer. Cancer Res 2007; 67:7106-7112.
42. Woods A, Johnstone SR, Dickerson K et al. LKB1 is the upstream kinase in the AMP-activated protein kinase cascade. Curr Biol 2003; 13:2004-2008.
43. Shaw RJ, Kosmatka M, Bardeesy N et al. Inaugural article: The tumor suppressor LKB1 kinase directly activates AMP-activated kinase and regulates apoptosis in response to energy stress. Proc Natl Acad Sci USA 2004; 101:3329-3335.
44. Inoki K, Li Y, Xu T et al. Rheb GTPase is a direct target of TSC2 GAP activity and regulates mTOR signaling. Genes & Dev 2003; 17:1829-1834.
45. Hawley SA, Boudeau J, Reid JL et al. Complexes between the LKB1 tumor suppressor, STRAD/and MO25/are upstream kinases in the AMP-activated protein kinase cascade. J Biol 2: 28. growth by directly phosphorylating Tsc2. Nat Cell Biol 2003; 4:658-665.
46. Corradetti MN, Inoki K, Bardeesy N et al. Regulation of the TSC pathway by LKB1: Evidence of a molecular link between tuberous sclerosis complex and Peutz-Jeghers syndrome. Genes Dev 2004; 18:1533-1538.
47. Sofer A, Lei K, Johannessen CM et al. Regulation of mTOR and cell growth in response to energy stress by REDD1. Mol Cell Biol 2005; 25:5834-5845.
48. Hara K, Maruki Y, Long X et al. Raptor, a binding partner of target of rapamycin (TOR), mediates TOR action. Cell 2002; 110:177-189.
49. Oshiro N, Yoshino K, Hidayat S et al. Dissociation of raptor from mTOR is a mechanism of rapamycin-induced inhibition of mTOR function. Genes Cells 2004; 9:359-366.
50. Kim DH, Sarbassov DD, Ali SM et al. mTOR interacts with raptor to form a nutrient-sensitive complex that signals to the cell growth machinery. Cell 2002; 110:163-175.
51. Cunningham JT, Rodgers JT, Arlow DH et al. mTOR controls mitochondrial oxidative function through a YY1-PGC-1alpha transcriptional complex. Nature 2007; 450:736-740.
52. Jacinto E, Loewith R, Schmidt A et al. Mammalian TOR complex 2 controls the actin cytoskeleton and is rapamycin insensitive. Nat Cell Biol 2004; 6:1122-1128.
53. Robb VA, Astrinidis A, Henske EP. Frequent hyperphosphorylation of ribosomal protein S6 in lymphangioleiomyomatosis-associated angiomyolipomas. Mod Pathol 2006; 19:839-840.
54. Loeb LA. Mutator phenotype may be required for multistage carcinogenesis. Cancer Res 1991; 51:3075-3079.
55. Chevillard S, Radicella JP, Levalois C et al. Mutations in OGG1, a gene involved in the repair of oxidative DNA damage, are found in human lung and kidney tumours. Oncogene 1998; 16:3083-3086.
56. Sakumi K, Tominaga Y, Furuichi M et al. Oggl knockout-associated lung tumorigenesis and its suppression by Mthl gene disruption. Cancer Res 2003; 63:902-905.
57. Kunisada M, Sakumi K, Tominaga Y et al. 8-Oxoguanine formation induced by chronic UVB exposure makes Oggl knockout mice susceptible to skin carcinogenesis. Cancer Res 2005; 65:6006-6010.
58. Liao J, Seril DN, Lu GG et al. Increased susceptibility of chronic ulcerative colitis-induced carcinoma development in DNA repair enzyme Oggl deficient mice. Mol Carcinogenesis 2008; 47:638-646.
59. Lubinski J, Hadaczek P, Podolski J et al. Common regions of deletions in chromosome regions 3p12 and 3p14.2 in primary clear cell renal carcinomas. Cancer Res 1994; 54:3710-3.
60. Shinmura K, Kohno T, Kasai H et al. Infrequent mutations of the hOGG1 gene, that is involved in the excision of 8-hydroxyguanine in damaged DNA, in human gastric cancer. Jpn J Cancer Res 1998; 89:825-828.

61. Sugimura H, Kohno T, Wakai K et al. hOGG1 Ser326Cys polymorphism and lung cancer susceptibility. Cancer Epidemiol Biomarkers Prev 1999; 8:669-674.
62. Audebert M, Chevillard S, Levalois C et al. Alterations of DNA repair gene OGG1 in human clear cell carcinomas of the kidney. Cancer Res 2000; 60:4740-4744.
63. Habib SL, Daniel E, Abboud HE et al. Genetic polymorphisms in OGG1 and their association with angiomyolipoma, a benign kidney tumor in patients with tuberous sclerosis. Cancer Biol Ther 2008; 7:23-27.
64. Farinati F, Cardin R, Bortolami M et al. Oxidative DNA damage in gastric cancer: CagA status and OGG1 gene polymorphism. Int J Cancer 2008; 123:51-55.
65. Jiao X, Huang J, Wu S et al. hOGG1 Ser326Cys polymorphism and susceptibility to gallbladder cancer in a Chinese population. Int J Cancer 2007; 121:501-505.
66. Xing DY, Tan W, Song N et al. Ser326Cys polymorphism in hOGG1 gene and risk of esophageal cancer in a Chinese population. Int J Cancer 2001; 95:140-143.
67. Elahi A, Zheng Z, Park J et al. The human OGG1 DNA repair enzyme and its association with orolaryngeal cancer risk. Carcinogenesis 2002; 23:1229-12s34.
68. Habib SL, Phan MN, Patel SK et al. Reduced constitutive 8-oxoguanine-DNA glycocylase expression and impaired induction following oxidative DNA damage in the tuberin deficient Eker rat. Carcinogenesis 2003; 24:573-582.
69. Habib SL, Simone S, Barnes JJ et al. Tuberin haploinsufficiency is associated with the loss of OGG1 in rat kidney tumors. Mol Cancer 2008; 7:10-14.
70. Habib SL, Riley DJ, Bhandari B et al. Tuberin regulates the DNA repair enzyme OGG1. Am J Physiol Renal Physiol 2008; 294:F281-F290.

Chapter 9

Hereditary Photodermatoses

Dennis H. Oh and Graciela Spivak*

Abstract

Photodermatoses are defined as the abnormal reactions of the skin to photons, usually those of wavelengths found in sunlight. These reactions can be caused by a wide variety of reasons, including defects in repair of light-induced DNA lesions, the interaction of certain chemicals or medications with sunlight to produce toxic mediators and photo-induced immune reactions. In this chapter we will describe photodermatoses that are associated with hereditary conditions. These can be subdivided into several groups: dermatoses caused by abnormal metabolic conditions, idiopathic photodermatoses, defects in cancer suppressor genes not directly involved in DNA repair but that predispose to photodistributed tumors and photosensitivity due to abnormalities in DNA repair pathways. Special emphasis will be placed on the relatively recently described UV-sensitive syndrome.

Introduction

Sunlight, the electromagnetic radiation emitted by the sun, comprises a wide spectrum of wavelengths, from 100 to 10^6 nanometers (nm). The types of radiation can be divided into five groups according to wavelength: Ultraviolet C (UVC) from 100 to 280 nm, ultraviolet B (UVB) from 280 to 315 nm, ultraviolet A (UVA) from 315 to 400 nm, visible from 400 to 700 nm and longer wavelengths (infrared, microwave, radiowave regions). Sunlight is essential for plant and animal life on Earth. In particular, it is necessary for photosynthesis in plants and for the production of vitamin D in mammals. However, sunlight can be harmful, causing premature aging of the skin and cancer in addition to photodermatoses. Biomolecules in cells can sustain light-induced damage; the types of lesions depend on the wavelength to which cells or organisms are exposed.

Although all UVC radiations from the sun are absorbed by the ozone layer in the atmosphere, a small fraction of UVB and most UVA reach the earth's surface. UVB and UVC can induce damage in DNA by direct excitation. The main DNA lesions caused by these wavelengths are cyclobutane pyrimidine dimers (CPD) and (6-4) pyrimidine-pyrimidone photoproducts (6-4PP). Mutations in genes that code for factors involved in the repair of CPD and 6-4PP result in enhanced photosensitivity, manifested as acute erythema, pigmentation anomalies, atrophy and dryness of the skin and, in some patients, a high incidence of cancer in sun-exposed areas.

UVA can cause acute reactions in photosensitive patients, but this type of radiation does not directly excite DNA; indirect photosensitized reactions, that result in the formation of reactive oxygen species (ROS), lead to oxidation of DNA bases, for example, 8-hydroxyguanine (8-oxo-7,8-dihydroguanine, 8-oxoG).[1,2]

To our knowledge, there are no human syndromes in which photosensitivity is caused by deficient repair of oxidized DNA bases; the special case of Cockayne syndrome will be discussed below. To address the question of whether light-induced oxidative damage is carcinogenic, investigators have utilized animal and cellular models; thus, inhibition of various steps in the base excision repair (BER)

*Corresponding Author: Graciela Spivak—Department of Biology, Stanford University, Stanford, California 94305, USA. Email: gspivak@stanford.edu

Diseases of DNA Repair, edited by Shamim I. Ahmad.

pathway causes hypersensitivity to UVA in cells[3] and laboratory mice with engineered mutations in genes coding for enzymes that initiate the BER pathway are slightly more prone to skin cancer upon exposure to UVB[4] but are not sun-sensitive.[5] NER-deficient human cell lines have been reported to be hypersensitive to oxidants;[6,7] however, analyses of the spectra of mutations in skin tumors from XP patients indicate that all the mutations were caused by CPD or 6-4PP.[8]

Pathologies associated with visible light and radiations in the higher wavelength regions have been reported, albeit rarely; the role of these types of radiation in skin disease, if any, are poorly understood.

Metabolic Photodermatoses

Porphyrias

The porphyrias are a group of photodermatoses characterized by an underlying defect in particular stages of heme biosynthesis. In most cases, the enzymatic deficiency results in accumulation of porphyrin derivatives that absorb strongly in the UVA/B and blue regions of the electromagnetic spectrum. This results in long-lived excited states that ultimately serve as Type II photodynamic agents by generating ROS that are the primary mediators of cellular and tissue damage. The mode of inheritance and clinical features vary with the type of porphyria (Table 1), although there is considerable overlap and correct diagnosis often requires additional biochemical characterization of porphyrin intermediates in urine and stool specimens. Not surprisingly, the two porphyrias that do not generate a tetrapyrole or porphyrin intermediate and thus do not result in accumulation of a chromophore with a long-lived excited state—ALA dehydratase-deficiency porphyria and acute intermittent porphyria—also exhibit no photosensitivity or notable cutaneous symptoms. Most cases of porphyria are those of porphyria cutanea tarda and are usually acquired rather than hereditary.

Hartnup Disease

A rare autosomal recessive condition, Hartnup disease usually first manifests in mid-childhood with cutaneous symptoms. Scaly, sharply bordered erythematous plaques appear predominantly on sun-exposed areas of the head and extremities. Neurological defects, commonly cerebellar ataxia and also nystagmus and tremor, occur years later. The underlying genetic defect is a mutation in the *SLC6A19* gene, which encodes a neutral amino acid transporter found in the intestinal brush border and proximal tubule of the kidneys.[9] The resulting impairment in tryptophan absorption and subsequent reduction in nicotinamide synthesis explains the resemblance of the cutaneous findings to those of pellagra, although the mechanism by which the defect causes photosensitivity is unclear.

Smith-Lemli-Opitz Syndrome (SLO)

This autosomal recessive syndrome was described relatively recently.[10] The defining clinical features are mental retardation associated with one or more of the following features: failure to thrive, dysmorphic facies, cleft palate, congenital heart disease, hypospadias and syndactyly of the second and third toes, as well as photosensitivity. Although the original reports did not mention photosensitivity, the disease is clinically heterogeneous and it now appears that about two-thirds of patients exhibit severe photosensitivity. The photosensitivity manifests as abnormal redness and itching, occasionally urticaria, of the sun-exposed skin following a brief delay of 20 minutes and then lasting over 24 hours before fading. Additionally, SLO is unusual because its action spectra are in the UVA region. SLO is biochemically characterized by excessive 7-dehydrocholesterol and deficiency of cholesterol, resulting from mutations in the *DHCR7* gene encoding dehydrocholesterol reductase, which catalyzes the conversion of 7-dehydrocholesterol to cholesterol.[11] Since 7-dehydrocholesterol itself possesses no significant UVA absorption, an alternative metabolite of 7-dehydrocholesterol, cholesta-5,7,9(11)-trien-3β-ol (9-DDHC), which strongly absorbs in the UVA region and sensitizes keratinocytes to UVA in association with ROS production, has been proposed as the pathogenic chromophore.[12]

Table 1. The porphyrias

Porphyria	Enzyme Defect	Inheritance	Clinical Features
Not photosensitive			
ALA Dehydratase Deficiency Porphyria	ALA dehydratase	Autosomal recessive	Extremely rare porphyria. Clinical features similar to AIP.
Acute Intermittent Porphyria (AIP)	Porphobilinogen deaminase	Autosomal dominant	Most frequent acute porphyria worldwide. Usually latent until precipitated by drugs or other external factors. Abdominal pain, peripheral neuropathy and other CNS or neuropsychiatric symptoms.
Photosensitive			
Congenital Erythropoeitic Porphyria (CEP) (Gunther's disease)	Uroporphryrinogen III cosynthase	Autosomal recessive	Rare. Onset of symptoms in early childhood. Discolored and fluorescent urine and teeth, exceedingly photosensitive (blistering, erythema, edema, photophobia), hypertrichosis, scarring, dyspigmentation.
Porphyria Cutanea Tarda (PCT)	Uroporphyrinogen decarboxylase	Autosomal dominant or acquired (80% of cases)	Most common porphyria overall. Hereditary form presents at any age. Skin fragility, blisters and crusting with scarring, milia on sun-exposed skin. Hypertrichosis, sclerodermoid changes.
Hepatoerythropoietic Porphyria (HEP)	Uroporphyrinogen decarboxylase	Autosomal recessive	Rare. Onset of symptoms at birth. Features of CEP. Extreme photosensitiviy (blistering, scarring, dyspigmentation). Hypertrichosis, sclerodermoid change. Hemolytic anemia.
Hereditary Coproporphyria (HCP)	Corproporphyrinogen oxidase	Autosomal dominant	Rare. Similar to AIP, but less severe, with abdominal pain and neurological symptoms. Photosesensitivity (blisters)
Variegate Porphyria (VP)	Protoporphyrinogen oxidase	Autosomal dominant	Rare. Onset of symptoms after puberty. Acute attacks similar to AIP. Photosensitivity similar to PCT
Erythropoietic protoporphyria (EP)	Ferrochelatase	Autosomal dominant >> Autosomal recessive	Onset in early childhood. Hepatic dysfunction, mild hemolytic anemia. Photosensitivity (sun-induced burning/tingling, followed by wheals and erythema, sometimes purpura. Diminishes with age). Scarring.

Hereditary Photodermatoses of Unknown Etiology or Pathogenesis

Kindler Syndrome

Originally described in 1954, Kindler syndrome was described as "congenital poikiloderma with traumatic bulla formation and progressive cutaneous atrophy".[13] In addition, sun-sensitivity was subsequently noted. The blistering tends to be acral and ectropion, keratoderma, atrophic skin elsewhere, telangiectasias, thin nails and large freckles and nevi are also characteristic features. The blistering and photosensitivity tend to improve with age, leaving a poikiloderma or atrophic skin with telangiectasias and variable reticulate erythema and dyspigmentation. For many years, the similarity of the blistering phenotype to dystrophic epidermolysis bullosa suggested that defects in Type VII collagen might be responsible, but mutations in the corresponding gene were never found.[14] In 2003, two groups identified mutations in the *FLJ20116* gene, which was renamed *KIND1* or *FERMT1*.[15,16] Its gene product, kindlin-1, is a homolog of the *Caenorhabditis elegans* protein UNC-112, which appears to be an actin-associated protein that couples the cytoskeleton to the extracellular matrix. Although the apparent structural role of kindlin-1 may satisfactorily explain certain cutaneous phenotypes such as the bullae, its relation to photosensitivity and the poikiloderma remain unclear.[17]

Actinic Prurigo

While there has been some debate as to whether actinic prurigo should be lumped with polymorphous light eruption, these entities are usually treated as distinct diagnoses. Actinic prurigo is characterized by severe itch, erythematous weeping, vesicles, papules, plaques and deeper nodules that appear hours after sun exposure. Affected areas include the lips and conjunctivae, but lesions on sun-exposed sites may also extend to sun-protected skin. Most, but not all, patients experience abatement of symptoms during the winter. Typically girls are affected starting in childhood, though as many of 30% of patients present in adulthood. Familial actinic prurigo is well known, particularly in Native American populations, with a dominant mode of inheritance.[18] Although there is a strong association with the HLA DR4 serotype in Caucasian populations, the genetic basis of actinic prurigo and the chromophores involved remain obscure.[19] One-third of affected individuals have no abnormal phototesting, while one-third may have a reduced minimal erythema dose to UVA and UVB together and one-third exhibit a reduced minimal erythema dose to UVA alone.

Polymorphous Light Eruption (PMLE)

While most cases of PMLE are sporadic, approximately one in six patients have a family history of the condition.[20] When a twin is affected, the other twin is only slightly more prone to PMLE if they are monozygotic rather than dizygotic twins, suggesting that even if a genetic basis for PMLE exists, there is a significant environmental influence. Action spectra for PMLE induction have been in either the UVA or the UVB regions, with a preponderance of sensitivity to UVA; however a significant fraction of patients are sensitive to both wavelength regions. Typically during the springtime and early summer, cutaneous lesions appear within minutes to hours following sun exposure and last several days. A hardening phenomenon occurs with further sun exposure and can be used prophylactically to minimize symptoms. True to its name, PMLE exhibits a variety of erythematous and itching lesions, including papules, plaques, blisters and edema; in any single individual the lesions are relatively monomorphic.

Solar Urticaria

A rare familial case has been mentioned,[21] but solar urticaria otherwise has not been associated with a genetic basis. Clinically, pruritis and hives appear within minutes of sun exposure. The responsible wavelengths are typically in the UVA/B regions and visible and even infrared wavelengths have been reported in rare cases.[22] Solar urticaria has generally been associated with an abnormal immune response mediated by immunoglobulin E, although the causal radiation-induced antigen has not been identified to date.

Chronic Actinic Dermatitis

Actinic reticuloid, persistent light reactor, photosensitive eczema and photosensitivity dermatitis have been lumped together as clinical variants under the rubric of chronic actinic dermatitis. Middle-aged to elderly men develop scaling, edema and more infiltrated papules and plaques (actinic reticuloid's morphology) in sun-exposed areas. Little is known about the origins of chronic actinic dermatitis other than the activating wavelengths, most commonly in the UVB region, though longer wavelengths including visible light have also been reported and this likely represents a photosensitization disorder.[23] The confusing nosological history and relative lack of clinical and basic data have limited our understanding of this disorder, which must currently be considered idiopathic, as there is no evidence of a genetic component.

Hydroa Vacciniforme

Usually sporadic, extremely rare familial cases of hydroa vacciniforme have been described.[24] Patients usually present within the first few years of life. Minutes to hours after sun exposure, a burning and pruritic eruption occurs, evolving from tender papules to vesicles and crust. Eventual resolution of the eruption leaves scars. Remission by puberty is the rule, although exceptional cases extending into adulthood have been reported.

Defects in Cancer Suppressor Genes

Basal Cell Nevus Syndrome (Gorlin Syndrome)

This rare autosomal dominant disorder is caused by mutations in the *PTCH* gene, a homolog of the *Patched (PTC)* gene of Drosophila. PTCH is a tumor-suppressor, involved in the sonic hedgehog signaling pathway and is a transmembrane protein. Roles in DNA maintenance, repair and/or replication have been suggested in view of the chromosomal instability associated with defective PTCH function.[25] Multiple organ systems may be affected in individuals with this disease, including the skeletal, genitourinary and central nervous systems. Although palmar/plantar pits are characteristic of the syndrome, the most dramatic cutaneous manifestation is the early appearance and high incidence of basal cell carcinomas, which are more prevalent in sun-exposed areas of the skin. The *PTCH* gene is frequently mutated in basal cell carcinomas that occur sporadically in normal individuals as well as in xeroderma pigmentosum (XP) patients.[26,27]

Familial Melanoma

Although melanomas comprise only 4% of all skin cancers, the mortality rates are much higher than those for basal cell and squamous cell carcinomas. However, if melanoma is detected and treated early, the 5-year survival rate is >96% for Stage I tumors. Melanoma is hereditary in about 10% of the cases.[28] The high correlation between atypical nevi and melanoma in some families led to the acronym FAMM, for Familial Atypical Mole-Melanoma syndrome (also designated B-K mole syndrome), although controversy remains as to the pathogenic relationship of atypical nevi to melanoma. Variants in the *MC1R* gene, encoding the melanocortin-1 receptor, are associated with impaired ability to produce melanin, photosensitivity in sunlight and a several-fold risk of melanoma and are considered low-risk alleles for familial melanoma. The *CDKN2A* gene on chromosome 9p21, is the most frequently mutated gene in familial melanomas. The gene codes for the p16 and $p14^{ARF}$ proteins, both suppressors of cell growth, acting through Rb and p53 respectively. Additional melanoma susceptibility factors are the *CDK4* gene on chromosome 12q13, involved in cell cycle regulation and a locus on chromosome 1p22.[29] Of course, XP is the other familial syndrome associated with an increased incidence of melanoma.

Human Syndromes Defective in DNA Repair

As mentioned above, the best characterized target of exposure to sunlight or to sources of UVC and UVB radiation is DNA; if the resulting lesions are not repaired, they might lead to mutagenesis and carcinogenesis, or to cell death. Mutations in genes that code for factors involved in the repair of CPD and 6-4PP through the nucleotide excision repair (NER) pathway can result in enhanced to extreme photosensitivity. This is the hallmark of a number of human hereditary conditions; the presence or the severity of other symptoms depend on the gene(s) affected, on the particular nucleotide changes within the gene and on the genetic make-up of each individual. NER can operate through two separate modes; the global genomic repair (GGR) pathway removes lesions from the entire genome, while the transcription-coupled repair (TCR) pathway repairs lesions on the template strand of actively transcribed genes.

In addition to NER and BER, photoproducts might be recognized by other repair pathways. For example, mismatch repair (MMR) complexes can bind bulky lesions, but the role of MMR appears to be related to suppression of mutagenesis and apoptosis, rather than repair. However, MMR-deficient individuals are not sun-sensitive. Most of the hereditary photosensitive disorders with defective DNA repair are related to NER deficiencies; the exceptions for which the defects are known include Bloom's and Rothmund-Thompson syndromes, as well as ataxia telangiectasia, described below.

Photodermatoses and Defects in DNA Helicases

To date, five homologs of *recQ* from *Escherichia coli*, which codes for a DNA helicase, have been identified in humans. In addition to Bloom's and Rothmund-Thompson syndromes (below), Werner's syndrome is also due to mutations in a RecQ-like helicase, the *WRN* gene. Werner's patients are characterized by premature aging and early onset of sarcomas and mesenchymal tumors, but photosensitivity has not been reported. Mutations in the other two human RecQ helicases, RECQ1 and RECQ5, have not yet been genetically linked to any disease.[30]

Bloom's Syndrome

This rare chromosome breakage syndrome primarily affects Ashkenazi Jews. It manifests as failure to thrive, stunted growth, small and narrow facies, sun-sensitive facial telangiectasias, immunodeficiency and increased risk of malignancies. Mutations in the *BLM* gene, which codes for a RecQL DNA helicase, are associated with the syndrome. Cells from Bloom's syndrome patients exhibit high numbers of chromosome aberrations and rearrangements, which reflect the high mutation rate associated with the loss of the BLM helicase.

Rothmund-Thomson Syndrome (RTS) or Poikiloderma Congenitale

RTS is a rare autosomal recessive disorder attributed to mutations of the *RECQL4* helicase gene. Key features include early photosensitivity and poikilodermatous skin changes, juvenile cataracts, skeletal dysplasias and a predisposition to osteosarcoma and skin cancer. The acute phase of the disease presents in early infancy as red patches on the cheeks, spreading later to other areas of the face, extremities or buttocks. About 30% of the patients are photosensitive.[31] Individuals mutated in the *RECQL4* gene can also develop RAPADILINO, a disease very similar to RTS except for the absence of photosensitivity and poikiloderma.[32]

Photodermatoses and Defects in DNA Damage Response Pathways

Ataxia telangiectasia (AT) has been the subject of a volume from this series.[33] The disease also known as Louis-Bar syndrome, results in ataxia, immune deficiency, elevated cancer incidence and premature aging. The autosomal recessive disease is caused by mutations in the *ATM* gene, a key factor in the cellular response to DNA damage, particularly double-strand breaks. Moreover, *ATM* regulates telomere length and this might correlate with the progeria observed in most patients. Although AT patients are not photosensitive, telangiectasia due to broken venous capillaries usually appears in childhood several years following ataxia and is more evident in sun-exposed areas of the skin, though typically sun-protected areas such as the flexural surfaces of the extremities and the chest are also affected. Other common cutaneous findings, contributing to the progeroid appearance, include subcutaneous fat atrophy and graying hair. Seborrehic dermatitis, ulcerative, non-infectious granulomatous lesions, dyspigmentation that is often segmental, poikiloderma, vitiligo, hirsutism and acanthosis nigricans have also been described.

Photodermatoses and Defects in NER

Xeroderma pigmentosum (XP) has been the subject of a volume from this series.[34] The disease is defined by the atrophic, dry, parchment-like texture of the skin and by highly prevalent and early development of tumors of the skin and other sun-exposed areas, such as the eyes and the tip of the tongue; as mentioned above, XP can be considered a category of familial melanoma. The disease comprises seven complementation groups, A to G, with defects in various steps of the NER pathway and an eighth group, XP variant, with a defective DNA polymerase, pol η, which has a role in DNA synthesis past UV-induced photoproducts. Some XP patients develop neurological and cognitive dysfunction.

Cockayne syndrome (CS) is a complex disease with a multitude of symptoms; it has also been the subject of a volume in this series.[35] There are two complementation groups of CS, CS-A and CS-B, with mutations in the *CSA* (*ERCC8*) and *CSB* (*ERCC6*) genes respectively. In very rare cases, patients with mutations in the *XPB*, *XPD* or *XPG* genes exhibit a combination of symptoms of CS and XP. CS patients present three major characteristics: failure to grow, progressive neurological dysfunction and leukodystrophy (progressive degeneration of the white matter). In addition, at least three of the following minor criteria have been suggested for a positive diagnosis: photosensitivity, demyelination of the peripheral nerves, pigmentary retinopathy and/or cataracts, sensorineural hearing loss, cachectic dwarfism with stooped posture and progeria with shortened lifespan. These are usually accompanied by numerous additional problems, including microcephaly, gait defects, contractures, spasticity, tremors, dental caries, basal ganglia calcifications, hypertension, osteoporosis and others.

Cells from CS patients are defective in the TCR subpathway of NER. Although a vast majority of the CS patients display photosensitivity, the disease is not caused by the TCR defect, as discussed below.

Cerebro oculo-facial-skeletal syndrome (COFS) shares some of the traits typical of CS, with additional findings. This syndrome will be described in Chapter 21 of this volume.

De Sanctis-Cacchione (DSC) syndrome was first named "xerodermic idiocy", because the patients exhibited symptoms of XP combined with mental deficiency, progressive neurologic deterioration, dwarfism and gonadal hypoplasia. The disease has been associated with mutations in XP genes, most often in *XPA*.[36] However, two siblings diagnosed with DSC had more severe neurological symptoms and less striking cutaneous manifestations than those usually seen in XP patients. Biochemical and complementation assays determined that these individuals were mutated in the *CSB* gene.[37-39]

Triothiodystrophy (TTD) will be discussed in Chapter 10 in this volume. We will briefly mention here that the sun-sensitive TTD patients carry mutations in genes involved with the NER pathway, namely *XPB*, *XPD* and *TTDA*; the proteins encoded by these genes are subunits of the transcription factor, TFIIH.

Other diseases associated with defective DNA repair. A puzzling case, presenting with numerous skin cancers but normal sun sensitivity and whose cells exhibit normal parameters of repair of photoproducts, has been described.[40] The only abnormality that has been detected in cells from this patient is reduced host cell reactivation (HCR) of UV-irradiated reporter plasmids. This activity could not be complemented by expression of any of the NER repair proteins, but expression of UVDE (an endonuclease that incises DNA containing CPD, 6-4PP and AP sites) restored the activity of the reporter gene; HCR of AP site-containing plasmids was normal. The authors hypothesize that the enhanced carcinogenesis in this patient is the result of novel UV-induced lesions that are recognized by UVDE but not by NER factors, or by a defect in a minor, as yet unknown, subpathway of NER.

UV-Sensitive Syndrome (UV^{S}S)

First described by Itoh and colleagues,[41] individuals with UV^{S}S are sun-sensitive and present pigmentation anomalies in sun-exposed areas of the skin. As with CS, skin or internal tumors have not been described in UV^{S}S patients. In high contrast with CS, no other pathologies have been associated with UV^{S}S. In this section we will describe in detail the cellular and molecular characteristics as well as the genetics of this rare disease; to date, only seven patients have been characterized.

Response of UV^{S}S Cells to UV Irradiation

The cellular responses of UV^{S}S and CS to treatment with UV light (254 nm) are identical: defective survival, impaired recovery of RNA synthesis, accumulation of p53 correlated with apoptotic response at low UV doses and proficient global repair of photoproducts. As these observations suggest, both types of cells are deficient in TCR of photoproducts (42 and references therein). These findings were puzzling because up to then CS had been defined as a "DNA repair deficient syndrome", but the results obtained with UV^{S}S cells suggested that defective TCR causes only sun sensitivity. This raises the question: what causes the variety and severity of the symptoms in CS?

Response of UV^{S}S and CS Cells to Oxidative DNA Damage

CS cells have been shown to be hypersensitive to treatment with agents that induce primarily oxidative DNA lesions in addition to single and/or double strand breaks. When UV^{S}S and CS cells were treated in parallel with hydrogen peroxide, potassium bromate or menadione, it became evident that UV^{S}S did not share the hypersensitivity of CS cells to oxidation, but instead like wild type cells.[43,44] These results suggest that CS cells might be defective in DNA repair (particularly TCR) of oxidative DNA lesions, or that these lesions were not efficiently bypassed during DNA or RNA synthesis in CS cells, and this might be the underlying cause of the disease.

Using a host cell reactivation assay in which undamaged cells were transfected with plasmids treated with agents that induced certain types of DNA lesions, Spivak et al demonstrated that expression of the UV-irradiated *lacZ* reporter gene was defective in both UV^{S}S and CS cells, as expected from their TCR defect.[43] When the plasmid contained the oxidized bases 8-oxo-guanine or thymine glycol, UV^{S}S cells exhibited responses similar to those of the wild type controls, while recovery of expression was defective in CS cells. Plasmids prepared to contain single strand breaks or abasic sites, which are intermediates in the BER pathway, yielded equal responses from all the cell lines; thus the CS defect is not associated with defective processing of the lesions through BER. Surprisingly, repair of thymine glycols was similar in the transcribed and nontranscribed strands of *lacZ* and there were no differences between wild type and repair-deficient cell lines. The conclusion drawn was that the system might not be sensitive enough to detect TCR in the portion of the transfected plasmids that had served as templates for transcription, or that the CS defect resides not in TCR but in transcriptional bypass of oxidative damage.

UV^{S}S Complementation Groups

There are three complementation groups of UV^{S}S, with mutations in the *CSA*, *CSB*, or in an unknown gene. In the initial description of the syndrome, the siblings Kps2 and Kps3 as well as the patient US^{s}1KO had been assigned to one complementation group.[41] The patients UV^{S}24TA and XP24KO were also assigned to the same group.[45] However, sequencing analysis revealed that US^{s}1KO cells carried a homologous mutation in the *CSB* gene.[46] The mutation found by Horibata et al implies that a severely truncated CSB protein, containing only the N-terminal 76 amino acids, would be generated in UV^{S}1KO cells, but Western blot analysis failed to reveal any CSB polypeptides in UV^{S}1KO cells. Thus it appeared that individuals with no detectable CSB protein develop UV^{S}S, whereas individuals with mutant CSB protein are more severely affected, with clinical outcomes such as Cockayne, DeSanctis-Cacchione or COFS syndromes.[47] An alternative hypothesis was proposed by Neuman et al, who reported that the domesticated PiggyBac-like transposon PGBD3 resides within intron 5 of the *CSB* gene and functions as an additional 3′ terminal exon, thus an alternatively spliced mRNA gives rise to a CSB-transposase fusion protein that is as abundant as CSB. Since most *CSB* mutations in patients have been found beyond intron 5, these authors suggested that CS could be caused by the unbalanced expression of the fusion protein.[48] These hypotheses were invalidated by the recent discovery of two CS patients with severe symptoms, who carried large deletions of the promoter and upstream exons in the *CSB* gene that prevented its expression and that of the fusion protein.[49] CS3AM is another UV^{S}S patient with the same *CSB* mutation as UV^{S}1KO.[50] A report of a patient with late onset symptoms of CS (at age 47), who has a mutation in *CSB* a few nucleotides downstream from the one in CS3AM or UV^{S}1KO and in whose cells the CSB protein could also not be detected,[47,51] is suggestive of a continuum between mild (as in UV^{S}S) and severe (as in CS) disease, depending on the location of the mutation(s) and on other factors that have not been identified to date. In support of this idea, individuals with the identical mutation in the *CSB* gene can present with different phenotypes.[52] Another possibility is that the UV^{S}S patients with mutations in *CSB*, might develop CS symptoms later in life; to our knowledge, the oldest UV^{S}S patient is about 40-years old.

The work of Nardo et al[44] demonstrates that certain mutations in the *CSA* gene can also lead to UV^{S}S. Thus, a picture emerges of dual roles for CSA and CSB: it is clear that functional CS proteins are required for TCR, but there appears to be a second activity that prevents developmental failure, premature aging and early death. Moreover, the activities involved in TCR of bulky DNA adducts and in resistance to oxidizing agents are uncoupled for both proteins. This concept is supported by the absence of CS-like phenotypes in XP-A individuals, who are defective in TCR and in global NER. The neurological degeneration observed in some XP-A patients is clinically distinct from that in CS individuals.

Another important conclusion drawn from the results described above is that CS is not caused by deficient TCR of bulky lesions, since UV^{S}S individuals with defective TCR present no developmental/neurological problems at all. The molecular abnormalities that lead to the progressive developmental problems and neurological degeneration typical of CS have not been determined. The hypotheses include problems with basal transcription, global repair of oxidative DNA lesions, activation of hormones involved in growth and metabolism, transcriptional bypass of base lesions and TCR of endogenously induced DNA lesions.[53]

Many of the factors involved in TCR have been characterized and various mechanisms have been suggested, but the lack of an in vitro TCR assay has hindered the detailed biochemical analyses of the pathway. The work of several research groups has defined plausible protein complexes involved in the various steps in TCR in humans: RNA polymerase II, CSB and XAB2 are required for assembly of the pre-incision complex, while CSA as a component of an E3 ubiquitin ligase complex, is required for post-incision events. Chromatin remodeling factors such as HMGN1 and p300 might be needed for displacing nucleosomes that become re-established behind the translocating transcription complex, so that the polymerase can regress, exposing the lesion to repair complexes. Following recognition of the lesion, the TCR pathway requires NER factors for incision, removal of the damaged oligonucleotide, repair synthesis and ligation. Additional factors are being identified, for example NEDD4 that has been implicated in DNA damage-induced ubiquitination of the RNA polymerase II subunit RPB1, marking it for proteosomal degradation.[54] Further studies, including the identification of the gene defective in the third complementation group of UV^{S}S, are necessary to elucidate the role of the respective proteins in DNA repair and transcription pathways.

Conclusion

The diseases described in this chapter provide a glimpse into the extremely complex reactions that occur in cells when they are exposed to various types of photons. The skin, which is the most spatially extensive organ and the main target of sunlight, responds to photons with a variety of inflammatory, pigmentation, premature aging and neoplastic pathologies. The correlations between defective biochemical pathways and the corresponding hereditary syndromes have yielded important insights into a number of hereditary photodermatoses, particularly those related to DNA damage and repair, which can serve as a foundation for the design of innovative therapeutic approaches. UV^{S}S is a particularly good example that illustrates the importance of precisely defining novel disease phenotypes, which can then guide identification of and be correlated with specific genetic and biochemical abnormalities. However, the number of photodermatoses, both hereditary and acquired, for which the molecular pathophysiology remains unclear, suggests that there are still rich opportunities to elucidate the photobiology of skin.

References

1. Kvam E, Tyrrell RM. Induction of oxidative DNA base damage in human skin cells by UV and near visible radiation. Carcinogenesis 1997; 18(12):2379.
2. Cadet J, Sage E, Douki T. Ultraviolet radiation-mediated damage to cellular DNA. Mutat Res 2005; 571(1-2):3.
3. Kim KJ, Chakrabarty I, Li GZ et al. Modulation of base excision repair alters cellular sensitivity to UVA1 but not to UVB1. Photochem Photobiol 2002; 75(5):507.
4. Kunisada M, Sakumi K, Tominaga Y et al. 8-Oxoguanine formation induced by chronic UVB exposure makes Ogg1 knockout mice susceptible to skin carcinogenesis. Cancer Res 2005; 65(14):6006.
5. Vens C, Dahmen-Mooren E, Verwijs-Janssen M et al. The role of DNA polymerase beta in determining sensitivity to ionizing radiation in human tumor cells. Nucleic Acids Res 2002; 30(13):2995.
6. D'Errico M, Parlanti E, Teson M et al. New functions of XPC in the protection of human skin cells from oxidative damage. EMBO J 2006; 25(18):4305.
7. Rünger T, Epe B, Moller K. Repair of ultraviolet B and singlet oxygen-induced DNA damage in xeroderma pigmentosum cells. J Invest Dermatol 1995; 104(1):68.
8. Dumaz N, Drougard C, Sarasin A et al. Specific UV-induced mutation spectrum in the p53 gene of skin tumors from DNA-repair-deficient xeroderma pigmentosum patients. Proc Natl Acad Sci USA 1993; 90(22):10529.
9. Azmanov DN, Kowalczuk S, Rodgers H et al. Further evidence for allelic heterogeneity in Hartnup disorder. Human Mut 2008; 29(10):1217.
10. Smith DW, Lemli L, Opitz JM. A newly recognized syndrome of multiple congenital anomalies. J Pediatr 1964; 64:210.
11. Yu H, Patel SB. Recent insights into the Smith-Lemli-Opitz syndrome. Clin Genet 2005; 68(5):383.
12. Chignell CF, Kukielczak BM, Sik RH et al. Ultraviolet A sensitivity in Smith-Lemli-Opitz syndrome: Possible involvement of cholesta-5,7,9(11)-trien-3 beta-ol. Free Radic Biol Med 2006; 41(2):339.
13. Kindler T. Congenital poikiloderma with traumatic bulla formation and progressive cutaneous atrophy. Br J Dermatol 1954; 66(3):104.
14. Yasukawa K, Sato-Matsumura KC, McMillan J et al. Exclusion of COL7A1 mutation in Kindler syndrome. J Am Acad Dermatol 2002; 46(3):447.
15. Siegel DH, Ashton GH, Penagos HG et al. Loss of kindlin-1, a human homolog of the Caenorhabditis elegans actin-extracellular-matrix linker protein UNC-112, causes Kindler syndrome. Am J Hum Genet 2003; 73(1):174.
16. Jobard F, Bouadjar B, Caux F et al. Identification of mutations in a new gene encoding a FERM family protein with a pleckstrin homology domain in Kindler syndrome. Hum Mol Genet 2003; 12(8):925.
17. White SJ, McLean WH. Kindler surprise: mutations in a novel actin-associated protein cause Kindler syndrome. J Dermatol Sci 2005; 38(3):169.
18. Schnell AH, Elston RC, Hull PR et al. Major gene segregation of actinic prurigo among North American Indians in Saskatchewan. Am J Med Genet 2000; 92(3):212.
19. Grabczynska SA, McGregor JM, Kondeatis E et al. Actinic prurigo and polymorphic light eruption: common pathogenesis and the importance of HLA-DR4/DRB1*0407. Br J Dermatol 1999; 140(2):232.
20. Morison WL, Stern RS. Polymorphous light eruption: a common reaction uncommonly recognized. Acta Derm Venereol 1982; 62(3):237.
21. Horkay I, Emri G, Varga V et al. Environmental dermatology in childhood: photosensitivity. Pediatric Health 2008; 2(6):749.
22. Botto NC, Warshaw EM. Solar urticaria. J Am Acad Dermatol 2008; 59(6):909.
23. Hawk JL. Chronic actinic dermatitis. Photodermatol Photoimmunol Photomed 2004; 20(6):312.
24. Gupta G, Mohamed M, Kemmett D. Familial hydroa vacciniforme. Br J Dermatol 1999; 140(1):124.
25. Shafei-Benaissa E, Savage JR, Babin P et al. The naevoid basal-cell carcinoma syndrome (Gorlin syndrome) is a chromosomal instability syndrome. Mutat Res 1998; 397(2):287.
26. Bodak N, Queille S, Avril MF et al. High levels of patched gene mutations in basal-cell carcinomas from patients with xeroderma pigmentosum. Proc Natl Acad Sci USA 1999; 96(9):5117.
27. Giglia G, Dumaz N, Drougard C et al. p53 mutations in skin and internal tumors of xeroderma pigmentosum patients belonging to the complementation group C. Cancer Res 1998; 58(19):4402.
28. Tsao H, Niendorf K. Genetic testing in hereditary melanoma. J Am Acad Dermatol 2004; 51(5):803.
29. de Snoo F, Grui N. Familial melanoma. In: Huret J-L, ed. Atlas Genet Cytogenet Oncol Haematol Poitiers 2005.
30. Brosh RM, Jr, Bohr VA. Human premature aging, DNA repair and RecQ helicases. Nucleic Acids Res 2007; 35(22):7527.
31. Hsu S, George S. Rothmund-Thompson syndrome. In: emedicine 2007.
32. Siitonen HA, Kopra O, Kaariainen H et al. Molecular defect of RAPADILINO syndrome expands the phenotype spectrum of RECQL diseases. Hum Mol Genet 2003; 12(21):2837.

33. Ahmad S, ed. Molecular Mechanisms of Ataxia Telangiectasia. Austin: Landes Bioscience, 2009.
34. Ahmad S, Hanaoka F, eds. Molecular Mechanisms of Xeroderma Pigmentosum. Austin/New York: Landes Bioscience/Springer Science+Business Media, 2008.
35. Ahmad S, ed. Molecular Mechanisms of Cockayne Syndrome. Austin: Landes Bioscience, 2009.
36. Kraemer KH, Lee MM, Scotto J. Xeroderma pigmentosum. Cutaneous, ocular and neurologic abnormalities in 830 published cases. Arch Dermatol 1987; 123(2):241.
37. Greenhaw GA, Hebert A, Duke-Woodside ME et al. Xeroderma pigmentosum and Cockayne syndrome: overlapping clinical and biochemical phenotypes. Am J Hum Genet 1992; 50(4):677.
38. Colella S, Nardo T, Botta E et al. Identical mutations in the CSB gene associated with either Cockayne syndrome or the DeSanctis-cacchione variant of xeroderma pigmentosum. Hum Mol Genet 2000; 9(8):1171.
39. Itoh T, Cleaver JE, Yamaizumi M. Cockayne syndrome complementation group B associated with xeroderma pigmentosum phenotype. Hum Genet 1996; 97(2):176.
40. Hashimoto S, Egawa K, Ihn H et al. A new disorder in UV-induced skin cancer with defective DNA repair distinct from xeroderma pigmentosum or Cockayne syndrome. J Invest Dermatol 2008; 128(3):694.
41. Itoh T, Fujiwara Y, Ono T et al. UVs syndrome, a new general category of photosensitive disorder with defective DNA repair, is distinct from xeroderma pigmentosum variant and rodent complementation group I. Am J Hum Genet 1995; 56(6):1267.
42. Spivak G. UV-sensitive syndrome. Mutat Res 2005; 577:162.
43. Spivak G, Hanawalt PC. Host cell reactivation of plasmids containing oxidative DNA lesions is defective in Cockayne syndrome but normal in UV-sensitive syndrome fibroblasts. DNA Repair (Amst) 2006; 5(1):13.
44. Nardo T, Oneda R, Spivak G et al. A UV-sensitive syndrome patient with a specific CSA mutation reveals separable roles for CSA in response to UV and oxidative DNA damage. Proc Natl Acad Sci USA 2009.
45. Itoh T, Linn S, Ono T et al. Reinvestigation of the classification of five cell strains of xeroderma pigmentosum group E with reclassification of three of them. J Invest Dermat 2000; 114(5):1022.
46. Horibata K, Iwamoto Y, Kuraoka I et al. Complete absence of Cockayne syndrome group B gene product gives rise to UV-sensitive syndrome but not Cockayne syndrome. Proc Natl Acad Sci USA 2004; 101:15410.
47. Spivak G. The many faces of Cockayne syndrome.(comment). Proc Natl Acad Sci USA 2004; 101(43):15273.
48. Newman JC, Bailey AD, Fan HY et al. An abundant evolutionarily conserved CSB-PiggyBac fusion protein expressed in Cockayne syndrome. PLoS Genet 2008; 4(3):e1000031.
49. Laugel V, Dalloz C, Stary A et al. Deletion of 5′ sequences of the CSB gene provides insight into the pathophysiology of Cockayne syndrome. Eur J Hum Genet 2008; 16(3):320.
50. Tanaka K, Itoh S. Transcription-coupled repair and its defect in cockayne syndrome. In: Ahmad S, ed. Molecular Mechanisms of Cockayne Syndrome. Austin: Landes. in press, 2009.
51. Hashimoto S, Suga T, Kudo E et al. Adult-onset neurological degeneration in a patient with Cockayne syndrome and a null mutation in the CSB gene. J Invest Dermatol 2008; 128(6):1597.
52. Colella S, Nardo T, Botta E et al. Identical mutations in the CSB gene associated with either Cockayne syndrome or the DeSanctis-cacchione variant of xeroderma pigmentosum. Hum Mol Genet 2000; 9(8):1171.
53. Hanawalt PC, Spivak G. Transcription-coupled DNA repair: two decades of progress and surprises. Nat Rev Mol Cell Biol 2008; 9(12):958.
54. Anindya R, Aygun O, Svejstrup JQ. Damage-induced ubiquitylation of human RNA polymerase II by the ubiquitin ligase Nedd4, but not Cockayne syndrome proteins or BRCA1. Molecular Cell 2007; 28(3):386.

CHAPTER 10

Trichothiodystrophy:
Photosensitive, TTD-P, TTD, Tay Syndrome

W. Clark Lambert,* Claude E. Gagna and Muriel W. Lambert

Abstract

Although the term, "trichothiodystrophy" (TTD) refers to the hair anomalies in this group of patients, this is a heterogeneous, multisystem disease in which any or every organ in the body may be affected.[1-5] Neuroectodermal derived tissues are particularly likely to be involved. This term was introduced by Price et al in 1980 to designate patients with sulfur-deficient brittle hair, which they recognized as a marker for this complex disease and designated it as a "neuroectodermal symptom complex".[6] Patients with TTD have brittle hair and nails (associated with reduced content of cysteine-rich matrix proteins), ichthyotic skin and physical and mental growth retardation.[7] Ichthyosis is usually apparent at birth but much less so after the first few weeks of life. Other frequently associated features include ocular cataracts, infections and maternal complications related to pregnancy. Atrophy of subcutaneous fat may also be present.

TTD occurs in a pattern of inheritance consistent with an autosomal recessive condition. The disease is extremely heterogeneous in severity and extent, with some patients showing no neurological deficiency.[8] Others show severe, multisystem disease. Many patients die at a young age, most commonly due to infectious disease.[1]

TTD is part of a more broadly defined group of diseases identified as IBIDS (**i**chthyosis, **b**rittle hair, **i**mpaired intelligence, **d**ecreased fertility and **s**hort stature). Photosensitive cases are also identified as PIBIDS (**p**hotosensitivity with IBIDS). Cases without manifest ichthyosis are also identified as PBIDS. These syndromes defy rigorous definition because of clinical variation between patients.[5] The original two cases were described by Tay in oriental siblings, whose parents were first cousins; thus the disease is also known as Tay syndrome.[9]

The hairs in patients with TTD have a distinctive, diagnostically useful appearance on polarized light microscopy consisting of alternating light and dark bands known as the "tiger tail" anomaly. Diagnosis may be confirmed by sulfur content analysis of hair shafts, which shows decreased sulfur and cysteine content.[5,7]

Approximately half of patients with TTD have photosensitivity, which correlates with a nucleotide excision repair (NER) defect. These patients are designated as having trichothiodystrophy—photosensitive (TTDP). Non-photosensitive patients are designated as having trichothiodystrophy—nonphotosensitive (TTDN). Skin cancer is very rare in sun-sensitive TTD.[5,10]

Clinical Manifestations

TTD is part of a complex and often clinically challenging group of diseases affecting skin, hair, nails and frequently displaying ichthyosis (increase, usually with malformation, of the superficial, cornified layer of skin). As noted above, not all patients within the formal definition of TTD are

*Corresponding Author: W. Clark Lambert—Department of Pathology and Laboratory Medicine, Room C520, Medical Science Building, UMDNJ-New Jersey Medical School, 185 South Orange Avenue, Newark, New Jersey 07103, USA. Email: lamberwc@umdnj.edu

Diseases of DNA Repair, edited by Shamim I. Ahmad. ©2010 Landes Bioscience and Springer Science+Business Media.

photosensitive. However, some of the other entities within this broad disease spectrum, outside this definition of TTD, are also photosensitive,[11] including a group of diseases falling within a syndrome known as *ichthyosiform follicularis atrichia with photophobia* (IFAP).[12] Like TTD, IFAP may also manifest neuropathological features.[13]

Faghri et al performed an extensive systematic literature review, finding 112 patients with trichothiodystrophy.[1] They ranged from 12 weeks to 47 years of age. In addition to hair abnormalities, there were developmental delay/intellectual impairment (86%), short stature (73%), ichthyosis (65%), abnormal characteristics at birth (55%), particularly low birth weight and length, ocular abnormalities (51%; primarily cataract), infections (46%), especially pulmonary infections, photosensitivity (42%), maternal pregnancy complications (28%) and defective DNA repair (37%). Hypogonadism and microcephaly are also characteristic. As seen in autosomal dominant ichthyosis vulgaris, flexural areas of the arms and legs may be spared. Decrease in subcutaneous fatty tissue is seen in some patients. In women, breast tissue may be completely absent despite normal nipple development. As seen in Cockayne syndrome, the face of TTD patients may have an aged appearance due to lack of subcutaneous fat. In the Faghri, et al series of 112 patients, there were 19 deaths in patients under 10 years of age, 13 of which were attributed to infection.[1] This represented a 20-fold higher rate of mortality than that of the general US population. The spectrum of clinical features varied widely, from mild disease with only hair involvement to severe disease with profound developmental defects, recurrent infections and a high mortality at a young age. A very rare temperature sensitive form of TTD, which is more severe when pyrexia (fever) is present, has been reported.[14,15] TTD has been described in whites, blacks and Asians.[3,9]

In patients with skin cancer or one or more precancerous entities such as actinic keratoses, xeroderma pigmentosum (XP) or the XP/TTD overlap syndrome should be considered. In patients with subcutaneous fat atrophy but limited or no epidermal, nail or hair changes, Cockayne syndrome should be considered.

Xeroderma Pigmentosum/Trichothiodystrophy Overlap Syndrome

Two patients with overlapping features of both XP and TTD have been reported; a 3-year-old girl with sun sensitivity and retarded mental and physical development and a 28 year-old woman with sun sensitivity, cutaneous dyspigmentation and skin cancer.[16] Both patients had a compound heterozygous mutation in the ERCC2/XPD gene. Cultured cells from the former patient showed barely detectable levels of NER associated with UVB radiation exposure. The other patient had a substantially higher rate of DNA repair following UVB exposure. In both patients, polarized light microscopy revealed a tiger-tail appearance of the hair. Analysis of the hair shafts from both patients showed levels of sulfur-containing proteins between those of normals and patients with TTD. Genetic association between XP and TTD has also been reported in three Italian families.[17,18]

Cockayne Syndrome/Trichothiodystrophy Overlap Syndrome

TTD shares some phenotypic features with other diseases, such as Cockayne syndrome (CS), having defective NER.[19] However, we are not aware of any patient with overlap between TTD and CS. This may be a matter of how carefully clinical information has been documented and of who has done it. Dermatologists may miss a neurological sign and neurologists may miss cutaneous impairments. These patients are best evaluated by a team of physicians in different specialties, preferably working together. We have reviewed XP/CS overlap syndrome patients elsewhere.[20,21]

Etiopathogenesis

The TTD phenotype has been identified to be caused by mutations in at least three different genes: (1) the overwhelming majority of patients have mutations in *ERCC2/XPD* (the Excision Repair Cross-Complementation group-2/Xeroderma Pigmentosum Complementation Group D gene);[22,23] (2) a single family has had, instead, biallelic mutations in the *ERCC3/XPB* (the Excision Repair Cross-Complementation group-3/Xeroderma Pigmentosum Complementation Group B gene).[24] These two genes encode the two helicase subunits of the transcription/DNA repair factor,

TFIIH (Transcription Factor for RNA polymerase II, subunit H, which is composed of at least 10 subunit proteins). One *ERCC2/XPD* mutation associated with TTD has been shown to be temperature sensitive.[15] Interestingly, it was found in a clinical situation in which the disease was only fully manifested when the patient was febrile.[14,15]

(3) A unique TTD complementation group has been identified by Stefanini et al,[25] in cells from a patient described by Jorizzo et al.[26] Stefanini et al showed that the cells from patients in this identified group could complement all known XP complementation groups and also that this complementation was not intragenic.[25] Cells from this new complementation group were originally identified as "TTD1BR". It was subsequently identified as "TTD2", with the other known TTD cases at the time (all due to mutations in *ERCC2/XPD*) identified as in "complementation group TTD1". Following the currently generally accepted revision in nomenclature proposed by Lehmann et al, this group is now known as TTD-A.[27] However, these alternative nomenclatures may be seen in reviewing the older literature. This new complementation group of TTD (TTD-A) has been found to be caused by a mutation in the gene, *GTF2H5/TTD-A*, encoding a small protein, the newly identified 10th subunit of TFIIH, TFB5.[28] The overall level of TFIIH in TTDA cells is also significantly depressed, presumably because TFB5, in addition to any other functions it may have, acts as a stabilization factor for TFIIH, protecting it from proteosomal degradation.[29] Both immunoblot and immunofluorescence analyses revealed a strong reduction in the TFIIH concentration and normal TFIIH, microinjected into TTD-A cells, has a markedly reduced half-life. TFIIH, from TTD cells in this complementation group, functions normally in both DNA repair and transcription assays. Thus its reduced level in the TTD cells appears to be responsible for the TTD phenotype. DNA repair appears to be more severely affected than transcription in these cells. Petrini has suggested that this was because a minimal level of TFIIH may be needed to enact the switch from transcription to repair when the transcription machinery encounters DNA damage in transcription coupled repair.[30]

Depending on the mutation and possibly other components of the genetic background in individual cases, mutations in *XP-B* or *XP-D* cause XP only, TTD only, or, for *XP-D*, an overlap syndrome incorporating elements of both diseases, known as the xeroderma pigmentosum/trichothiodystrophy overlap syndrome (XP/TTD syndrome). In their extensive review of reported cases, Faghri et al noted that complications of pregnancy and abnormal characteristics at birth were frequent features of TTD, suggestive of a role of DNA repair genes in normal fetal development.[1]

Laboratory Diagnosis

In patients in whom the diagnosis of TTD is suspected, a hair pull should be performed and the hairs examined under a polarizing microscope. The hairs in patients with TTD have a distinctive appearance on polarized light microscopy consisting of alternating light and dark bands known as the "tiger tail" anomaly. Diagnosis may then be confirmed by analysis of sulfur content of these hair shafts, which should show decreased sulfur and cysteine content.[5,7] This test is performed by a reference laboratory.

In photosensitive patients, biochemical testing includes obtaining fibroblast cultures for in vitro testing, although this is carried out at the research level at present. Cellular sensitivity to ultraviolet C (UVC) radiation is increased, as measured by vital dye exclusion, a metabolic assay such as a WST-1 test, or colony forming ability. Testing for unscheduled DNA synthesis (UDS) would be expected to produce a normal or a variable result, since the factor, TFIIH, plays a role in transcription coupled DNA repair, which comprises only about 15 per cent of total cellular DNA repair (which is responsible for UDS) following UVC exposure. Other DNA repair assays may also be carried out.

Most TTDP patients have been found to have a mutation(s) in the *ERCC2/XPD* gene, which should be probed first in attempting to establish a molecular diagnosis. If no mutation is found, *ERCC3/XPB* or *GTF2H5/TTD-A* may be probed. In all testing for TTD, it should be kept in mind that all expectations are based on a very small data base. One should expect the unexpected.

Animal Model

An animal model has been described that has some relationship to TTD, but the phenotype is far from an exact match, despite the elegance of the model.[31]

Conclusion

Trichothiodystrophy is a multifaceted disease characterized, in part, by brittle hair which shows a banded, "tiger tail" abnormality on polarized light microscopy and low sulfur content on biochemical analysis, either or both of which may be used to establish the diagnosis. Other frequently encountered characteristics include developmental delay/intellectual impairment, short stature, ichthyosis, abnormal characteristics at birth (often accompanied by prenatal complications), especially low birth weight and length, ocular complications, especially cataracts, and infections, especially pulmonary infections, which may cause death during childhood and which are primarily responsible for an increased mortality rate in affected children. The disease varies widely in severity, with some patients extremely affected, whereas others show only the hair anomaly. Approximately half of patients are sun-sensitive; the majority of these have a defective DNA nucleotide excision repair mechanism. The overwhelming majority of sun-sensitive patients with identified genetic defects have been found to have mutations in *ERCC2/XPD*; a single family has been found to have biallelic mutations in *ERCC3/XPB*. These two genes encode the two helicases in the transcription/DNA repair factor, TFIIH, which contains at least ten proteins. The smallest group of genetically defined patients with TTD has been found to have a defect in third gene, *GTF2H5/TTD-A*, which encodes a small protein, TFB5, which is also a component of TFIIH that functions to prevent its proteosomal degradation.

References

1. Faghri S, Tamura D, Kraemer KH et al. Trichothiodystrophy: a systematic review of 112 published cases characterizes a wide spectrum of clinical manifestations. J Med Genet 2008; 45:609-621.
2. Blomquist HK, Back O, Fagerlund M et al. Tay or IBIDS syndrome: a case with growth and mental retardation, congenital ichthyosis and brittle hair. Acta Paediatr Scand 1991; 80:1241-1245.
3. Braun-Falco O, Ring J, Butenandt O et al. Ichthyosis vulgaris, Minderwuchs, Haardysplasie, Zahnanomalien, Immundefekte, psychomotorische Retardation und Resorptionsstorungen. Kasuistischer Bericht ueber zwei Geschwister. Hautarzt 1981; 33:67-74.
4. Happle R, Traupe H, Grobe H et al. The Tay syndrome (congenital ichthyosis with trichothiodystrophy). Eur J Pediat 1984; 141:147-152.
5. Itin PH, Pittelkow MR. Trichothiodystrophy: review of sulfur-deficient brittle hair syndromes and association with the ectodermal dysplasias. J Am Acad Dermatol 1990; 22:705-717.
6. Price VH, Odom RB, Ward WH et al. Trichothiodystrophy. Sulfur-deficient brittle hair as a marker for a neuroectodermal symptom complex. Arch Dermatol 1980; 116:1375-1384.
7. Liang C, Morris A, Schlucker S et al. Structural and molecular hair abnormalities in trichothiodystrophy. J Invest Dermatol 2006; 126(10):2210-2216.
8. Peter C, Tomczok J, Hoting E et al. Trichothiodystrophy without associated neuroectodermal defects. Brit J Dermatol 1998; 139:137-140.
9. Tay CH. Ichthyosiform erythroderma, hair shaft abnormalities and mental and growth retardation: a new recessive disorder. Arch Dermatol 1971; 104:4-13.
10. Stary A, Sarasin A. The genetic basis of xeroderma pigmentosum and trichothiodystrophy syndromes. Cancer Surv 1996; 26:155-171.
11. Eisman S, Ngwanya RM, Pillay K et al. Whorled follicular keratosis, scarring alopecia in ichthyosis follicularis atrichia with photophobia syndrome. J Eur Acad Dermatol Venereol 2009; 23:842-843.
12. Sato-Matsumura KC, Matsumura T, Kumakiri T et al. Ichthyosiform follicularis with alopecia and photophobia in a mother and daughter. Br J Dermatol 2000; 142:157-162.
13. Keyvani K, Paulus W, Traupe H et al. Ichthyosis follicularis, alopecia and photophobia (IFAP) syndrome: Clinical and neuropathological observation in a 33 year-old man. Am J Clin Genet 1998; 78:371-377.
14. Kleijer WJ, Beemer FA, Boom BW. Intermittent hair loss in a child with PIBI(D)S syndrome and trichothiodystrophy with defective DNA repair-xeroderma pigmentosum group D. Am J Med Genet 1994; 52:227-230.
15. Vermeulen W, Rademakers S, Jaspers NGJ et al. A temperature-sensitive disorder in basal transcription and DNA repair in humans. Nature Genet 2001; 27:299-303.

16. Broughton BC, Berneburg M, Fawcett H et al. Two individuals with features of both xeroderma pigmentosum and trichothiodystrophy highlight the complexity of the clinical outcomes of mutations in the XPD gene. Hum Molec Genet 2001; 10:2539-2547.
17. Nuzzo F, Stefanini M, Colognola R et al. Association of two rare hereditary disorders, xeroderma pigmentosum and trichothiodystrophy, in three families from north-east Italy. [Abstract] Seventh International Congress on Human Genetics, Berlin, 1986.
18. Nuzzo F, Zei G, Stefanini M et al. Search for consanguinity within and among families of patients with trichothiodystrophy associated with xeroderma pigmentosum. J Med Genet 1990; 27:21-25.
19. Kraemer KH, Patronas NJ, Schiffmann R et al. Xeroderma pigmentosum, trichothiodystrophy and Cockayne syndrome: A complex genotype-phenotype relationship. Neuroscience 2007; 145(4):1388-1396.
20. Lambert WC, Gagna CE, Lambert MW. Xeroderma pigmentosum: Its overlap with trichothiodystrophy, Cockayne syndrome and other progerioid syndromes. In: Ahmad SI, Hanaoka F, Eds. Molecular Mechanisms of Xeroderma Pigmentosum. Springer Science/Landes Bioscience, Austin: Texas, 2008;28-237. Adv Exp Med Biol 2008; 637:128-137.
21. Lambert WC, Neilan EG, Gagna CE et al. Cockayne syndrome: Its overlap with xeroderma pigmentosum and other progerioid syndromes. In: Ahmad SI, Ed. Molecular Mechanisms of Cockayne Syndrome. Landes Bioscience, Austin: Texas, 2009;81-89.
22. Stefanini M, Giliani S, Nardo T et al. DNA repair investigations in nine Italian patients affected by trichothiodystrophy. Mutat Res 1992; 273:119-125.
23. Stefanini M, Lagomarsini P, Arlett CF et al. Xeroderma pigmentosum (complementation group D) mutation is present in patients affected by trichothiodystrophy with photosensitivity. Hum Genet 1986; 74:107-112.
24. Weeda G, Eveno E, Donker I et al. A mutation in the XPB/ERCC3 DNA repair transcription gene associated with trichothiodystrophy. Am J Hum Genet 1997; 60:320-329.
25. Stefanini M, Vermeulen W, Weeda G et al. A new nucleotide-excision-repair gene associated with the disorder trichothiodystrophy. Am J Hum Genet 1993; 53:817-821.
26. Jorizzo JL, Atherton DJ, Crounse RG et al. Ichthyosis, brittle hair, impaired intelligence, decreased fertility and short stature (IBIDS syndrome). Brit J Dermatol 1982; 106:705-710.
27. Lehmann AR, Bootsma D, Clarkson SG et al. Nomenclature of human DNA repair genes. Mutat Res 1994; 315:41-42.
28. Giglia-Mari G, Coin F, Ranish JA et al. A new, tenth subunit of TFIIH is responsible for the DNA repair syndrome trichothiodystrophy group A. Nature Genet 2004; 36:714-719.
29. Vermeulen W, Bergmann E, Auriol J et al. Sublimiting concentration of TFIIH transcription/DNA repair factor causes TTD-A trichothiodystrophy disorder. Nature Genet 2000; 26:307-313.
30. Petrini JHJ. When more is better. Nature Genet 2000; 26:257-258.
31. de Boer J, Andressoo JO, de Wit J et al. Premature aging in mice deficient in DNA repair and transcription. Science 2002; 296:1276-1279.

CHAPTER 11

Cornelia de Lange Syndrome

Jinglan Liu* and Gareth Baynam*

Abstract

Cornelia de Lange syndrome (CdLS) (OMIM # 122470, #300590 and #610759) is an autosomal dominant disorder that is classically characterized by typical facial features, growth and mental retardation, upper limb defects, hirsutism, gastrointestinal and other visceral system involvement. Heterozygous mutations in the cohesin regulator, *NIPBL*, or the cohesin structural components *SMC1A* and *SMC3*, have been identified in approximately 65% of individuals with CdLS. Cohesin regulates sister chromatid cohesion during the mitotis and meiosis. In addition, cohesin has been demonstrated to play a critical role in the regulation of gene expression. Furthermore, multiple proteins in the cohesin pathway are also involved in additional fundamental biological events such as double strand DNA break repair, chromatin remodeling and maintaining genomic stability. Here, we will discuss the biology of cohesin and its associated factors, with emphasis on the clinical manifestations of CdLS and mechanistic studies of the CdLS related proteins.

Clinical Characteristics

Introduction

Cornelia de Lange syndrome (CdLS) (OMIM #122470, #300590 and #610759) was initially described by Vrolik in 1849 in a child with severe oligodactyly.[1] In 1916, Brachmann[2] described an individual with an overlapping phenotype and in the 1930s Cornelia de Lange, a Dutch pediatrician reported two unrelated girls with similar features and named the condition "degeneration typus amstelodamensis".[3,4] In honor of her formal characterization the eponym Cornelia de Lange syndrome is widely used.

CdLS is a dominantly inherited multisystem developmental disorder with locus heterogeneity. The facies are striking and may be one of the most useful diagnostic signs. The facial gestalt includes synophrys, long eyelashes, a depressed nasal root with an up-turned nasal tip and anteverted nares, a long philtrum, a thin upper lip, small widely spaced teeth, brachymicrocephaly, low-set and posteriorly rotated ears. Other characteristic findings include hirsutism; abnormalities of the upper extremities ranging from small hands and subtle phalangeal and metacarpal changes to severe forms of oligodactyly and forearm truncation primarily involving the ulnar structures; gastroesophageal dysfunction and retardation of growth and neurodevelopment. (Fig. 1A-D).[5,6] Other common findings include ptosis, myopia, intestinal malrotation, cryptorchidism, hypospadias, pyloric stenosis, congenital diaphragmatic hernias, cardiac septal defects, seizures and hearing loss. Mental retardation is typically severe; IQs range from less than 30 to 102 with an average of 53. Many patients also demonstrate autistic traits and other behavioural abnormalities including self-destructive tendencies, avoidance or rejection of social interactions and physical contact.[5] The estimated prevalence is 1 in 10, 000;[7] this

*Corresponding Authors: Jinglan Liu—Division of Human and Molecular Genetics, The Children's Hospital of Philadelphia, Philadelphia, Pennsylvania, USA; and Gareth Baynam—School of Paediatrics and Child Health, The University of Western Australia, Perth, Australia. Emails: liujin@email.chop.edu and gareth.baynam@health.wa.gov.au

Diseases of DNA Repair, edited by Shamim I. Ahmad.

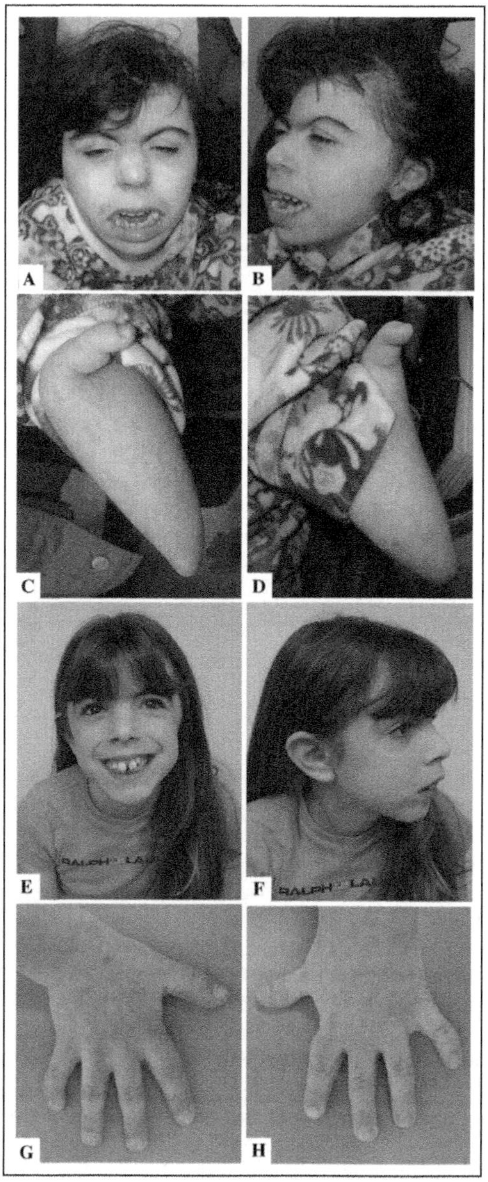

Figure 1. Typical phenotypic characteristics of CdLS. Panels (A-D) show manifestations in a severely affected 17 year-old girl. A,B) Note characteristic facial features (arched eyebrows, synophrys, long eyelashes, ptosis, short nose, long philtrum, thin upper lip, posteriorly angulated ears). C,D) Note severe oligodactyly with single radial digit, hypoplasia of ulnar structures and pterygia of antecubital region bilaterally. Panels E-H shows the characteristic mild phenotype in an 8 year-old girl. E,F) Note the characteristic facial features noted above (without the ptosis) but with much milder expression. G,H). Note the small hands with 5th finger clinodactyly and proximally placed thumbs bilaterally. Reprinted, with permission, from the Annual Review of Genomics and Human Genetics, Volume 9 (c) 2008 by Annual Reviews. www.annualreviews.org.

may be an underestimate as the advent of molecular testing has broadened the reported phenotype.[8] There is marked variable expressivity and a milder phenotype has been described[6,9-12] (Fig. 1E,F). With increasing recognition of milder phenotypes further isolated and familial cases have been identified.[13-15] The mild phenotype is characterised by lesser psychomotor and growth retardation, a lower incidence of major malformations and milder limb anomalies[12]; it is estimated to account for approximately 20-30% of the CdLS population.[12]

CdLS in Adolescents and Adults

Kline et al[16] presented a review of 59 North American adolescents and adults, ascertained through a multidisciplinary clinic with a mean age of 17 and 9 months (range 11-50 years) years. The majority of individuals were diagnosed in childhood. Almost all had growth below the 5th centile and approximately three quarters had microcephaly. The typical facial gestalt persisted whilst the facies coarsened and lengthened and premature graying was common. Seizures occurred in one quarter and typically commenced in childhood. Mental retardation was present in majority of the cases; 43% severe, 8% moderate to severe, 16% moderate and 16% were mild. Approximately 80% had gastrointestinal reflux, with approximately 50% requiring fundoplication; a similar number reported constipation. Reflux persisted or worsened, with some having complications including Barret oespohagus. Submucous clefts occurred in 14% and were often detected late. Approximately two-thirds had sensorineural hearing loss and one-third had chronic sinusitis. Myopia was common; cataracts and retinal detachment were noted in some. One quarter had leg length discrepancy, 39% had scoliosis and reduced bone density was noted with occasional fracture. Most females had delayed or irregular menstrual cycle. Behavioural and psychiatric disturbances often deteriorated. 70% had oppositional defiant disorder. Approximately 60% had a tendency for self-injury, one-third had aggression or a roaming type behavior and approximately two-third had sleep disturbance. Approximately half had mutations identified in *NIPBL* or *SMC1A*; whilst no clear genotype-phenotype conclusions could be drawn. Those with missense mutations were generally milder when compared to those with other mutations. The ascertainment bias in this cohort may have resulted in skewing towards the more severe end of the spectrum. A further study of the behavioural phenotype revealed an increased frequency of severe autism and compulsive behaviours in individuals with CdLS compared to those with comparable levels of mental retardation.[17] Other adults with mild presentations have been ascertained following the diagnosis of a more severely affected child.[18,19] With increasing availability of molecular diagnosis the adult spectrum will likely broaden.

Prenatal Presentation and Recurrence Risks

A variety of findings have been reported throughout gestation. Low levels of maternal pregnancy-associated plasma protein-A have been reported in the first and second trimesters.[20-22] Abnormal ultrasonic findings included increased nuchal translucency in the first trimester[23]; growth restriction, diaphragmatic hernia and other major organ malformations, characteristic facial profile and abnormalities of the extremities in the second trimester[24-27]; and intrauterine growth retardation and macrocephaly in the third trimester. Diaphragmatic hernia recurrently identified prenatally[27-31] and CdLS may be particularly considered in association with upper limb anomalies.[27]

The empirical recurrence risk has been estimated to be 1.5%.[5] In known familial cases the recurrence risk is dependent on the mode of transmission—either autosomal dominant or X-linked dominant. Paternal gonadal mosaicism for a *NIPBL* mutation has also been reported.[32]

Molecular Genetics of CdLS

Roles of NIPBL, SMC1A and SMC3 in CdLS

Most CdLS patients have normal chromosomes except a few specific chromosomal rearrangements that have been reported over the years.[33,34] The dup 3q syndrome has long been considered as a phenocopy of CdLS due to clinical overlap[35-39] and a critical region has been defined at 3q26.3-q27,[40-42] although the two entities can easily be differentiated on clinical examination. Several genes at these previously reported candidate regions were screened as potential CdLS disease genes including: *CHRD* and *SOX2* (3q27),[43] *SHOT* (3q25-q26),[44,45] *GSC* (14q32),[43] *NAALADL2* (3q26.3) (which was found to be disrupted by a t(3;17))[46] and *NLGN1*, a gene involved in synaptogenesis in the central nervous system; gene dosage of which has been implicated in the mental retardation associated with the dup(3q) syndrome (35). Interestingly, none of these genes were found to be mutated among CdLS patients and the *CDL1* locus at 3q26.3 did not segregate in at least half of familial cases studied, questioning the veracity of this region as a CdLS locus.[47] A genome wide linkage exclusion analysis, performed on 12 families, was able to narrow down candidate gene loci to 5 genomic regions, one of which mapped to 5p13.1 and corresponded to a locus involved in a de novo balanced t(5;13)(p13.1;q11.2) in a patient with CdLS. Screening of candidate gene identified a novel gene named *Nipped-B Like* (*NIPBL*) after its *Drosophila* homolog, *nipped-b*, heterozygous mutations in which were found to cause CdLS.[48-49] NIPBL is a regulator of the cohesin complex. For the first time this finding implicated the cohesin complex and its regulators to a developmental disorder. Shortly afterwards mutations in cohesin components, *SMC1A* and *SMC3* were also found to cause CdLS.[8,50]

Molecular Genetics of NIPBL, SMCA1 and SMC3

Human *NIPBL* is located on chromosome 5p13, its full length transcript is approximately 10 Kb with various smaller alternative transcripts and two protein isoforms reported being conserved in vertebrates.[48,49] The NIPBL protein is referred as delangin. *SMC1A* is located at Xp11.22-p11.21 and has been reported to escape X-inactivation in humans.[51] *SMC3* is located at 10q25. *NIPBL* transcripts are ubiquitously expressed in humans but with the highest expression found in heart and skeletal muscle.[48,49] In situ hybridization on whole mount mouse embryos demonstrated *NIPBL* expression in developing limbs, craniofacial bones and muscles, the craniofacial mesenchyme surrounding the cochlear canal, the respiratory and gastrointestinal systems, heart, renal tubules and genitourinary system concordant with the clinically involved systems in CdLS.[48] Mouse *Nipbl* has been reported to be expressed in limb buds, branchial arch and craniofacial mesenchyme robustly, although other tissues and organs also show its expression at various levels.[49] *Xenopus tropicalis* delangin is found throughout the ectoderm and in the mesoderm during gastrulation.[52] UniGene cluster and EST expression studies revealed both *SMC1A* and *SMC3* are universally expressed with some increased expression in certain tissues (http://smd-www.stanford.edu/cgi-bin/source).

Mutational analysis by sequencing of *NIPBL* exons and flanking intronic regions has been performed among various ethnic groups[53-57] as has *SMC1A* mutation analysis.[8,58] As expected, a wide variety of pathogenic mutations have been identified, but overall only approximately 50% of CdLS probands are found to carry an *NIPBL* mutation, most of which are point mutations, small insertions or deletions in coding regions or splice junctions. These mutations are assumed to produce either malfunctioning full length or truncated NIPBL proteins, consistent with haploinsufficiency.[53] In rare cases, large genomic rearrangements have been found as have alterations in the upstream noncoding region of the gene. For example, one affected girl and her mildly affected father have heterozygous deletion-insertion mutation in *NIPBL* 5'-UTR[19] and multiplex ligation-dependent probe amplification (MLPA) has revealed a 5.2 Kb deletion encompassing exons 41-42 of *NIPBL* in another patient.[59] Among *NIPBL* mutation-negative CdLS probands, approximately 8% (4% of total CdLS population) have been found to carry an *SMC1A* mutation.[8] One mildly affected adult male with CdLS was found to carry a de novo *SMC3* mutation. At least 35% of clinically diagnosed patients with CdLS are not found to carry a mutation in any of these identified genes, suggesting the existence of other CdLS genes or potential alternative mechanisms of alteration in these known genes may be involved during pathogenesis.

Genotype/phenotype correlations are beginning to emerge but are mostly gross observations that individuals with missense mutations, or nonidentifiable mutations, tend to present a milder phenotype than those with truncating mutations and individuals with *SMC1A* mutations also tend to be milder and rarely, if ever, manifest structural organ or limb defects.[8,53] At the cellular level, B-lymphoblastoid cells from CdLS patients, also manifest some phenotypic features; these include (1) 41% of metaphase spreads from *NIPBL* mutation positive probands, show precocious sister chromatid separation (PSCS) while the same phenotype only appeared in 9% of control samples[60] and (2) both fibroblast and B-lymphoblastoid cells, derived from CdLS patients, either with or without detectable *NIPBL* mutations, are more sensitive to DNA crosslinking agent, mitomycin C (MMC) and demonstrate decreased ability to repair double strand DNA breaks at G_2 phase after X-ray exposure.[61] Rare CdLS associated malignant cases such as Wilm's tumor have been reported, but in general cancer is not a predominant manifestation.[62]

Cohesin Biology

The primary biological role, identified for cohesin is to control sister chromatid segregation during both, mitosis and meiosis cell cycles. Four evolutionarily conserved subunits form the core structural component of cohesin complex; two SMC (structural maintenance of chromosomes) proteins SMC1A and SMC3, a kleisin protein RAD21 (also called MCD1 or SCC1) and SCC3 (also called SA1/SA2 or STAG1/STAG2). Homologous genes to cohesin and its regulators were identified in several eukaryotic model systems including *Saccharomyces cerevisiae, Schizosaccharomyces pombe, Drosophila melanogaster, Xenopus laevis* and human.[63]

The protein structure of SMC1A and SMC3 is similar; each spans 1000-1500 amino acids and both contain globular domains at the N- and C-terminal ends and a globular hinge domain in the middle that separates the alpha helical structure. N and C termini fold at the hinge domain to form an antiparallel coiled-coil by bringing together the two halves of the alpha helix, which also subsequently form the head domain. Three highly conserved motifs, Walker A, Walker B and a signature motif were identified in the ATPase domain in cohesin head formed by the N and C termini.[64,65] SMC1A and SMC3 dimerize at the hinge domains forming a V-shape structure most likely through a hydrophobic interaction. The N terminus of kleisin RAD21 subunit interacts with SMC3 head and the C terminus interacts with SMC1A head and ATP is required for the RAD21 and SMC1 association.[66] RAD21 was therefore suggested as a central regulator of cohesin function because it simultaneously crosslinks the SMC heterodimer while locating closely to the ATPase active sites in cohesin heads and binds to the fourth known cohesin subunit, SCC3.[67] The maturely formed cohesin is a four subunit, ring like complex with globular hinge and head domains separated by ~45 nm of coiled coil (Fig. 2).

Cohesin binds to chromatin during G1/S phase in budding yeast; in vertebrates, binding happens during telophase of the preceding cell division. Additionally, cohesin binds at G2/M phase when double strand DNA breaks are created.[68] The mechanisms by which sister chromatid cohesion is established is unclear, although it appears in yeast that the cohesin loading complex Scc2/Scc4 (or adherin) and the acetyl-transferase Eco1/Ctf7 (the human homolog of which, ESCO2, which results in Roberts syndrome when mutated) are essential for this process.[63,69] During or after DNA replication, replication factor C and DNA helicase are also required to establish cohesion (reviewed in ref. 70). In yeast, cohesin-associated regions (CARs) are scattered on average at 15-kb intervals on chromosome arms at transcriptional convergent sites which are usually AT-rich.[71,72] Centromeres recruit more cohesin than other chromosomal sites.[73,74] Although no consensus DNA sequence for cohesin binding has been illustrated, cohesin binding is enriched at heterochromatin[75] and the chromatin surrounding a DNA double-strand break (DSB).[76] Removal of cohesin starts from chromosome arms in prophase and is completely finished by anaphase. Cohesins begin to disassociate from the arms during prophase while the pericentric cohesin is protected until the onset of anaphase. In vertebrates, both cohesin subunit, SA2 and RAD21 are substrates of Polo-like kinase 1 (Plk1) and the phosphorylation of SA2 as well as a functioning Wapl is particularly required for prophase removal.[77-79] At the same time, Shugoshin/MEIS-322/Sgo1[80] and sororin[81] are able to protect centromere cohesins. An evolutionally conserved protein Pds5/BimD/Spo76 interacting with Wapl, sororin and Eco1 is also involved in maintaining pericentric cohesion in prophase.[82] Inactivation of cohesin and the complete dissolution of cohesin at anaphase onset enable faithful sister chromatid segregation. Securin and separase were identified as regulators of this event. Securin binds and inhibits the protease activity of separase (Esp1) before anaphase.[83] At the beginning of anaphase, securin is degraded by APC (anaphase-promoting complex) and separase is activated.[84] The active separase

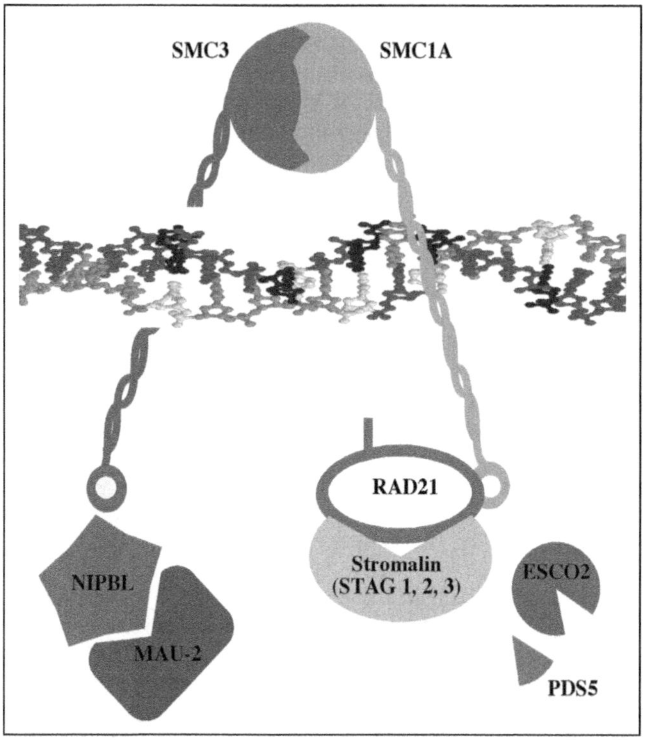

Figure 2. Schematic illustration of a single "open" cohesin molecule on a DNA strand in the "embracing" model. SMC1A and SMC3 (mutations in which have been identified in CdLS) are attached at their hinge domains. Coiled-coil arms connect the hinge domain to the head domain that contains ATP-binding cassettes crucial in the dynamic opening and closing of the ring structure mediated by the RAD21 and stromalin proteins. The cohesin regulatory proteins NIPBL (and interacting protein MAU-2) and ESCO2 (and putative interacting factor PDS5) involved in CdLS and RBS-SC phocomelia respectively are also indicated. Reprinted, with permission, from the Annual Review of Genomics and Human Genetics, Volume 9 (c) 2008 by Annual Reviews www.annualreviews.org.

cleaves RAD21 and the cohesin complex is further degraded.[83] Mechanisms other than cohesin cleavage to ensure the complete inactivation of all cohesion have also been proposed.[85]

The human heterodimer NIPBL/SCC4 complex is evolutionarily conserved and required for loading of cohesin onto chromatin in both, mitosis and meiosis and to all chromosome regions such as heterochromatin, CARs, centromeres, DSBs, etc.[76,86] Our understanding is incomplete as to how NIPBL loads cohesin onto DNA, as no such study has been reported to date. Through sequence homology analysis, the yeast homolog of NIPBL, Scc2 has been suggested to be a kinase[87] and mutations in Scc2 have greatly reduced Rad21 phosphorylation.[88] NIPBL/SCC4 seems to be involved in all cohesin activities including SMC ATPase activation, hinge dimerization, chromatin binding and chromatin remodeling.[89] In addition to NIPBL/SCC4, certain histone modifications and local chromosome structure are also targeted by cohesin.[73,75,90] Distinct protein complexes such as MRX, CEN complexes and Rep, which are distinct from NIPBL/SCC4 can also recruit cohesin to specific chromosome regions.[76,91,92]

In summary, cells have elaborated mechanisms to regulate chromatin binding of cohesin temporally and spatially, most of these mechanisms are not fully known as yet. This complexity of regulation enables cohesin to perform diverse biological functions and mutations in genes, encoding the associated major players are expected to contribute to multiple human diseases. Cohesin-independent mechanisms are also suggested to exist, as complete un-pairing of sister chromatids can not be achieved by simply knocking down cohesin. Condensin complexes, origin recognition complexes (ORCs), centromere complexes and DNA catenation, each has been reported to have a role in mediating cohesin-independent sister chromatid cohesion.[93-96]

Cohesin and Its Accessory Proteins in Transcriptional Regulation

The Nipped-B protein, the homolog of cohesin loading protein Scc2 in yeast, was identified in *Drosophila* through a genetic screening for proteins that facilitated expression of the *cut* homeobox gene in the developing wing margin that is driven by a distant transcriptional enhancer located more than 80 Kb upstream of the transcription start site. Nipped-B alleviates the *gypsy* insulator function by assisting long distance promoter-enhancer interactions.[97] *Cut* regulates *Drosophila* wing and limb development. *Nipped-B* null mutations are lethal to the fly, while heterozygous mutated *Nipped-B* results in reduced expression of *cut* protein with notch wing phenotypes.[97] In *Drosophila*, heterozygous *Nipped-B* mutants do not show cohesion defects while homozygous *Nipped-B* mutants show only the defects right before death at the second instar stage, indicating sister chromatid cohesion is independent from cohesin regulated gene expression. Cohesin binds to chromosomes throughout interphase when gene expression also takes place. Cohesin binds to the *cut* regulatory sequences in cultured cells and to the *cut* locus in *Drosophila* salivary gland chromosomes.[98] Reducing cohesin increases *cut* expression while reducing Nipped-B diminishes its expression in the emergent wing margin.[98,99] Enhancer-promoter communication at the *cut* gene locus may be interfered with by cohesin binding, thereby inhibiting gene expression. Nipped-B is able to reverse these suppression effects by dynamically removing the bound cohesin.[100] Scc4 (mau-2 in other species) interacts with Scc2 and is also essential for sister chromatid cohesion in yeast. In *Drosophila*, Nipped-B was shown to interact with the Mau-2.[101,102] In *C. elegans*, cells with RNAi knocked down of *mau-2* have abnormal axonal migration without sister chromatid cohesion defects, while in human cultured cells, knocked down of human *MAU-2* by RNAi does cause cohesion defects.[102] *Mau-2* mutations in *C. elegans* may have a similar effect as heterozygous *Nipped-B* mutations in *Drosophila* which may reduce gene expression and activity without obvious effect on cohesion.[97,99] Compared to Nipped-B, Mau-2 expression alterations seem less critical, as partial knockdown of *Mau-2* by RNAi in *Drosophila* had no effects on either *cut* gene expression or sister chromatid cohesion.[99,102] Recently, homozygous *Drosophila* mutants of Smc1 or Scc3 demonstrated defective axon pruning in the postmitotic mushroom body. The ecdysone steroid hormone receptor (EcR), an important factor for axon pruning, was found to have reduced protein expression in the mutants.[103] Over-expression of EcR in the postmitotic neurons in those knockout flies could rescue the pruning defect, indicating that reduced EcR expression is primarily responsible for the pruning defect.[103] The block of pruning could also be induced by knocking out another cohesion subunit Rad21 in the same neurons, indicating a complete cohesin complex is essential for pruning.[104] As all the above observations occur in the postmitotic cells, chromosome segregation and cell cycle regulation are clearly not involved. In addition cholinergic neurons, lacking of Rad21, caused abnormal larva locomotion without obvious effects on mitosis.[104] In other model organisms, such as zebra fish, the functions of Rad21 and Smc3 are needed for proper expression of *runx1* and *runx3* genes in early embryonic development.[105] In humans, the Scc3 homolog STAG2 was shown to be a transcriptional co-activator in the NF-kappa B signaling pathway.[106] Another cohesion accessory protein *Pds5B* gene, is conserved from fungi to man, dynamically interacts with cohesin and is involved in establishment and/or maintenance of sister chromatid cohesion.[98] In *Drosophila*, there is a single *pds5* gene and heterozygous *pds5* mutations alter *cut* gene expression without effects on cohesion. Homozygous mutants have both, gene expression and cohesion defects.[98] In mammals, there are two forms of Pds5 proteins. Homozygous *Pds5B* knock out mice show CdLS-like developmental abnormalities without cohesion defects, suggesting that changes in gene expression also likely underlie the Pds5B function in mouse development.[107] Genome-wide chromatin immunoprecipitation experiments in human and mouse cells have revealed colocalization of cohesin and CTCF, a zinc-finger protein with enhancer blocking/boundary activities.[108-110]

Knockdown of either cohesin or CTCF would influence gene expression levels in whole genome and knockdown of cohesin alone resulted in dys-regulated gene expression of CTCF targets.[108,109] Cohesin was therefore suggested to regulate gene expression through multiple mechanisms which are fairly similar to the regulatory mechanisms of CTCF.[108,109,111] All the above observations suggest that cohesin and its accessory proteins can influence both gene expression and sister chromatid cohesion during development, but the latter may require more dramatic protein changes. Thus, gene expression appears to be more sensitive to subtle dosage alterations of the cohesin apparatus.

Genome wide chromatin immunoprecipitation (ChIP-on-chip) assays to examine the binding sites of Nipped-B, RNA polymerase II (Pol II) and cohesin subunit SMC1A in 3 different *Drosophila* cell lines has discovered that Nipped-B colocalizes with cohesin which supports the idea that it dynamically regulates cohesin binding.[112] The preferential association of cohesin with transcribed regions suggests additional mechanisms by which cohesin binding might affect transcription and vice versa. The same study has also suggested that cohesin could interfere with both transcriptional activation and elongation, because cohesin binds to active *Abd-B* gene in Sg4 cells and some cohesin and Pol II peaks overlap within both, the *Abd-B* transcription unit and the regulatory region. Genes with distant regulatory elements, such as *cut* and *Ubx*, may be more sensitive to Nipped-B dosage because of the combined effects on activation and elongation. However, genome wide mapping of the cohesin and cohesin loading protein binding sites in yeast and *Drosophila* are conflicting. The reason for this could be that gene regulation through distant DNA elements does not play a major role in yeast and higher organisms may possess additional mechanisms for cohesin to bind in the genome and to regulate gene expression.[113]

Increasingly accumulated data suggest that the dys-regulated gene expression is in the etiology of CdLS. CdLS probands, with missense *NIPBL* mutations usually are mildly affected while severely affected probands have protein truncating alleles[53]; in *Drosophila*, the expression of *cut* and *Ubx* are less affected in mutants with missense *Nipped-B* alleles similar to some CdLS-causing mutations than those with truncating or null alleles.[114] Thus the milder effects of *NIPBL* missense mutations in humans could reflect milder effects on gene expression. Moreover, all identified CdLS probands which have heterozygous *SMC1A* or *SMC3* mutations, are mildly affected. The mutations in *SMC1A* or *SMC3* are either missense or small in-frame deletions and cell lines from probands with cohesin subunit mutations lack cohesion defects and produce normal levels of SMC1A.[115] More interestingly, the level of *NIPBL* mRNA was only reduced by 30% in both CdLS proband cell lines and heterozygous *NIPBL* knockout mice as well as in *Nipped-B* mutant flies.[115] In summary, the slightly impaired NIPBL expression (<30%) or a single amino acid changes in SMC1A or SMC3 can lead to enormous effects on human development without considerable defects in sister chromatid cohesion. These findings further support effects of cohesin on gene expression and is likely the main contributor to the etiology of CdLS. This indicates that in addition to altered cohesin activity, abnormal gene expression could also be mediated through NIPBL or cohesin mutations, affecting the changes in the dynamics of the binding of cohesin to chromosomes.[116]

Cohesin and Chromatin Remodeling

Cohesin also forms stable associations with chromatin remodeling complexes in vivo.[90] A human ISWI (SNF2h)-containing complex was copurified with cohesin and NuRD. The RAD21 subunit of the cohesin complex directly interacts with the ATPase subunit of SNF2h. Loading RAD21 onto chromatin might involve ATPase activity of SNF2h.[90] Alu sequences specifically bind to RAD21, SNF2h and Mi2 demonstrating that Alu repeats may also act as cohesin binding sites. Modifications of histone tails may be associated to SNF2h/cohesin complex as well. Cohesin was shown to bind to AT-rich noncoding regions which are the bases of chromosome loops attached by chromatin fiber and flanked by genes, transcribed convergently in yeast.[117] Yeast Sir2 protein recruits cohesin to multiple tandem rDNA arrays and suppresses chromatin recombination.[118] Additional support for NIPBL (and cohesin's) role in chromatin remodeling and epigenetic regulation came from the demonstration that NIPBL binds directly to the chromoshadow domain (CSD) of hetechromatin protein 1α (HP1α) with high affinity. The HP1α protein regulates epigenetic gene silencing by promoting and maintaining chromatin condensation. The HP1α chromodomain also binds to the methylated histone H3.[119]

Other Cohesinopathies

Since *NIPBL* was first found mutated in ~50% of CdLS probands,[48,49] the term "cohesinopathy" was coined in 2004 to describe any human disorder that is due to mutations in cohesin or its accessory proteins with manifestations that may or may not involve sister chromatid cohesion defects. At present, in addition to CdLS, there are three other multi-system genetic disorders that fit the above criteria, with all of them demonstrating abnormal sister chromatid pairing: (1) Roberts-SC phocomelia syndrome (RBS OMIM #268300; SC OMIM #269000) caused by mutations in the cohesin regulator *ESCO2* (establishment of cohesion 1 homolog 2),[120] (2) α-Thalassemia/ Mental Retardation Syndrome, X-Linked (ATRX OMIM #301040) caused by mutations in the *ATRX* gene on the X chromosome and (3) Rothmund—Thomson syndrome (RTS, also known as poikiloderma congenital) (OMIM #268400) caused by mutations in helicase gene RECQL4.[121] Recently, there has been mounting evidence that links mutations in genes involved in cohesion pathways to many forms of human cancer. For example, the *WAPL* protein is over-expressed in cervical cancers.[122] *SMC1A, NIPBL, SMC3* and *STAG3,* all have been suggested to play a major role in human colorectal cancers.[123] Failed sister chromatid cohesin has been suggested to be a common pathway leading to chromosome instability in human cancers.[123]

Conclusion

CdLS is a dominant multi-system genetic disorder caused by heterozygous mutations in *NIPBL, SMC1A* and *SMC3* which are key proteins in sister chromatid cohesion pathway. Although dys-regulated gene expression is most likely to be the underlying molecular mechanism, however, sister chromatid cohesion deficiency can not be excluded. Slightly reduced DNA repair capability has also demonstrated in CdLS. Identification of additional underlying molecular pathogenic causes are needed in the remaining ~35% of probands without identifiable gene mutations. As the first cohesinopathy ever identified, CdLS will serve as an excellent human disease model to study cohesin related biological mechanisms.

Acknowledgements

The authors are grateful to Dr. Ian D. Krantz at the Children's hospital of Philadelphia for helpful discussion and suggestions.

References

1. Oostra RJ, Baljet B, Hennekam RC. Brachmann-de Lange syndrome "avant la lettre". Am J Med Genet 1994; 52:267-268.
2. Brachmann W. Ein fall von symmetrischer monodactylie durch ulnadefekt, mit symmetrischer flughautbildung in den ellenbeugen, sowie anderen abnormalitaten (zwerghaftigkeit, halsrippen, behaarung). Jahrbuch Kinderheilkd 1916; 84.
3. de Lange C. Sur un type nouveau de dégénération (typus Amstelodamensis) [On a new type of degeneration (type Amsterdam)]. Arch Méd Enfants 1933; 36.
4. de Knecht-van Eekelen A, Hennekam RC. Historical study: Cornelia C. de Lange (1871-1950)—a pioneer in clinical genetics. Am J Med Genet 1994; 52:257-266.
5. Jackson L, Kline AD, Barr MA et al. de Lange syndrome: a clinical review of 310 individuals. Am J Med Genet 1993; 47:940-946.
6. Ireland M, Donnai D, Burn J. Brachmann-de Lange syndrome. Delineation of the clinical phenotype. Am J Med Genet 1993; 47:959-964.
7. Opitz JM. The Brachmann-de Lange syndrome. Am J Med Genet 1985; 22:89-102.
8. Deardorff MA, Kaur M, Yaeger D et al. Mutations in cohesin complex members SMC3 and SMC1A cause a mild variant of cornelia de Lange syndrome with predominant mental retardation. Am J Hum Genet 2007; 80:485-494.
9. Van Allen MI, Filippi G, Siegel-Bartelt J et al. Clinical variability within Brachmann-de Lange syndrome: a proposed classification system. Am J Med Genet 1993; 47:947-958.
10. Selicorni A, Lalatta F, Livini E et al. Variability of the Brachmann-de Lange syndrome. Am J Med Genet 1993; 47:977-982.
11. Saul RA, Rogers RC, Phelan MC et al. Brachmann-de Lange syndrome: diagnostic difficulties posed by the mild phenotype. Am J Med Genet 1993; 47:999-1002.

12. Allanson JE, Hennekam RC, Ireland M. De Lange syndrome: subjective and objective comparison of the classical and mild phenotypes. J Med Genet 1997; 34:645-650.
13. Greenberg F, Robinson LK. Mild Brachmann-de Lange syndrome: changes of phenotype with age. Am J Med Genet 1989; 32:90-92.
14. Chodirker BN, Chudley AE. Male-to-male transmission of mild Brachmann-de Lange syndrome. Am J Med Genet 1994; 52:331-333.
15. Russell KL, Ming JE, Patel K et al. Dominant paternal transmission of Cornelia de Lange syndrome: a new case and review of 25 previously reported familial recurrences. Am J Med Genet 2001; 104:267-276.
16. Kline AD, Grados M, Sponseller P et al. Natural history of aging in Cornelia de Lange syndrome. Am J Med Genet 2007; 145C:248-260.
17. Moss JF, Oliver C, Berg K et al. Prevalence of autism spectrum phenomenology in Cornelia de Lange and Cri du Chat syndromes. Am J Ment Retard 2008; 113:278-291.
18. Borck G, Redon R, Sanlaville D et al. NIPBL mutations and genetic heterogeneity in Cornelia de Lange syndrome. J Med Genet 2004; 41:e128.
19. Borck G, Zarhrate M, Cluzeau C et al. Father-to-daughter transmission of Cornelia de Lange syndrome caused by a mutation in the 5' untranslated region of the NIPBL Gene. Hum Mutat 2006; 27:731-735.
20. Aitken DA, Ireland M, Berry E et al. Second-trimester pregnancy associated plasma protein-A levels are reduced in Cornelia de Lange syndrome pregnancies. Prenatal Diagnosis 1999; 19:706-710.
21. Arbuzova S, Nikolenko M, Krantz D et al. Low first-trimester pregnancy-associated plasma protein-A and Cornelia de Lange syndrome. Prenatal Diagnosis 2003; 23:864.
22. Westergaard JG, Chemnitz J, Teisner B et al. Pregnancy-associated plasma protein A: a possible marker in the classification and prenatal diagnosis of Cornelia de Lange syndrome. Prenatal Diagnosis 1983; 3:225-232.
23. Huang WH, Porto M. Abnormal first-trimester fetal nuchal translucency and Cornelia De Lange syndrome. Obstet Gynecol 2002; 99:956-958.
24. Kliewer MA, Kahler SG, Hertzberg BS et al. Fetal biometry in the Brachmann-de Lange syndrome. Am J Med Genet 1993; 47:1035-1041.
25. Lalatta F, Russo S, Gentilin B et al. Prenatal/neonatal pathology in two cases of Cornelia de Lange syndrome harboring novel mutations of NIPBL. Genet Med 2007; 9:188-194.
26. Le Vaillant C, Quere MP, David A et al. Prenatal diagnosis of a 'minor' form of Brachmann-de Lange syndrome by three-dimensional sonography and three-dimensional computed tomography. Fetal Diagn Ther 2004; 19:155-159.
27. Marino T, Wheeler PG, Simpson LL et al. Fetal diaphragmatic hernia and upper limb anomalies suggest Brachmann-de Lange syndrome. Prenatal Diagnosis 2002; 22:144-147.
28. Bozner P, Blackburn W, Cooley NR Jr. Bilateral ulnar agenesis: case report and review of the literature. Pediatr Pathol Lab Med 1995; 15:895-913.
29. Jelsema RD, Isada NB, Kazzi NJ et al. Prenatal diagnosis of congenital diaphragmatic hernia not amenable to prenatal or neonatal repair: Brachmann-de Lange syndrome. Am J Med Genet 1993; 47:1022-1023.
30. Stoll C, Alembik Y, Dott B et al. Associated malformations in cases with congenital diaphragmatic hernia. Genetic counseling (Geneva, Switzerland) 2008; 19:331-339.
31. Baynam G, Goldblatt J, Walpole I. Deletion of 8p23.1 with features of Cornelia de Lange syndrome and congenital diaphragmatic hernia and a review of deletions of 8p23.1 to 8pter? A further locus for Cornelia de Lange syndrome. Am J Med Genet 2008; 146A:1565-1570.
32. Niu DM, Huang JY, Li HY et al. Paternal gonadal mosaicism of NIPBL mutation in a father of siblings with Cornelia de Lange syndrome. Prenatal Diagnosis 2006; 26:1054-1057.
33. Kousseff BG, Newkirk P, Root AW. Brachmann-de Lange syndrome. 1994 update. Arch Pediatr Adolesc Med 1994; 148:749-755.
34. DeScipio C, Kaur M, Yaeger D et al. Chromosome rearrangements in cornelia de Lange syndrome (CdLS): report of a der(3)t(3;12)(p25.3;p13.3) in two half sibs with features of CdLS and review of reported CdLS cases with chromosome rearrangements. Am J Med Genet A 2005; 137:276-282.
35. Meins M, Hagh JK, Gerresheim F et al. Novel case of dup(3q) syndrome due to a de novo interstitial duplication 3q24-q26.31 with minimal overlap to the dup(3q) critical region. Am J Med Genet A 2005; 132:84-89.
36. McKenney RR, Elder FF, Garcia J et al. Brachmann-de Lange syndrome: autosomal dominant inheritance and male-to-male transmission. Am J Med Genet 1996; 66:449-452.
37. Sciorra LJ, Bahng K, Lee ML. Trisomy in the distal end of the long arm of chromosome 3. A condition clinically similar to the Cornelia de Lange syndrome. Am J Dis Child 1979; 133:727-730.
38. Wilson GN, Hieber VC, Schmickel RD. The association of chromosome 3 duplication and the Cornelia de Lange syndrome. J Pediatr 1978; 93:783-788.
39. Holder SE, Grimsley LM, Palmer RW et al. Partial trisomy 3q causing mild Cornelia de Lange phenotype. J Med Genet 1994; 31:150-152.

40. Aqua MS, Rizzu P, Lindsay EA et al. Duplication 3q syndrome: molecular delineation of the critical region. Am J Med Genet 1995; 55:33-37.
41. Ireland M, English C, Cross I et al. Partial trisomy 3q and the mild Cornelia de Lange syndrome phenotype. J Med Genet 1995; 32:837-838.
42. Rizzu P, Haddad BR, Vallcorba I et al. Delineation of a duplication map of chromosome 3q: a new case confirms the exclusion of 3q25-q26.2 from the duplication 3q syndrome critical region. Am J Med Genet 1997; 68:428-432.
43. Smith M, Herrell S, Lusher M et al. Genomic organisation of the human chordin gene and mutation screening of candidate Cornelia de Lange syndrome genes. Hum Genet 1999; 105:104-111.
44. Blaschke RJ, Monaghan AP, Schiller S et al. SHOT, a SHOX-related homeobox gene, is implicated in craniofacial, brain, heart and limb development. Proc Natl Acad Sci USA 1998; 95:2406-2411.
45. Ozkinay F, Cogulu O, Gunduz C et al. A case of Brachman de Lange syndrome with cerebellar vermis hypoplasia. Clin Dysmorphol 1998; 7:303-305.
46. Tonkin ET, Smith M, Eichhorn P et al. A giant novel gene undergoing extensive alternative splicing is severed by a Cornelia de Lange-associated translocation breakpoint at 3q26.3. Hum Genet 2004; 115:139-148.
47. Krantz ID, Tonkin E, Smith M et al. Exclusion of linkage to the CDL1 gene region on chromosome 3q26.3 in some familial cases of Cornelia de Lange syndrome. Am J Med Genet 2001; 101:120-129.
48. Krantz ID, McCallum J, DeScipio C et al. Cornelia de Lange syndrome is caused by mutations in NIPBL, the human homolog of Drosophila melanogaster Nipped-B. Nat Genet 2004; 36:631-635.
49. Tonkin ET, Wang TJ, Lisgo S et al. NIPBL, encoding a homolog of fungal Scc2-type sister chromatid cohesion proteins and fly Nipped-B, is mutated in Cornelia de Lange syndrome. Nat Genet 2004; 36:636-641.
50. Musio A, Selicorni A, Focarelli ML et al. X-linked Cornelia de Lange syndrome owing to SMC1L1 mutations. Nat Genet 2006; 38:528-530.
51. Brown CJ, Miller AP, Carrel L et al. The DXS423E gene in Xp11.21 escapes X chromosome inactivation. Hum Mol Genet 1995; 4:251-255.
52. Strachan T. Cornelia de Lange Syndrome and the link between chromosomal function, DNA repair and developmental gene regulation. Curr Opin Genet Dev 2005; 15:258-264.
53. Gillis LA, McCallum J, Kaur M et al. NIPBL mutational analysis in 120 individuals with Cornelia de Lange syndrome and evaluation of genotype-phenotype correlations. Am J Hum Genet 2004; 75:610-623.
54. Bhuiyan ZA, Klein M, Hammond P et al. Genotype-phenotype correlations of 39 patients with Cornelia De Lange syndrome: the Dutch experience. J Med Genet 2006; 43:568-575.
55. Yan J, Saifi GM, Wierzba TH et al. Mutational and genotype-phenotype correlation analyses in 28 Polish patients with Cornelia de Lange syndrome. Am J Med Genet A 2006; 140:1531-1541.
56. Schoumans J, Wincent J, Barbaro M et al. Comprehensive mutational analysis of a cohort of Swedish Cornelia de Lange syndrome patients. Eur J Hum Genet 2007; 15:143-149.
57. Selicorni A, Russo S, Gervasini C et al. Clinical score of 62 Italian patients with Cornelia de Lange syndrome and correlations with the presence and type of NIPBL mutation. Clin Genet 2007; 72:98-108.
58. Borck G, Zarhrate M, Bonnefont JP et al. Incidence and clinical features of X-linked Cornelia de Lange syndrome due to SMC1L1 mutations. Hum Mutat 2007; 28:205-206.
59. Bhuiyan ZA, Stewart H, Redeker EJ et al. Large genomic rearrangements in NIPBL are infrequent in Cornelia de Lange syndrome. Eur J Hum Genet 2007; 15:505-508.
60. Kaur M, DeScipio C, McCallum J et al. Precocious sister chromatid separation (PSCS) in Cornelia de Lange syndrome. Am J Med Genet A 2005; 138:27-31.
61. Vrouwe MG, Elghalbzouri-Maghrani E, Meijers M et al. Increased DNA damage sensitivity of Cornelia de Lange syndrome cells: evidence for impaired recombinational repair. Hum Mol Genet 2007; 16:1478-1487.
62. Charles AK, Porter HJ, Sams V et al. Nephrogenic rests and renal abnormalities in Brachmann-de Lange syndrome. Pediatr Pathol Lab Med 1997; 17:209-219.
63. Nasmyth K, Haering CH. The structure and function of SMC and kleisin complexes. Annu Rev Biochem 2005; 74:595-648.
64. Hopfner KP, Karcher A, Shin DS et al. Structural biology of Rad50 ATPase: ATP-driven conformational control in DNA double-strand break repair and the ABC-ATPase superfamily. Cell 2000; 101:789-800.
65. Walker JE, Saraste M, Runswick MJ et al. Distantly related sequences in the alpha- and beta-subunits of ATP synthase, myosin, kinases and other ATP-requiring enzymes and a common nucleotide binding fold. EMBO J 1982; 1:945-951.
66. Gruber S, Haering CH, Nasmyth K. Chromosomal cohesin forms a ring. Cell 2003; 112:765-777.
67. Haering CH, Lowe J, Hochwagen A et al. Molecular architecture of SMC proteins and the yeast cohesin complex. Mol Cell 2002; 9:773-788.

68. Haering CH, Schoffnegger D, Nishino T et al. Structure and stability of cohesin's Smc1-kleisin interaction. Mol Cell 2004; 15:951-964.
69. Hirano T. At the heart of the chromosome: SMC proteins in action. Nat Rev Mol Cell Biol 2006; 7:311-322.
70. Skibbens RV. Unzipped and loaded: the role of DNA helicases and RFC clamp-loading complexes in sister chromatid cohesion. J Cell Biol 2005; 169:841-846.
71. Glynn EF, Megee PC, Yu HG et al. Genome-wide mapping of the cohesin complex in the yeast Saccharomyces cerevisiae. PLoS Biol 2004; 2:E259.
72. Lengronne A, Katou Y, Mori S et al. Cohesin relocation from sites of chromosomal loading to places of convergent transcription. Nature 2004; 430:573-578.
73. Partridge JF, Scott KS, Bannister AJ et al. cis-acting DNA from fission yeast centromeres mediates histone H3 methylation and recruitment of silencing factors and cohesin to an ectopic site. Curr Biol 2002; 12:1652-1660.
74. Bernard P, Maure JF, Partridge JF et al. Requirement of heterochromatin for cohesion at centromeres. Science 2001; 294:2539-2542.
75. Chang CR, Wu CS, Hom Y et al. Targeting of cohesin by transcriptionally silent chromatin. Genes Dev 2005; 19:3031-3042.
76. Unal E, Arbel-Eden A, Sattler U et al. DNA damage response pathway uses histone modification to assemble a double-strand break-specific cohesin domain. Mol Cell 2004; 16:991-1002.
77. Sumara I, Vorlaufer E, Stukenberg PT et al. The dissociation of cohesin from chromosomes in prophase is regulated by Polo-like kinase. Mol Cell 2002; 9:515-525.
78. Kueng S, Hegemann B, Peters BH et al. Wapl controls the dynamic association of cohesin with chromatin. Cell 2006; 127:955-967.
79. Hauf S, Roitinger E, Koch B et al. Dissociation of cohesin from chromosome arms and loss of arm cohesion during early mitosis depends on phosphorylation of SA2. PLoS Biol 2005; 3:e69.
80. McGuinness BE, Hirota T, Kudo NR et al. Shugoshin prevents dissociation of cohesin from centromeres during mitosis in vertebrate cells. PLoS Biol 2005; 3:e86.
81. Schmitz J, Watrin E, Lenart P et al. Sororin is required for stable binding of cohesin to chromatin and for sister chromatid cohesion in interphase. Curr Biol 2007; 17:630-636.
82. Losada A, Yokochi T, Hirano T. Functional contribution of Pds5 to cohesin-mediated cohesion in human cells and Xenopus egg extracts. J Cell Sci 2005; 118:2133-2141.
83. Uhlmann F, Lottspeich F, Nasmyth K. Sister-chromatid separation at anaphase onset is promoted by cleavage of the cohesin subunit Scc1. Nature 1999; 400:37-42.
84. Ciosk R, Zachariae W, Michaelis C et al. An ESP1/PDS1 complex regulates loss of sister chromatid cohesion at the metaphase to anaphase transition in yeast. Cell 1998; 93:1067-1076.
85. Guacci V. Sister chromatid cohesion: the cohesin cleavage model does not ring true. Genes Cells 2007; 12:693-708.
86. Watanabe Y. Sister chromatid cohesion along arms and at centromeres. Trends Genet 2005; 21:405-412.
87. Jones S, Sgouros J. The cohesin complex: sequence homologies, interaction networks and shared motifs. Genome Biol 2001; 2:RESEARCH0009.
88. Tomonaga T, Nagao K, Kawasaki Y et al. Characterization of fission yeast cohesin: essential anaphase proteolysis of Rad21 phosphorylated in the S phase. Genes Dev 2000; 14:2757-2770.
89. Onn I, Heidinger-Pauli JM, Guacci V et al. Sister chromatid cohesion: a simple concept with a complex reality. Annu Rev Cell Dev Biol 2008; 24:105-129.
90. Hakimi MA, Bochar DA, Schmiesing JA et al. A chromatin remodelling complex that loads cohesin onto human chromosomes. Nature 2002; 418:994-998.
91. Weber SA, Gerton JL, Polancic JE et al. The kinetochore is an enhancer of pericentric cohesin binding. PLoS Biol 2004; 2:E260.
92. Yang XM, Mehta S, Uzri D et al. Mutations in a partitioning protein and altered chromatin structure at the partitioning locus prevent cohesin recruitment by the Saccharomyces cerevisiae plasmid and cause plasmid missegregation. Mol Cell Biol 2004; 24:5290-5303.
93. Lam WW, Peterson EA, Yeung M et al. Condensin is required for chromosome arm cohesion during mitosis. Genes Dev 2006; 20:2973-2984.
94. Shimada K, Gasser SM. The origin recognition complex functions in sister-chromatid cohesion in Saccharomyces cerevisiae. Cell 2007; 128:85-99.
95. Monje-Casas F, Prabhu VR, Lee BH et al. Kinetochore orientation during meiosis is controlled by Aurora B and the monopolin complex. Cell 2007; 128:477-490.
96. Diaz-Martinez LA, Gimenez-Abian JF, Azuma Y et al. PIASgamma is required for faithful chromosome segregation in human cells. PLoS ONE 2006; 1:e53.

97. Rollins RA, Morcillo P, Dorsett D. Nipped-B, a Drosophila homologue of chromosomal adherins, participates in activation by remote enhancers in the cut and Ultrabithorax genes. Genetics 1999; 152:577-593.
98. Dorsett D, Eissenberg JC, Misulovin Z et al. Effects of sister chromatid cohesion proteins on cut gene expression during wing development in Drosophila. Development 2005; 132:4743-4753.
99. Rollins RA, Korom M, Aulner N et al. Drosophila nipped-B protein supports sister chromatid cohesion and opposes the stromalin/Scc3 cohesion factor to facilitate long-range activation of the cut gene. Mol Cell Biol 2004; 24:3100-3111.
100. Dorsett D. Adherin: key to the cohesin ring and cornelia de Lange syndrome. Curr Biol 2004; 14:R834-836.
101. Giot L, Bader JS, Brouwer C et al. A protein interaction map of Drosophila melanogaster. Science 2003; 302:1727-1736.
102. Seitan VC, Banks P, Laval S et al. Metazoan Scc4 homologs link sister chromatid cohesion to cell and axon migration guidance. PLoS Biol 2006; 4:e242.
103. Schuldiner O, Berdnik D, Levy JM et al. piggyBac-based mosaic screen identifies a postmitotic function for cohesin in regulating developmental axon pruning. Dev Cell 2008; 14:227-238.
104. Pauli A, Althoff F, Oliveira RA et al. Cell-type-specific TEV protease cleavage reveals cohesin functions in Drosophila neurons. Dev Cell 2008; 14:239-251.
105. Horsfield JA, Anagnostou SH, Hu JK et al. Cohesin-dependent regulation of Runx genes. Development 2007; 134:2639-2649.
106. Lara-Pezzi E, Pezzi N, Prieto I et al. Evidence of a transcriptional co-activator function of cohesin STAG/SA/Scc3. J Biol Chem 2004; 279:6553-6559.
107. Zhang B, Jain S, Song H et al. Mice lacking sister chromatid cohesion protein PDS5B exhibit developmental abnormalities reminiscent of Cornelia de Lange syndrome. Development 2007; 134:3191-3201.
108. Wendt KS, Yoshida K, Itoh T et al. Cohesin mediates transcriptional insulation by CCCTC-binding factor. Nature 2008; 451:796-801.
109. Parelho V, Hadjur S, Spivakov M et al. Cohesins functionally associate with CTCF on mammalian chromosome arms. Cell 2008; 132:422-433.
110. Stedman W, Kang H, Lin S et al. Cohesins localize with CTCF at the KSHV latency control region and at cellular c-myc and H19/Igf2 insulators. EMBO J 2008; 27:654-666.
111. Yusufzai TM, Tagami H, Nakatani Y et al. CTCF tethers an insulator to subnuclear sites, suggesting shared insulator mechanisms across species. Mol Cell 2004; 13:291-298.
112. Misulovin Z, Schwartz YB, Li XY et al. Association of cohesin and Nipped-B with transcriptionally active regions of the Drosophila melanogaster genome. Chromosoma 2007.
113. Misulovin Z, Schwartz YB, Li XY et al. Association of cohesin and Nipped-B with transcriptionally active regions of the Drosophila melanogaster genome. Chromosoma 2008; 117:89-102.
114. Gause M, Webber HA, Misulovin Z et al. Functional links between Drosophila Nipped-B and cohesin in somatic and meiotic cells. Chromosoma 2008; 117:51-66.
115. Dorsett D, Krantz ID. On the molecular etiology of Cornelia de Lange syndrome. Ann N Y Acad Sci 2009; 1151:22-37.
116. Revenkova E, Focarelli ML, Susani L et al. Cornelia de Lange syndrome mutations in SMC1A or SMC3 affect binding to DNA. Hum Mol Genet 2009; 18:418-427.
117. Filipski J, Mucha M. Structure, function and DNA composition of Saccharomyces cerevisiae chromatin loops. Gene 2002; 300:63-68.
118. Kobayashi T, Horiuchi T, Tongaonkar P et al. SIR2 regulates recombination between different rDNA repeats, but not recombination within individual rRNA genes in yeast. Cell 2004; 117:441-453.
119. Lechner MS, Schultz DC, Negorev D et al. The mammalian heterochromatin protein 1 binds diverse nuclear proteins through a common motif that targets the chromoshadow domain. Biochem Biophys Res Commun 2005; 331:929-937.
120. Vega H, Waisfisz Q, Gordillo M et al. Roberts syndrome is caused by mutations in ESCO2, a human homolog of yeast ECO1 that is essential for the establishment of sister chromatid cohesion. Nat Genet 2005; 37:468-470.
121. Kitao S, Shimamoto A, Goto M et al. Mutations in RECQL4 cause a subset of cases of Rothmund-Thomson syndrome. Nat Genet 1999; 22:82-84.
122. Oikawa K, Ohbayashi T, Kiyono T et al. Expression of a novel human gene, human wings apart-like (hWAPL), is associated with cervical carcinogenesis and tumor progression. Cancer Res 2004; 64:3545-3549.
123. Barber TD, McManus K, Yuen KW et al. Chromatid cohesion defects may underlie chromosome instability in human colorectal cancers. Proc Natl Acad Sci USA 2008; 105:3443-3448.
124. Liu J, Krantz ID. Cohesin and human disease. Annu Rev Genomics Hum Genet 2008; 9:303-320.

Chapter 12

Rectal Cancer and Importance of Chemoradiation in the Treatment

Sergio Huerta*

Abstract

Neoadjuvant chemoradiation is used as a preoperative standard treatment for the majority of patients with Stage II/III rectal cancers and, in attempt to facilitate surgical intervention, reduce the rate of local recurrence and improve overall outcomes in patients suffering from this malignancy. However, this modality results in a wide range of clinical responses. In many cases, a pathological complete response is achieved, while in others, receiving the same form of treatment, the tumor continues to grow. The specific phenotype of the tumor plays a major role in rendering tumor cells a survival advantage to the cytotoxic effects of chemoradiation. Several factors participating in proliferation, cell cycle, apoptosis and hypoxia have been investigated under a variety of conditions in pre-irradiated tissues and post-irradiated tumors. Mutations in the genes of these pathways have shown to lead to resistance to chemoradiation. This chapter describes colon cancer with emphasis on the molecular mechanisms that may lead to chemoradio-resistance of cells with cytotoxically induced DNA injury in rectal cancer.

Introduction

Colon and rectal cancers are typically grouped and staged together. However, their preoperative and intraoperative management is vastly different. Due to the pelvic location of the rectum and its proximity to the anal sphincter and bladder as well as to sympathetic and parasympathetic nerves, patients with rectal cancers present a substantial surgical challenge compared to individuals suffering from colon cancer. The pelvic location of the rectum creates the reduced ability to obtain clear radial margins at the time of surgical resection.

Neoadjuvant modalities have been developed to downgrade rectal tumors and facilitate surgical intervention as well as decrease the rate of local recurrence.[1] In spite of current aggressive multimodality treatments, recurrence rate remains high (5-15%). The expected 5-year survival in patients who develop recurrent disease is a dismal 9%. Efforts to improve local control and survival in rectal cancer are the focus of multiple current clinical and preclinical research efforts.

Preoperative chemoradiation results in a significantly wide range of clinical response. From one side of the spectrum, 9 to 37% of patients are able to achieve a pathological complete response (pCR) after surgical intervention. At the other end of this wide range, up to 9% of patients do not respond to this treatment modality. In other cases, tumor continues to grow in spite of neoadjuvant treatment.[2] The wide variations in the response of treatment are due to the tumor size, differentiation and the molecular phenotype of rectal cancer cells. The relapse rate of up to 15% in early rectal cancers and the 90% mortality in patients with recurrence mandates a refinement of surgical techniques and optimization of the therapeutic ratios of radiotherapy and chemotherapy. In the

*Sergio Huerta—University of Texas Southwestern Medical Center, 4500 Lancaster Road, Dallas, Texas 75216, USA. Email: sergio.huerta@utsouthwestern.edu

Diseases of DNA Repair, edited by Shamim I. Ahmad.

following discussion, we address the possible mechanisms leading to resistance to the cytotoxic effects induced by chemoradiation in rectal cancer.

Rectal Cancer: Metastasis and Survival Rates

Rectal cancer is one of the most common cancers amongst other human cancers. Over 40,000 Americans are expected to be diagnosed with rectal cancer in 2009 with an expected mortality of approximately 7,000 individuals in the same year.[3] Tumors located in the rectum, in terms of neoadjuvant management, are defined as cancers located within 12 cm from the anal verge by rigid proctoscopy according to the guidelines by the National Comprehensive Cancer Network (NCCN).[4] Current sharp radial resection of the tumor [total mesorectal excision (TME)] is curative for small tumors [T1 or T2, N0 (Stage I)]. However, the risk of locoregional recurrence, distant metastasis and death increase with tumors extending through the muscularis propria [> T3, any N, any M (Stages II-IV)].

One retrospective study[1] and one randomized control trial[5] show that patients with Stage II and III benefit from neoadjuvant modalities. Following induction chemoradiation and resection that includes the mesentery which encircles the rectum (the mesorectum) [a total mesorectal excision, TME] has emerged as the gold-standard operation for the management of rectal cancer treated for cure. In the current era of a multidisciplinary approach for the management of rectal cancer, the distal margin for low rectal cancer does not require the historical 2.0-cm free margins and the goal is to perform a surgery without tumor involvement in the surgical margin distally.[6] In distal rectal cancer in which the anal sphincter cannot be successfully cleared of margins or in patients with anal sphincter dysfunction, an abdominoperineal resection is the surgical treatment of choice.

Preoperative chemoradiation can downgrade tumors and facilitate surgical intervention, thereby theoretically decreasing the rate of locoregional recurrence because of a better ability to obtain radial margins.[1] Unfortunately, induction chemoradiation has an unpredictable tumor response. A Cochrane study[2] of 19 randomized control clinical trials, comparing preoperative radiation to surgery alone, demonstrated a clear improvement in local recurrence and a marginal improvement in overall mortality; the most effective dose was >30Gy. However, the magnitude of benefit was heterogeneous across the trials,[2] and early complications included wound and perineal infections. Long-term complications were more serious such as venous thrombosis, femoral or pelvic fractures, intestinal obstruction, postoperative fistulas, as well as rectal and sexual dysfunction (Fig. 1).

Independent of the possible benefits from neoadjuvant chemoradiation, in terms of reduction of locoregional recurrence as well as a survival advantage, a major objective for this preoperative modality is to reduce the size of the tumor. This reduction might permit anal sphincter preservation as well as allowing the performance of a low anterior resection compared to an abdominoperineal resection.[1] An objective parameter to determine the efficacy of the neoadjuvant regimen in patients,

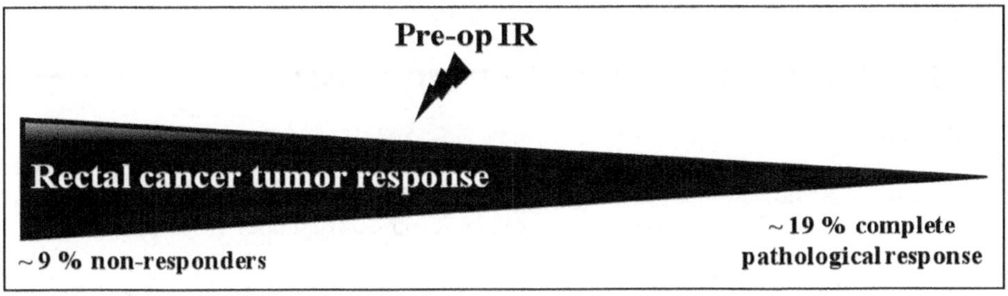

Figure 1. The wide response to preoperative ionizing radiation (IR) is depicted by the narrowing of the arrow. There is a 9% to 37% complete pathological response depending on the study. Tumor size, grade, and the phenotype of the rectal cancer cells as well as the amount of IR administered and the mode of administration accounted for the wide range in tumor response.

affected with rectal cancer, is pathological complete response (complete obliteration of the tumor following preoperative chemoradiation; pCR). Results from a randomized Phase II trial, assessing combined chemoradiation for rectal cancer showed 28% of patients with complete pathological response and 78% with tumor downgrading.[7]

The presensitizing agents, including 5-fluorouracil (5-FU) and leucovorin were the only available agents prior to the year 2000. In less than a decade, the available modalities utilized as radiosensitizers have substantially increased, which include: irinotecan, oxaliplatin, capecitibine and bevacizumab. Several of these modalities are currently under studies. However, even with these agents, there is still a wide response to ionizing radiation. A review of selected studies, utilizing various agents in the neoadjuvant setting with patients with T3N0 rectal cancers, has shown a response rate between 9.0% to 37%.[8]

Neoadjuvant chemoradiation, in successfully eradicating larger tumors (T3 and T4), is limited by normal tissue tolerance, tumor sensitivity (i.e., tumor phenotype) and microscopic spread beyond the borders of the tumor. In addition, the local recurrence remains high (5-15%) with a 5-year survival of 9% in this cohort of patients. The wide heterogeneity in tumor response is primarily the result of the individual phenotype of the lesions. How the cells are able to escape the cytotoxic effects of chemoradiation and the DNA damage induced by drugs, in some cases more than others, is dictated by the molecular phenotype of the cell and its ability to repair the DNA damage induced by a given agent. These observations point to an understanding of the molecular events leading to radio-resistance, such that those specific molecules can be targeted in order to obtain an improved response to radiation therapy and to reduce the side effects. If radiosensitive tumors can be identified, a selective and individualized form of chemoradiation could be instituted.

Mechanisms of Cell Death By Ionizing Radiation

Ionizing radiation results in DNA damage and exerts its therapeutic activity primarily by inducing DNA strand breaks of both types: single strand breaks (SSBs) and double strand breaks (DSBs).[9] The DSBs are more difficult to repair than SSBs and thus apoptosis typically ensues following therapeutic application of chemoradiation to a tumor. In addition to strand breaks, radiation induces DNA damage of the following types: (1) DNA base damage; (2) DNA to protein cross-links; and (3) install replication forks. In addition, membrane damage leading to signal transduction may affect: (1) gene expression of cell cycle regulators; (2) growth factor production; and (3) oxidative stress pathway activation. Post-radiation induced cell death ensues via: (1) apoptosis, (2) necrosis, or (3) mitotic catastrophe. It is not clear which type of cell death predominates over other, but cells response differently to ionizing radiation depending on their phenotype.[10] The classical pathway of radiation-induced cell death begins with DNA damage, which is identified by a DNA damage sensor. This induces to DNA repair systems, a transcriptional response and check point DNA damage. If the cells are unable to successfully repair the DNA damage, induced by IR, the cells undergo programmed cell death or apoptosis.[9]

Mechanisms of Resistance to Radiation in Rectal Cancer

Cell Cycle

Several mediators of the cell cycle have been implicated in colon carcinogenesis, including cyclin D1, the retinoblastoma protein, p53, as well as the cylcin dependent kinase inhibitors (CDKIs), p21 and p27. With regards to the cell cycle, p53, p21 and p27 have been extensively studied in rectal cancer.

A study undertaken by Lebe et al evaluated the prognostic significance of p53, p21 and p27 in paraffin-embedded tissues from patients with rectal cancers (Stage I-IV). Examination of 45 formalin-fixed, paraffin-embedded rectal cancer tissues, by immunohistochemistry (IHC) with antibodies specific for p53, p21, p27, demonstrated no clinicopathological association between the expression of these cell cycle mediators and tumor aggressiveness, metastatic potential, or survival in this cohort of patients.[11] The only positive finding of this study was that p27 expression in the rectal tumor correlated with hepatic metastasis.[11] Whether p27 expression represented a mutated form of the protein was not clear in this study.

The mechanisms that lead to radiation response of tissues is dependent on several factors including: (1) proliferation kinetics of the cells (proliferation vs apoptosis); (2) Pressure of oxygenation of the tissue to be irradiated (i.e., hypoxia); (3) Intrinsic cell's radiosensitivity (i.e., susceptibility of the cell to DNA damage); (4) DNA repair mechanisms.[12]

Proliferation Markers and Mitotic Index (Ki-67)

Pre-irradiated biopsy samples from rectal cancer patients have shown high to moderate Ki-67 and PCNA immunostaining.[12] In evaluating the post-irradiated tumors, all indices for proliferation have shown to decrease following treatment with irradiation and such response was associated with a decrease in tumor size by 50%. Additionally, the five year disease free survival was associated with high PCNA and high mitotic count.[12] Another two studies have shown that high Ki-67 stain correlated with a positive response to IR.[13,14] Nevertheless, the majority of analyses have consistently shown that proliferating nuclear antigen labeling index are poor indicators of a response to IR.[8]

p53, p21 and p27 and Apoptotic Index (AI)

The results from preclinical analyses have consistently indicated that wild-type p53 is essential for apoptotic cell death induced by IR.[8] This notion was first reported in vitro in mouse thymocyte cells, which showed that p53 was required for these cells to undergo IR-induced apoptosis.[15] Importantly, p53 mutant and wild-type cells were sensitive to other apoptotic cell death when induced by chemotherapeutic agents other than IR. This indicated that p53 is required for IR cell death in thymocytes, but not for every form of apoptosis.[15] The required form of p53 wild-type in cell cycle arrest following gamma irradiation of colon cancer cells has also been documented.[16] Several other studies have replicated these observations in vitro and in vivo colorectal cancer.[17,18] However, the observations of a positive correlation between p53 and a good response to IR have not been universal as some studies have indicated that the status of p53 does not play a role in radiation response.[8] Furthermore, some investigators have suggested that p53 mutations may render cells a more radiosensitive phenotype owning to a reduction in p53 DNA dependent pathway mechanisms.[19]

The half-life of wild-type p53 is short and may not be detected by a single point in time by IHC.[20] Conformational changes of the p53 protein, resulting from mutations, could end up in protein stability and a longer half-life, which may allow for an increased ability for detection of this protein by IHC.[8] Consequently, multiple observations by IHC in rectal cancer have failed to establish a significant relationship between the p53 status and spontaneous apoptosis or to tumor response to IR.

Phenotypic differences that confer variable response rates to IR in patients with rectal cancer have also been shown to be dependent on the mutation(s) in p53. While multiple mutations are localized in exons 5 to 8 of the p53 gene, it is the mutations of codon 288 in exon 8 that seem to lead to worse prognosis in rectal cancer.[21] Thus, these specific mutations may be the reason to show a more resistant phenotype in patients with rectal cancer. If, however, a mutation in exon 5 is present, in certain studies, it may not show resistance to apoptotic cell death following IR. Thus, p53 has been demonstrated to play a crucial role in cellular response to radiation in vitro and the wild-type p53 renders a radiosensitive phenotype in cells.

The importance of the p53-p21 axis in colon cancer was demonstrated in DLD-1 colorectal cancer cells. DLD-1 cells, with mutations in the p53 gene, expressed low levels of p21 protein following DNA damage with irradiation or treatment with chemotherapeutic agents.[22] Additionally, colon cancer-p21 deficient cells are resistant to irradiation-induced cell death.[23]

Low levels or absence of p21 have directly been correlated with poor prognosis in patients with colorectal cancer.[8] However, these observations have been challenged by other studies, which showed no correlation between p21 expression and colorectal cancer.[11,24] For instance, some studies have shown that the levels of p21 correlated positively with a good pathological response[25] and p21 expression was substantially reduced in radioresistant tumor cells.[26] Other studies have evaluated the prognostic significance of p21 and demonstrated that increased p21 expression, after treatment, was associated with an improved 5-year survival.[27] While other investigators showed that p21 was not useful in the post irradiated tumor samples in determining a good response to IR.[28] Furthermore, A dual role for p21 has been suggested depending upon its wild-type vs mutational status. In wild-type p21 tumors, radiation-induced cell cycle arrest may prevent cells from programmed cell death.[8] In this dual role, $p21^{+/+}$ tumors, which lead to cell cycle arrest, may render cells a radio-resistant phenotype; while $p21^{-/-}$ tumors undergo apoptosis and they regress following treatment with IR. [29]

In support of this hypothesis is loss of p21 causing a substantial increase in radiation-induced apoptosis and p21-negative tumor xenografts responding better to IR.[22,30] Radioresistant HCT116/$p21^{+/+}$ colon cancer cells were rendered a radiosensitive phenotype following treatment with an antisense oligodeoxynucleotide (ODN) against p21. The resistant apoptotic phenotype of cells (treated with ODN and 10 Gy IR) and HCT116/$p21^{+/+}$ xenografts [treated with i.p administration of (ODN) and 15 Gy IR] was similar to that of HCT116/$p21^{-/-}$ cells and established tumors in nude mice.[29] Thus, growth arrest was replaced by cellular apoptosis by p21 inhibition.[29] While, this compelling evidence in preclinical studies suggests that a positive p21 status in pre-irradiated tumor biopsies would predict poor response to IR, this has not been observed in clinical studies. Whether the stability of detectable p21 by IHC represents a wild-type form or a mutated status of this protein (similar to the observation in p53), remains to be determined.[8]

Lack of immunoreactivity with antibodies against p53 and p27 was associated with poor response to IR. The responsible mechanism was postulated to be the result of apoptosis induced by these proteins as level of p53 and p21 increased in tumor samples compared to biopsies following chemoradiation, which was thought to be the result of protein induction from DNA damage. Several other studies involving p27 have evaluated its prognostic potential in resected specimens following IR treatment. For instance, a decreased level of p27 or its absence was an independent predictor of poor survival in patients with rectal cancer who did not receive neoadjuvant chemoradiation.[8] Similarly, high p27 protein expression correlated with improved outcomes and its loss with decreased survival and metastatic disease in patients with colorectal cancers.[8]

Other studies have demonstrated contrasting results. For example, a study showed that the median 5-year survival was significantly reduced in patients expressing p27 in the residual tumor following IR treatment.[31] It was concluded that p27 expression may also allow cancer cells to escape apoptosis. Additionally, tissue microarray studies of early rectal cancers (T1-T2, N0), treated exclusively by surgical intervention with analysis for p53, MDM2, p21, bcl-2 and p27, demonstrated that none of these tumor markers were important predictors of outcome with the exception of a trend towards metastatic potential in tumor biopsies negative for p27.[32]

Thus, several studies addressing cell cycle regulators as predictors for response to IR or as markers to determine outcome have had contradicting results. Positive CDKI status in pre-irradiated samples appears to be a good predictor of radiosensitivity in patient with rectal cancers. However, preclinical studies showed that $p21^{+/+}$ cells had a more radioresistant phenotype as a result of cell cycle arrest. Inhibition of p21 promoted apoptosis and radiosensitivity in cells and xenografts. The major problem of the clinical studies is the relatively low number of samples analysed and also the small percentage of positivity showed by the tissue samples for any given protein, hence it is difficult to predict with certainty the response to IR for those tumors that are not positive for these protein markers.

Apoptosis

Mutations in pro-apoptotic genes or over activation of anti-apoptotic proto-oncogenes are known to lead to resistance to radiation therapy in rectal cancer.[33,34] Murine models of solid tumors demonstrated that their response to IR correlated with the ability of the cells to undergo apoptosis. The best correlation exists with the apoptotic index of pretreated tumor cells.[35]

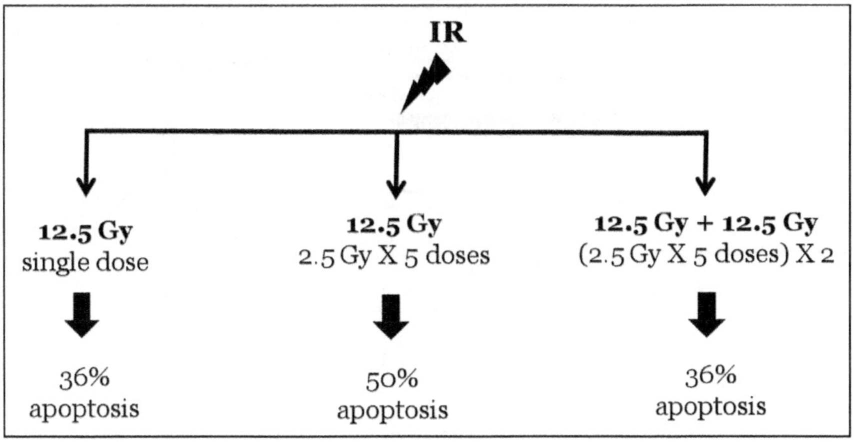

Figure 2. Murine models of solid tumors were subjected to various doses of IR in a single vs. multiple doses. Apoptosis was examined following treatment with IR. The highest response to apoptosis was observed with solid tumors receiving low dose IR in divided doses. [Data from Meyn et al.][35]

These studies also demonstrated the following findings: (1) not all tumors were susceptible to IR-induced apoptosis; (2) the apoptotic response was observed within few hours after treatment and this was followed by a decline in AI; (3) low doses of IR were more effective in achieving apoptosis and (4) tumors that achieved significant regression still had cells that were resistant to die by apoptosis following treatment. In these studies, spontaneous apoptosis in the untreated tumors had a better correlation to tumor regression compared to apoptosis in the irradiated tumors.[36] Another important aspect of this study was that the highest rate of apoptosis was achieved by smaller doses of IR given at intervals (Fig. 2).[36]

The prognostic significance of AI to determine disease free survival and recurrence was evaluated in 160 tumors of patients with rectal cancer at the time of curative surgery. High AI occurred in 25% of tumors and AI did not demonstrate prognostic significance. However, p53, Bcl-2, tumor stage and gender were independent predictors of recurrence. Similarly, all variables with the exception of Bcl-2 predicted disease free survival.[37]

In a clinical study of patients from the health care region in the Swedish Rectal Cancer Trial,[38] Adell investigated the outcome in this population with regards to apoptosis in patients who did and did not receive short-term preoperative radiotherapy. The AI in pre-irradiated tumor biopsies was 0.3% compared to 1.1% in tumor samples of patients who had undergone surgical intervention over two days following neoadjuvant radiotherapy. The main finding of this study was that tumors with high AI had low rate of local recurrence after surgery (with or without preoperative radiotherapy) compared to tumors with low AI. Patients who received neoadjuvant chemoradiation did have a decrease rate of local recurrence. However, there was no change in tumor size even if the rate of apoptosis was increased in pre-irradiated vs post-irradiated samples.

A pivotal study that demonstrated the role of apoptosis in rectal cancer was conducted by Rodel's group, who demonstrated that spontaneous apoptosis in the pretreatment biopsies was a good predictor of pathological response, in accordance to the reports by Meyn's observations.[35] The apoptotic index in pre-irradiated tumor biopsies predicted relapse-free survival. AI was related to Ki-67, but not to p53 or Bcl-2.[14]

The prognostic value of intrinsic, compared to radiation-induced apoptosis was evaluated in 45 irradiated patients with rectal cancers and this was compared with 45 non-irradiated patients. Non-irradiated patients, with high intrinsic apoptosis, had low recurrence rates and also an increase in cancer specific survival, while IR increased apoptosis by 2-fold, this increase did

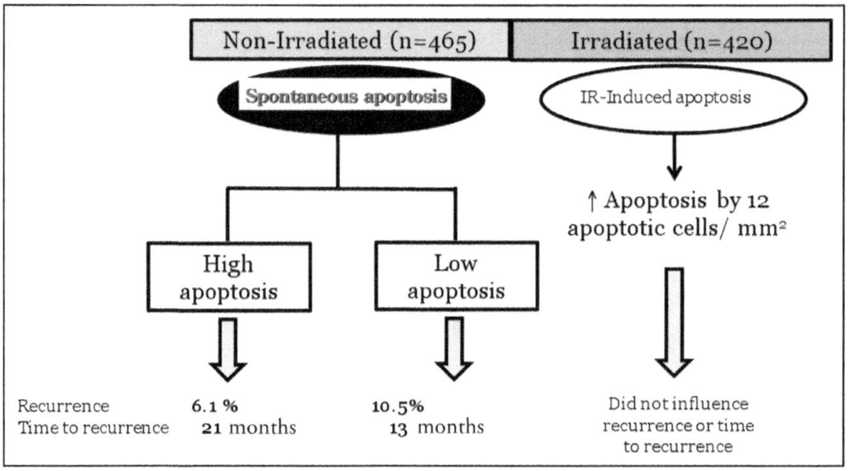

Figure 3. Rectal cancer pateints with low intrinsic apoptosis had a much higher rate of local recurrance and recurrences occured earlier compared to the cohort that had high spontaneous apoptosis in pre-irradiated samples. Ionizing radiation indcued apoptosis, but this has no effect on prognosis. [Data from de Bruin et al.][40]

not correlate either with local control or survival.[39] Finally, the role of both intrinsic apoptosis and radiation-induced apoptosis as markers for prognosis in rectal cancer was examined by tissue microarray in 1198 tumor samples from the Dutch Total Mesorectal Excision trial. The rate of recurrence in patients, who received irradiation, was 5% compared to 10% in patient who did not. Non irradiated patients with high apoptosis rate had a decrease in local recurrence by 1.7-fold. While there was an increase in apoptosis in the irradiated tumors, this effect did correlate with prognosis (Fig. 3).[40]

NFκB

NFκB activation induced by IR exposure, was prevented by transfection of colorectal cancer cells WiDR, KM1214 and HT-29 with a mutated form of a regulatory molecule of the NFκB complex (AdCMV) IκBα or pretreatment with the proteosome inhibitor PS-341. NFκB inhibition was associated with an increase in apoptosis, with a maximal response 48 hours after irradiation of cells. Additionally, clonogenic assays demonstrated that NFκB inhibition was associated with a reduction in cell growth.[41]

Cusack's et al has demonstrated that by inhibiting radiation-induced NFκB, apoptosis had increased and cell growth decreased in colorectal cancer cells. Radiosensitivity was demonstrated in colorectal cancer LOVO, WiDR and KM12L4 cells pretreated with the proteosome inhibitor PS-341 or AdCMV IκBα. Colorectal LOVO xenografts, pretreated with a single fraction i.p. administration of the proteosome inhibitor PS-341(1 mg/kg) followed by 6 Gy IR (four hours after PS-341 treatment) had an 84% reduction in tumor growth.[41]

Inhibitors of Apoptosis (IAPs: Survivin)

Cell lines with different sensitivities to IR demonstrated higher spontaneous and IR-induced apoptosis in radiosensitive SW48 colorectal cancer cells compared to the radioresistant SW480 colorectal cancer cells. SW480 cells showed higher levels of survivin at base line 48 hours after treatment with IR; while, radiosensitive SW48 cells had low levels of survivin at base line and the levels of this protein did not increase with IR treatment. Intermediate IR responsive HCT15 cell lines were in between these two cell lines, both in spontaneous apoptosis and survivin expression.[42] Survivin expression was investigated by quantitative RT-PCR in all of these cell lines.

Transcriptional activity increased more substantially in SW480 cells, both at 24 hours and 48 hours at 2 Gy and 8 Gy. Furthermore intermediate in HCT-15 cells and did not change in SW48 cells treated with the same doses of IR during the same period of time. Consequently, there was a dose-dependent increase both of survivin mRNA expression and protein levels in radioresistant SW480 and HCT-15 cells. These observations indicate that survivin may act as a constitutive radioresistant factor in colorectal cancer cells.[43]

Studies on short interfering RNA (siRNA) inhibition of survivin demonstrated an increase in apoptosis and reduced survival in radioresistant SW480 and HCT-15 colorectal cancer cells following IR-treatment.[44] The cologenic assays data showed that survivin siRNA decreased cell viability, induced cell cycle arrest at G2-M, increased DNA double-strand breaks and resulted in an increase in radiosensitivity. Furthermore, these experiments also revealed that increased survivin levels were associated with a reduction in apoptosis in 59 tumors examined from patients with rectal cancers treated with chemoradiotherapy and this was related to a higher risk of tumor recurrence.[44]

Similarly, inhibition of survivin by specific synthetic single-stranded DNA antisense oligonucleotide (ASOs) showed substantial radiosensitization of SW480 cells as well as established SW480 xenografts. SW480 cells, transfected with ASO, had higher spontaneous and IR-induced apoptosis and an increase of cell accumulation at the G2/M phase as well as an increase in DNA double strand breaks (DSBs).[45] These studies indicate that survivin inhibition improves IR-induced cell death by mechanism beyond caspase-mediated apoptosis. Alternative pathways of cell death may include (1) mitotic arrest, (2) cell-cycle redistribution and (3) impairment in DSBs DNA repair.[45] Finally, survivin mRNA expression was examined in 20 patients by oligonucleotide microarray chips. Survivin mRNA levels were 4.2-fold higher in tumor tissue compared to normal mucosa. High surviving mRNA levels significantly correlated with an increased risk of tumor relapse.[45]

Conclusion

The results and discussion presented in this chapter point that the phenotype of the cell is paramount in determining the cellular response to the DNA damage induced by the cytotoxic effects of ionizing radiation. Rectal cancer is an important disease of DNA repair. Mutations of p53 prevent the cell from undergoing apoptosis and escape the DNA damage induced by ionizing radiation. A similar role of cyclin dependent kinase inhibitors has been shown to play in response to ionizing radiation. Several reports addressing cell cycle regulators as predictors for response to IR have demonstrated some inconsistencies. Arguments with regard to p53 lean towards an important role of wild-type p53 in pre-irradiated samples and a good response to IR. Positive CDKI status in pre-irradiated samples by IHC also seems to be a reasonable predictor of radiosensitivity in patients with rectal cancers. However, the results obtained have been different in preclinical studies where p21-positive cells had a more radioresistant phenotype, which was likely due to cell cycle arrest. Apoptosis is an important mechanism by which IR exerts its therapeutic response. Faulty apoptosis is known to render resistance to radiation therapy in rectal cancer. In vitro and in vivo data with regards to the NFκB-IAP axis of radioresistance, have shown consistently that the NFκB survival pathway leads to over expression of survivin and radioresistance. Ex vivo data support the role of survivin in raidoresistance. A complete understanding in DNA repair mechanisms, especially those involved in DSBs repair is crucial in the development of therapies that take advantage of these mechanisms in the quest to radiosensitize tumors that become resistant to conventional therapeutic interventions.

References

1. Rodel C, Sauer R. Radiotherapy and concurrent radiochemotherapy for rectal cancer. Surg Oncol 2004; 13:93-101.
2. Wong RK, Tandan V, De SS et al. Pre-operative radiotherapy and curative surgery for the management of localized rectal carcinoma. Cochrane Database Syst Rev 2007; CD002102.
3. Jemal A, Siegel R, Ward E et al. Cancer statistics, 2008. CA Cancer J Clin 2008; 58:71-96.
4. Engstrom PF, Arnoletti JP, Benson AB, III et al. Rectal cancer. J Natl Compr Canc Netw 2007; 5:940-981.

5. Kapiteijn E, Marijnen CA, Nagtegaal ID et al. Preoperative radiotherapy combined with total mesorectal excision for resectable rectal cancer. N Engl J Med 2001; 345:638-646.
6. Nagtegaal ID, Marijnen CA, Kranenbarg EK et al. Circumferential margin involvement is still an important predictor of local recurrence in rectal carcinoma: not one millimeter but two millimeters is the limit. Am J Surg Pathol 2002; 26:350-357.
7. Mohiuddin M, Winter K, Mitchell E et al. Randomized phase II study of neoadjuvant combined-modality chemoradiation for distal rectal cancer: Radiation Therapy Oncology Group Trial 0012. J Clin Oncol 2006; 24:650-655.
8. Huerta S, Gao X, Saha D. Mechanisms of resistance to ionizing radiation in rectal cancer. Expert Rev Mol Diagn 2009; 9:469-480.
9. Sancar A, Lindsey-Boltz LA, Unsal-Kacmaz K et al. Molecular mechanisms of mammalian DNA repair and the DNA damage checkpoints. Annu Rev Biochem 2004; 73:39-85.
10. Elshaikh M, Ljungman M, Ten HR et al. Advances in radiation oncology. Annu Rev Med 2006; 57:19-31.
11. Lebe B, Sarioglu S, Sokmen S et al. The clinical significance of p53, p21 and p27 expressions in rectal carcinoma. Appl Immunohistochem Mol Morphol 2005; 13:38-44.
12. Willett CG, Warland G, Hagan MP et al. Tumor proliferation in rectal cancer following preoperative irradiation. J Clin Oncol 1995; 13:1417-1424.
13. Willett CG, Warland G, Cheek R et al. Proliferating cell nuclear antigen and mitotic activity in rectal cancer: predictor of response to preoperative irradiation. J Clin Oncol 1994; 12:679-682.
14. Rodel C, Grabenbauer GG, Papadopoulos T et al. Apoptosis as a cellular predictor for histopathologic response to neoadjuvant radiochemotherapy in patients with rectal cancer. Int J Radiat Oncol Biol Phys 2002; 52:294-303.
15. Lowe SW, Schmitt EM, Smith SW et al. p53 is required for radiation-induced apoptosis in mouse thymocytes. Nature 1993; 362:847-849.
16. Bracey TS, Miller JC, Preece A et al. Gamma-radiation-induced apoptosis in human colorectal adenoma and carcinoma cell lines can occur in the absence of wild type p53. Oncogene 1995; 10:2391-2396.
17. Merritt AJ, Potten CS, Kemp CJ et al. The role of p53 in spontaneous and radiation-induced apoptosis in the gastrointestinal tract of normal and p53-deficient mice. Cancer Res 1994; 54:614-617.
18. Spitz FR, Nguyen D, Skibber JM et al. Adenoviral-mediated wild-type p53 gene expression sensitizes colorectal cancer cells to ionizing radiation. Clin Cancer Res 1996; 2:1665-1671.
19. Ribeiro JC, Barnetson AR, Fisher RJ et al. Relationship between radiation response and p53 status in human bladder cancer cells. Int J Radiat Biol 1997; 72:11-20.
20. Lane DP, Benchimol S. p53: oncogene or anti-oncogene? Genes Dev 1990; 4:1-8.
21. Vidaurreta M, Maestro ML, Sanz-Casla MT et al. Colorectal carcinoma prognosis can be predicted by alterations in gene p53 exons 5 and 8. Int J Colorectal Dis 2008; 23:581-586.
22. Waldman T, Lengauer C, Kinzler KW et al. Uncoupling of S phase and mitosis induced by anticancer agents in cells lacking p21. Nature 1996; 381:713-716.
23. Brugarolas J, Chandrasekaran C, Gordon JI et al. Radiation-induced cell cycle arrest compromised by p21 deficiency. Nature 1995; 377:552-557.
24. Slebos RJ, Baas IO, Clement M et al. Clinical and pathological associations with p53 tumour-suppressor gene mutations and expression of p21WAF1/Cip1 in colorectal carcinoma. Br J Cancer 1996; 74:165-171.
25. Qiu H, Sirivongs P, Rothenberger M et al. Molecular prognostic factors in rectal cancer treated by radiation and surgery. Dis Colon Rectum 2000; 43:451-459.
26. Palazzo JP, Kafka NJ, Grasso L et al. The role of p53, p21WAF1/C1PI and bcl-2 in radioresistant colorectal carcinoma. Hum Pathol 1997; 28:1189-1195.
27. Sogawa N, Takiguchi N, Koda K et al. Value of expression of p21WAF1/CIP1 as a prognostic factor in advanced middle and lower rectal cancer patients treated with preoperative radio-chemotherapy. Int J Oncol 2002; 21:787-793.
28. Lin LC, Lee HH, Hwang WS et al. p53 and p27 as predictors of clinical outcome for rectal-cancer patients receiving neoadjuvant therapy. Surg Oncol 2006; 15:211-216.
29. Tian H, Wittmack EK, Jorgensen TJ. p21WAF1/CIP1 antisense therapy radiosensitizes human colon cancer by converting growth arrest to apoptosis. Cancer Res 2000; 60:679-684.
30. Waldman T, Zhang Y, Dillehay L et al. Cell-cycle arrest versus cell death in cancer therapy. Nat Med 1997; 3:1034-1036.
31. Moore HG, Shia J, Klimstra DS et al. Expression of p27 in residual rectal cancer after preoperative chemoradiation predicts long-term outcome. Ann Surg Oncol 2004; 11:955-961.
32. Hoos A, Nissan A, Stojadinovic A et al. Tissue microarray molecular profiling of early, node-negative adenocarcinoma of the rectum: a comprehensive analysis. Clin Cancer Res 2002; 8:3841-3849.

33. Huerta S, Goulet EJ, Huerta-Yepez S et al. Screening and detection of apoptosis. J Surg Res 2007; 139:143-156.
34. Huerta S. Recent advances in the molecular diagnosis and prognosis of colorectal cancer. Expert Rev Mol Diagn 2008; 8:277-288.
35. Meyn RE, Stephens LC, Milas L. Programmed cell death and radioresistance. Cancer Metastasis Rev 1996; 15:119-131.
36. Meyn RE, Stephens LC, Ang KK et al. Heterogeneity in the development of apoptosis in irradiated murine tumours of different histologies. Int J Radiat Biol 1993; 64:583-591.
37. Schwandner O, Schiedeck TH, Bruch HP et al. Apoptosis in rectal cancer: prognostic significance in comparison with clinical histopathologic and immunohistochemical variables. Dis Colon Rectum 2000; 43:1227-1236.
38. [Anonymus]. Improved survival with preoperative radiotherapy in resectable rectal cancer. Swedish Rectal Cancer Trial. N Engl J Med 1997; 336:980-987.
39. Marijnen CA, Nagtegaal ID, Mulder-Stapel AA et al. High intrinsic apoptosis, but not radiation-induced apoptosis, predicts better survival in rectal carcinoma patients. Int J Radiat Oncol Biol Phys 2003; 57:434-443.
40. de Bruin EC, van de Velde CJ, van de Pas S et al. Prognostic value of apoptosis in rectal cancer patients of the dutch total mesorectal excision trial: radiotherapy is redundant in intrinsically high-apoptotic tumors. Clin Cancer Res 2006; 12:6432-6436.
41. Russo SM, Tepper JE, Baldwin AS, Jr. et al. Enhancement of radiosensitivity by proteasome inhibition: implications for a role of NF-kappaB. Int J Radiat Oncol Biol Phys 2001; 50:183-193.
42. Rodel C, Haas J, Groth A et al. Spontaneous and radiation-induced apoptosis in colorectal carcinoma cells with different intrinsic radiosensitivities: survivin as a radioresistance factor. Int J Radiat Oncol Biol Phys 2003; 55:1341-1347.
43. Rodel F, Hoffmann J, Grabenbauer GG et al. High survivin expression is associated with reduced apoptosis in rectal cancer and may predict disease-free survival after preoperative radiochemotherapy and surgical resection. Strahlenther Onkol 2002; 178:426-435.
44. Rodel F, Hoffmann J, Distel L et al. Survivin as a radioresistance factor and prognostic and therapeutic target for radiotherapy in rectal cancer. Cancer Res 2005; 65:4881-4887.
45. Rodel F, Frey B, Leitmann W et al. Survivin antisense oligonucleotides effectively radiosensitize colorectal cancer cells in both tissue culture and murine xenograft models. Int J Radiat Oncol Biol Phys 2008; 71:247-255.

Chapter 13

Familial Cutaneous Melanoma

Johan Hansson*

Abstract

Approximately 5-10 % of all cutaneous melanomas occur in families with hereditary melanoma predisposition. Worldwide, approximately 20-40% of kindreds with familial melanoma harbor germline mutations in the *CDKN2A* gene, located on chromosome 9p21, which encodes two different proteins, p16INK4 and p14ARF, both involved in regulation of cell cycle progression and induction of senescence. In different populations several recurring *CDKN2A* founder mutations have been described. The risk of melanoma in *CDKN2A* mutations carriers varies between populations and is higher in regions with high sun exposure and high incidence of melanoma in the general population. Some *CDKN2A* mutations have been associated not only with melanoma but also with increased risk of other malignancies—most notably pancreatic carcinoma. A much smaller number of families have germline mutations in the *CDK4* gene on chromosome 12q14, encoding a cyclin dependent kinase which normally interacts with p16INK4A. The management of families with hereditary melanoma is discussed.

Introduction

The incidence of cutaneous malignant melanoma (CMM) is rapidly increasing in white skinned populations across the world.[1] Therefore improved preventive strategies are needed. One essential task is to define high-risk groups for CMM, who may be enrolled in preventive programs. A particular high risk group is represented by members of families with hereditary CMM. In this chapter the clinical characteristics, genetic aspects and guidelines for management of familial melanoma (FM) will be presented.

Risk Factors for Melanoma

The major environmental risk factor for melanoma is exposure to ultraviolet (UV) irradiation, both long wavelength UVA (320-400 nm) and intermediate wavelength UVB (290-320 nm). The most consistent relationship between increased risks of CMM is seen with intermittent sun exposure, particularly in early life and a history of sunburns.[2] The presence of large numbers of melanocytic nevi and atypical moles/dysplastic nevi (DN) are additional risk factors for CMM.[3] Other phenotypic risk factors include skin type with an inability to tan, large number of freckles, red or blond hair color, blue/grey eye color, actinic skin damage and a history of a previous premalignant or skin cancer lesions (melanoma or nonmelanoma).[4] In addition to environmental and phenotypic risk factors, genetic makeup also plays important roles.

Familial Melanoma—The Clinical Picture

It is estimated that 5-10% of all cases of CMM occur in families with FM.[5,6] In population-based studies 1-13% of melanoma cases reported were in individuals with one first-degree

*Johan Hansson—Department of Oncology-Pathology, Karolinska Institutet, Karolinska University Hospital Solna S-171 76, Stockholm, Sweden. Email: johan.hansson@ki.se

Diseases of DNA Repair, edited by Shamim I. Ahmad. ©2010 Landes Bioscience and Springer Science+Business Media.

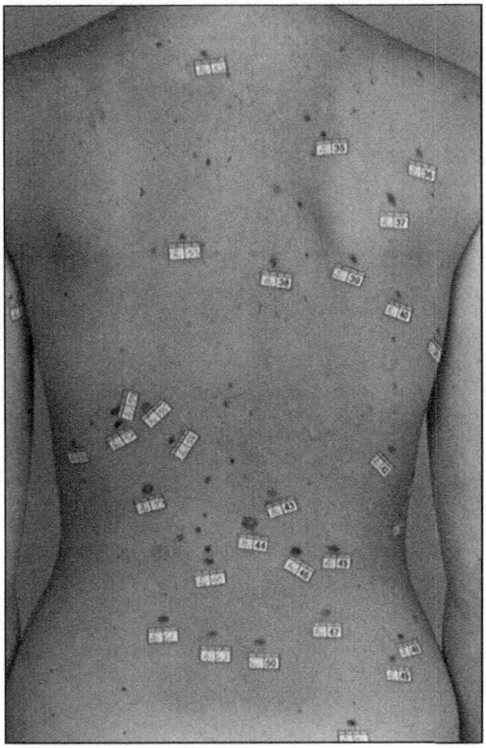

Figure 1. The FM-DNS phenotype. The back of a young woman belonging to a kindred with familial melanoma, who exhibits numerous dysplastic nevi. Reprinted from: Hansson J. Familial melanoma. Surg Clin North Am 2008; 88(4):897-916; ©2008 with permission from Elsevier.

relative suffering from CMM.[7,8] Thus, according to a report from the Genetic Epidemiology of Melanoma Group (GEM), first degree relatives of individuals with CMM have increased risk of developing melanoma, with a cumulative risk of 6-7% at age 80.[9]

Families with increased melanoma risk have been recognized for nearly two centuries. The first description of CMM appeared in 1820 in which Norris reported a family in United Kingdom in which two members had CMM and several relatives had large moles.[10] In 1978 Clark reported six melanoma prone families where both, patients and relatives, had large "funny-looking" nevi, which were designated as potential precursors of CMM (Figs. 1,2).[11] It was named the "dysplastic nevus syndrome" (DNS).[8] At the same time Lynch reported the same syndrome, which he named the "familial atypical multiple mole melanoma" (FAMMM) syndrome.[12] The syndrome is also called the "atypical mole syndrome" (AMS).[13]

For the clinical diagnosis of DNS the established *ABCD*(*E*) criteria are useful. Nevi are judged as dysplastic nevus (DN) if they have an *A*symmetrical form, irregular *B*orders, multi-*C*olor pigmentation and a *D*iameter of ≥6 mm.[14] A further criterion of DN is *E*levation of the nevus, i.e., the simultaneous presence of macular and papular components.[15] It should be noted that these criteria are also signs of early CMM. In a trained hand dermatoscopy (epiluminescence microscopy) is a useful technique to increase the diagnostic accuracy.[16]

Melanoma is diagnosed at an earlier age in FM patients, they have thinner tumors and a higher frequency of multiple primary melanomas than patients with sporadic melanoma.[17] In some families there is also an increased risk of pancreatic carcinoma (see below). In the original descriptions of melanoma families there was an association between melanoma risk and the DNS phenotype

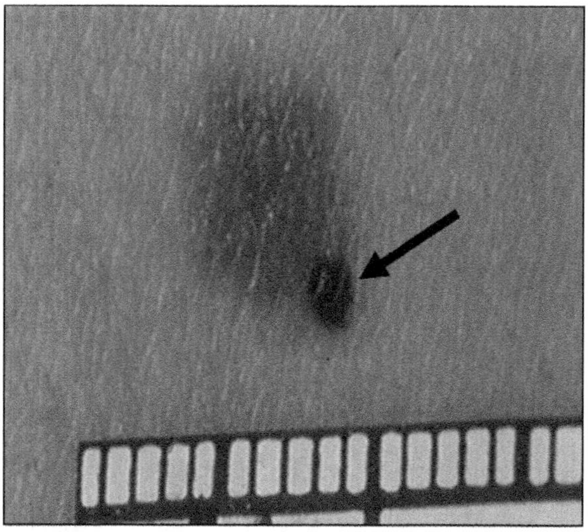

Figure 2. Development of an early melanoma in a dysplastic nevus. During follow-up of this member of a kindred with FM a dark pigmentation arose in the lower part of this dysplastic nevus (arrow). Histopathological examination of the excised nevus showed that the darkly pigmented area corresponds to an early CMM: a superficial spreading melanoma with a tumor thickness of 0.5 mm (T1a, according to the AJCC classification). Reprinted from: Hansson J. Familial melanoma. Surg Clin North Am 2008; 88(4):897-916; ©2008 with permission from Elsevier.

and DN were considered as precursor lesions of CMM[11,18] However, it has now become clear that there can be variation between families as to whether they exhibit the DNS phenotype or not. Moreover, although DN may be precursor lesions of CMM (Fig. 2),[19] CMM can also develop in individuals without DNS.[20]

Molecular Genetics of Familial Cutaneous Melanoma

Since 1997 the Melanoma Genetics Consortium, GenoMEL, has been active. GenoMEL is comprised of researchers worldwide, working on the genetics of familial CMM. The mission of GenoMEL is to identify melanoma susceptibility genes, to evaluate gene-environment interactions and to assess the risk of CMM and other cancers related to variations in these genes. Information on GenoMEL and its activities can be obtained at: http://www.genomel.org.

High Risk Melanoma Genes

CDKN2A

Studies have shown that affected members in some FM families harbor germline mutations in the *CDKN2A* gene on chromosome 9p21.[21,22] The *CDKN2A* gene encodes two unrelated tumor suppressor proteins, which play key roles in cell cycle regulation (Figs. 3,4).[2] The $p16^{INK4}$ protein, encoded by exons 1α, 2 and 3 of *CDKN2A,* negatively regulates cell cycle progression by inhibiting the cyclin dependent kinases *CDK4* and CDK6. Inhibition of the kinases prevents phosphorylation of the retinoblastoma protein, pRb and thereby entry into the S-phase of the cell cycle. A large proportion of melanomas contain BRAF gene mutations; the product of the BRAF activates the mitogen-activated protein kinase (MAPK) pathway and this is important for cellular growth, survival and migration.[24] Interestingly, BRAF mutations are also very frequent in nevi.[25] It has been suggested that $p16^{INK4}$ functions as a barrier to melanoma development by causing senescence in nevus cells containing activating BRAF mutations.[26,27] Germline loss of one *CDKN2A* allele would thus weaken this protective mechanism against melanoma development.[28]

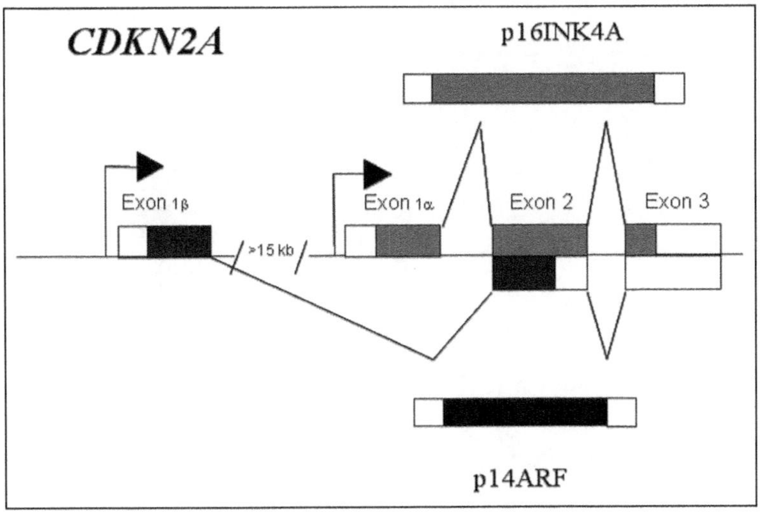

Figure 3. The *CDKN2A* locus on chromosome 9p21. This gene encodes two different proteins: p16^{INK4A} and p14^{ARF}, respectively. p16^{INK4A} is encoded by exons 1α, 2 and 3, whereas the p14^{ARF} protein is encoded by alternative splicing of an alternative exon 1β to exon 2. The two proteins have quite different amino acid sequences since they are translated in different reading frames. Reprinted from: Hansson J. Familial melanoma. Surg Clin North Am 2008; 88(4):897-916; ©2008 with permission from Elsevier.

The second *CDKN2A* product, protein p14^{ARF}, is encoded by splicing of an alternative exon 1β to exon 2 of the gene. This protein is translated in an *A*lternative *R*eading *F*rame (hence ARF) and lacks amino acid homology to p16^{INK4}. p14^{ARF} blocks HDM2-mediated degradation of p53. p53 is suggested to be involved in a second p16^{INK4}-independent senescence barrier in melanocytes.[29] Loss of p14^{ARF} function may have the same effect as loss of p53 and the frequent loss of p14^{ARF} in melanoma tumors, through deletions or other mutations may explain the relatively low frequency of TP53 mutations in melanoma.[23,30,31] In addition p14^{ARF} has been implicated in sumoylation of several binding proteins as well as inhibiting the transcriptional activator E2F-1 and promoting its degradation.[32,33] Mutations in *CDKN2A* thus have the capacity to target negative regulators in two key signaling pathways, the pRb and p53 pathways, both of which have central roles in cell cycle regulation and act as senescence barriers.

A large number of different germline *CDKN2A* mutations have been identified in FM families as well in patients with multiple primary CMM.[34-37] The mutations are scattered throughout the gene without any specific hot spot. Most mutations have been reported in exons 1α and 2, consistent with the inactivation of the p16^{INK4} protein as the main predisposing factor for CMM. However, alterations, affecting exon 1β only, have also been shown in melanoma families where also neural system tumors occur.[38-42] Therefore it seems that mutations affecting either protein may be involved in development of FM.

Globally it has been estimated that approximately 20-40% of FM families are associated with germline *CDKN2A* mutations.[43] GenoMEL carried out a large study of 466 families with at least 3 melanoma patients.[44] Overall 41% of families had *CDKN2A* mutations, mostly involving p16^{INK4A}, while mutations in *CDK4* and in exon 1β of *CDKN2A*, exclusively affecting p14^{ARF}, occurred at equally low frequency (2-3%). Thus, although p16^{INK4A} is most frequently mutated protein, the results are consistent with a role of both p16^{INK4A} and p16ARF in melanoma suppression. In total 66 different *CDKN2A* mutations have been reported, of which 43 are unique while the remainder occurred in two or more kindreds. There were striking differences in mutation

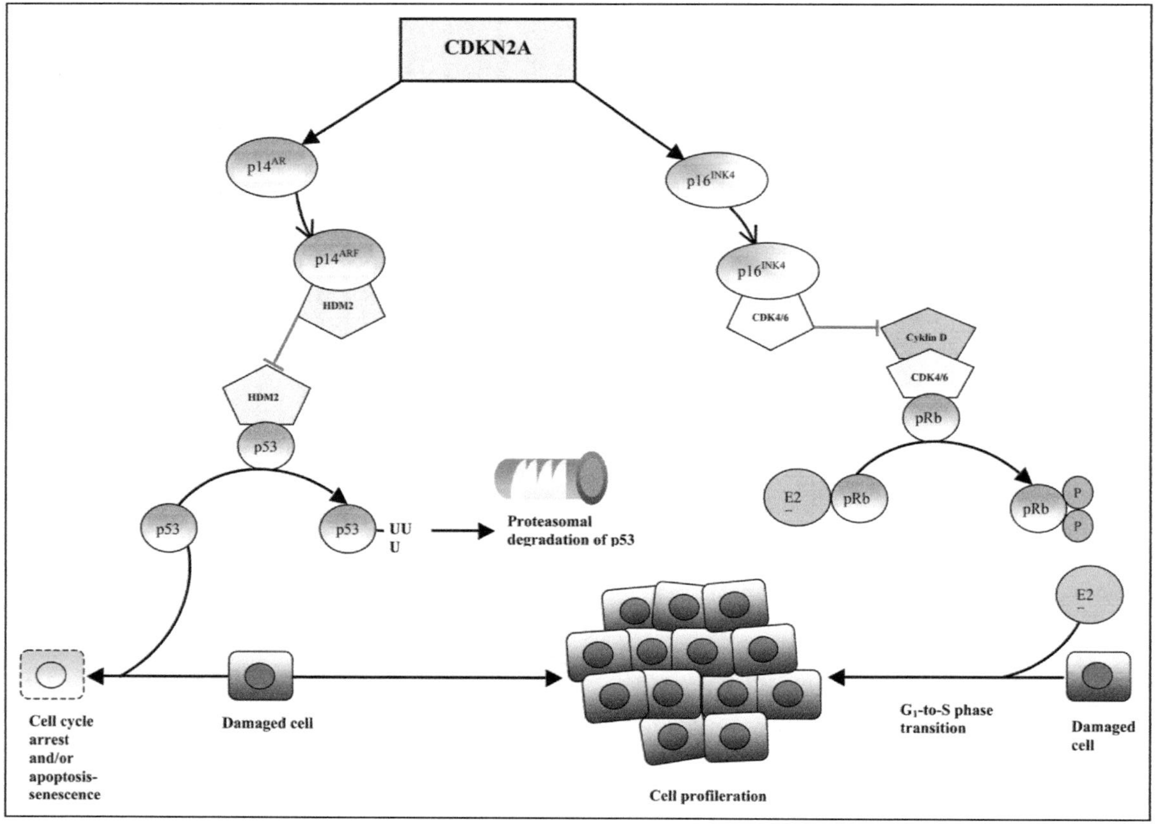

Figure 4. The two *CDKN2A* gene products: p16^{INK4A} and p14^{ARF} act as negative regulators in the CDK4/cyclin D/Rb and the p53 pathways respectively. Both proteins thus have roles as inhibitors of unrestricted proliferation and promote senescence (modified from Chudnovsky Y. et al J Clin Invest 2005; 115:813-824).

patterns across geographical areas. Specific founder *CDKN2A* mutations have been found in several countries and the proportion of families with such mutations differed significantly among geographical regions (P = 0.0009). Thus, single founder *CDKN2A* mutations are predominant in Sweden (p.R112_L113insR, 92% of familial mutations),[45] the Netherlands (c.225_243del19, 90% of familial mutations)[46] and Iceland (p.G89D).[47] France, Spain and Italy have almost the same mutation frequency (p.G101W).[48] Similarly, Australia and United Kingdom share the same most common mutations (p.M53I, c.IVS2-105A>G, p.R24P and p.L32P).

In another GenoMEL study of 385 families with 3 or more melanoma cases, frequency of germline *CDKN2A* mutations in different continents were analysed.[47] Overall, 39% of families had *CDKN2A* mutations, ranging from 20% (32/162) in Australia, 45% (29/65) in North America and 57% (89/157) in Europe. The lower frequency of *CDKN2A* mutation in areas with high CMM incidence, such as Australia, may be explained by a higher frequency of clustering of sporadic cases, or cases associated with low-penetrance genes and/or high solar UV exposure.

CDK4

Germline mutations in the *CDK4* gene on chromosome 12q14, encoding a cyclin dependent kinase, which normally interacts with $p16^{INK4A}$, were first reported in 1996.[49] To date, less than 15 families with germline *CDK4* mutations have been identified worldwide.[50-52] In every case the mutation affected codon 24 where two different mutations have been described: p.R24C, p.R24H. These mutations abolish binding of CDK4 protein to $p16^{INK4A}$ and causes *CDK4* to act as dominant oncogene.[23,30,31] GenoMel studies have confirmed that germline *CDK4* mutations are rare (about 2%) compared to *CDKN2A* mutations.[44] The phenotype of the affected families seems to be similar to those with *CDKN2A* mutations.

Candidate Loci for Novel Genes Predisposing to Familial CMM

GenoMEL has carried out a study on microsatellite markers in 82 melanoma families, mostly from Australia and found a significant linkage to markers on chromosome 1p22,[53] while Loss of Heterozygocity (LOH) studies indicated that a tumor suppressor gene may be present at this locus. Despite considerable efforts, so far no susceptibility gene has been identified.[54] More recently a linkage analysis of three Danish kindreds, carrying both ocular and cutaneous melanoma, resulted in a possible linkage to markers on chromosome 9q21.32, but the putative gene responsible for the syndrome has not been identified.[55]

Risk of Melanoma and Other Cancers in Melanoma Families with Germline *CDKN2A* Mutations

For genetic counseling it is important to estimate the penetrance of a certain germline gene mutation, i.e.,the risk of developing the disease among carriers of the mutation. The penetrance of germline *CDKN2A* mutations for melanoma development in members of kindreds with FM has been the subject of a large collaborative GenoMEL study of members of 80 kindreds with FM from different parts of the world.[67] Overall, *CDKN2A* mutation penetrance was estimated to be 0.30 (95% confidence interval (CI) = 0.12 to 0.62) by age 50 years and 0.67 (95% CI = 0.31 to 0.96) by age 80 years. There was a significant increase in melanoma penetrance in individuals living in a location with a high incidence rate of melanoma ($P = 0.003$). Thus, by age 50 years *CDKN2A* mutation penetrance reached 0.13 in Europe, 0.50 in the United States and 0.32 in Australia; by age 80 years it was 0.58 in Europe, 0.76 in the United States and 0.91 in Australia. Thus, the same factors that affect population incidence of melanoma may also modulate *CDKN2A* penetrance, strongly supporting an interaction between germline *CDKN2A* mutations and environmental UV exposure.

As reported in a recent publication from the GEM study, the risk of melanoma in carriers of germline *CDKN2A* mutations in the general population is lower.[56] The risk of melanoma in carriers of germline *CDKN2A* mutations was 14% by age 50, 24 % by age 70 and 28% by age 80. Thus, carriers of *CDKN2A* germline mutations in the general population seem to have a considerably lower melanoma risk than those who belong to kindreds with FM. This is most likely due to the influence of other unknown melanoma predisposing factors in FM, which interact with and increase the penetrance of *CDKN2A* mutations.

Apart from cutaneous melanoma, an increased risk for pancreatic carcinoma (PC) has been documented in several families, including families carrying the Dutch p16 Leiden mutation, the Mediterranean p.G101W and the Swedish p.R112_L113insR founder mutations.[46,57-59] A recent study of 22 families, positive for the p16-Leiden founder mutation, showed an increased risk of cancer other than melanoma and nonmelanoma skin cancer, with a relative risk of 4.4, predominantly attributable to increased risk for pancreatic cancer with a relative risk of 46.6.[60] Although the risk of pancreatic cancer is lower than that of melanoma, a study of Dutch melanoma families showed nearly equal mortality rates due to melanoma and pancreatic cancer in these families.[61]

In a small number of families CMM and neural system tumors (NST) have been associated with large deletions of *CDKN2A*/ARF and/or mutations that affect $p14^{ARF}$.[38-42] In GenoMEL study, however, there was a marginally significant association of NST with mutations affecting the $p14^{ARF}$ transcript ($p = 0.05$), thus giving some support for the impact of germline $p14^{ARF}$ alterations and NST as well as melanoma.[44] Furthermore, there was no significant association between *CDKN2A* germline mutations in families and the occurrence of uveal melanoma ($p = 0.25$).

Genetic Testing in Familial Melanoma

Genetic testing is widely used to identify individuals with hereditary predisposition for colorectal and breast/ovarian cancers, but genetic testing for *CDKN2A* mutations for melanoma is not routinely performed. However, *CDKN2A* mutation testing is becoming increasingly available. Members of melanoma families should be counselled for the advantages and disadvantages of genetic testing.[62] Arguments against genetic testing include: (i) many melanoma families still lack identifiable germline mutations in known high-risk genes, therefore a negative test result is uninformative; (ii) the penetrance of germline mutations is still insufficiently characterised, making the implications of a positive test results imprecise; (iii) there is an increased risk of CMM also in family members without mutations in *CDKN2A*. A negative result could lead to false reassurance with a negative impact on preventive activities, although there is no evidence for this from other familial cancers.[63,64]

Arguments in favour of gene testing have been given: (1) provided the limitations of the test are explained to the family, the information presented may be valuable to them; (2) a positive test result may improve the compliance of the family to enter in preventive programs; (3) a negative test result of a member of a FM family, where a relative has died from CMM, will be reassuring to the individual.

Gene testing for melanoma should only be offered by a certified genetic counselling service. Gene testing for germline *CDKN2A* mutation should only be considered where there is reasonable likelihood of finding reliable results. It is not possible to define exact criteria for FM testing, due to the variability between different populations for the penetrance of *CDKN2A* mutations.[65] However, the current data indicate that, in moderate to high melanoma incidence areas, individuals fulfilling the following criteria are appropriate candidates for testing:

- Individuals with 3 or more primary invasive melanomas.
- Families with at least one invasive melanoma and two cases of melanoma and/or pancreatic cancer among first- or second-degree relatives on the same side of the family.

Management of Familial Melanoma

There is a long-standing consensus that members of FM kindreds should be invited to participate in preventive programs and a consensus statement on the management of FM has been published by GenoMEL.[66]

To identify patients with FM it is important to question every newly diagnosed CMM patient for family history of melanoma and other malignancies. Diagnoses of CMM and other cancers should be verified, preferably through histopathology reports, as well as age at diagnosis. When a

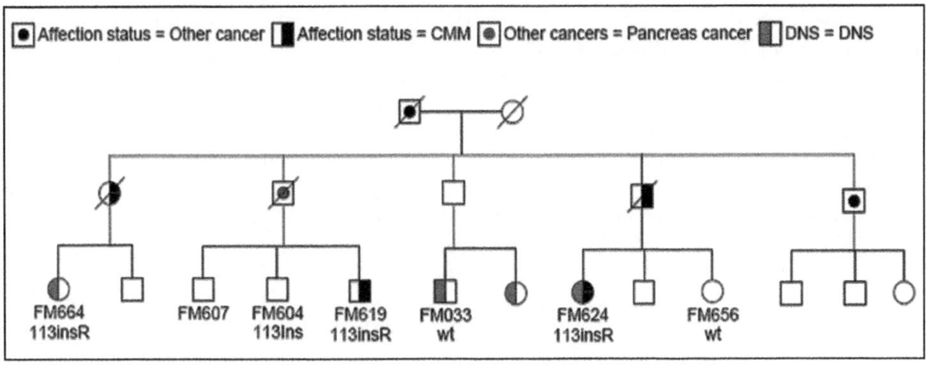

Figure 5. Pedigree of a familial melanoma kindred with both CMM and pancreatic carcinoma in which affected members carry the Swedish *CDKN2A* founder mutation. Reprinted from: Hansson J. Familial melanoma. Surg Clin North Am 2008; 88(4):897-916; ©2008 with permission from Elsevier.

family history of melanoma has been established a careful extended pedigree of the family should be established in collaboration with the proband of the family (Fig. 5).

Due to the limitations of testing for germline *CDKN2A* mutations, it is recommended that members of melanoma families are offered preventive measures regardless of *CDKN2A* mutation status. At least all first degree relatives of CMM patients should be invited to participate in the preventive program.

Primary Prevention

An integral part of primary prevention is educating the family members for protection from the harmful effects of sun exposure. Efforts should aim to reduce sun exposure to all members of melanoma families, particularly in their early life. Parents should be educated for sun protective measures specially for children,[68,69] including the use of sun-protective clothing, hats and sunglasses, avoidance of sun exposure during peak UV levels and absolute avoidance of sunburns. The issue of use of sunscreens remains controversial and hence may be considered as a supplement to other sun protective measures.[70]

Secondary Prevention of CMM

Since many melanomas develop from precursor lesions such as DN and since melanomas which are detected and treated early have an excellent prognosis, there is a clear role for monitoring of pigmented skin lesions in members of melanoma families. Family members should be instructed for self examination of skin and also be given the opportunity to participate in regular screening programmes. Information regarding the significance of change in shape and size of pigmented lesions should be given and instruction on the ABCD(E) rules may also be useful.[14,15,74] However, it should be noted that these criteria do not apply for every melanomas, since considerable fraction of early melanomas have a diameter less than 6 mm.[75] Moreover, since a proportion of CMM tumors arise apparently de novo and not by progression of a precursor lesion, the individuals must also be informed to be watchful for novel skin lesions.[73]

Commencing at age 10 years, members of kindreds with familial CMM should have a baseline whole-skin examination, including scalp and external genitals. The examination should focus on detection and characterization of nevi and any suspected melanoma lesions. Photographs of the entire skin as well as close-up pictures of DN are very useful for follow-up. Family members should be followed with skin examinations approximately every 6 months, at least until the nevi are stable and the individual is judged competent in self-surveillance. Subsequently, the individual should be examined annually or have prompt access to the health provider as necessary. Dermoscopy can be helpful during skin examinations.[16,71,72]

Any changing nevus should be considered for excision for histopathological examination. There is, however, no justification for prophylactic removal of nevi, since the probability of progression to melanoma is low for every individual lesion.[73]

There are reports which indicate that preventive measures may result in early diagnosis of CMM and tumours with small thickness can be detected during follow-up.[76-78] In a report on long-term follow-up of 844 members of 33 kindreds with FM, of which 19 had germline mutations in either *CDKN2A* or *CDK4*, 86 new CMMs were identified.[79] Of these, 72 were classified as early lesions with an average thickness of 0.3 mm. Similarly, in an analysis of 2,080 family members of 280 Swedish FM kindreds who were followed between 1987 and 2001, 41 CMMs were detected during the follow-up. Of these 15 (37%) were in situ tumors and 27 of the 41 CMMs (66%) lacked vertical growth phase and metastatic ability.[80]

Pancreatic Carcinoma Surveillance

At present there are no methods to detect pancreatic cancer at a curable stage.[81-83] Endoscopic retrograde cholangiopanceratography (ERCP) is considered the golden standard for visualization of pancreatic carcinoma. However, due to the risk of serious complications such as bleeding, intestinal perforation or pancreatitis, ERCP cannot be used in routine surveillance. Endoscopic ultrasound (EUS) is a novel technique which may be able to detect early PC and precursor lesions.[81] More recently MRI, combined with cholangiopanceratography (MR/MRCP), has been proposed as a screening tool. The benefit of such screening programs need to be investigated in prospective studies and there are a number of research programs in the US and Europe addressing this.

Conclusion

Considerable advances have been made in our understanding of familial melanoma, and guidelines for management of melanoma families and genetic testing for germline *CDKN2A* mutations have been developed. However, more research is important to further characterize the interaction between inherited mutations, and environment factors as well as individual phenotype. Also, gene-gene interactions between *CDKN2A* and risk modifying genes need to be further explored. Moreover, the genetic background for the majority of kindreds with familial melanoma is still unknown. Further investigations aimed at identifying novel melanoma predisposing genes in families without mutations in known high risk genes should be prioritized.

Acknowledgements

This work was supported by the Swedish Cancer Society, The Swedish Medical Research Council, The Radiumhemmet Research Funds, The NIH (RO1-CA-083115-06) and the European Commission (6th Framework Program, GenoMEL).

References

1. Jemal A, Siegel R, Ward E et al. Cancer statistics, 7. CA Cancer J Clin 2007; 57:43-66.
2. Gandini S, Sera F, Cattaruzza MS et al. Meta-analysis of risk factors for cutaneous melanoma: II. Sun exposure. Eur J Cancer 2005; 41:45-60.
3. Gandini S, Sera F, Cattaruzza MS et al. Meta-analysis of risk factors for cutaneous melanoma: I. Common and atypical naevi. Eur J Cancer 2005; 41:28-44.
4. Gandini S, Sera F, Cattaruzza MS et al. Meta-analysis of risk factors for cutaneous melanoma: III. Family history, actinic damage and phenotypic factors. Eur J Cancer 2005; 41:2040-2059.
5. Platz A, Ringborg U, Hansson J. Hereditary cutaneous melanoma. Semin Cancer Biol 2000; 10:319-326.
6. Florell SR, Boucher KM, Garibotti G et al. Population-based analysis of prognostic factors and survival in familial melanoma. J Clin Oncol 2005; 23:7168-7177.

7. Ford D, Bliss JM, Swerdlow AJ et al. Risk of cutaneous melanoma associated with a family history of the disease. The International Melanoma Analysis Group (IMAGE). Int J Cancer 1995; 62:377-381.
8. Greene MH, Clark WH Jr, Tucker MA et al. Precursor naevi in cutaneous malignant melanoma: a proposed nomenclature. Lancet 1980; 2:1024.
9. Begg CB, Hummer A, Mujumdar U et al. Familial aggregation of melanoma risks in a large population-based sample of melanoma cases. Cancer Causes Control 2004; 15:957-965.
10. Norris W. Case of fungoid disease. Edinburgh Med Surg J 1820; 16:562-565.
11. Clark WH Jr, Reimer RR, Greene M et al. Origin of familial malignant melanomas from heritable melanocytic lesions. 'The B-K mole syndrome'. Arch Dermatol 1978; 114:732-738.
12. Lynch HT, Frichot BC 3rd, Lynch JF. Familial atypical multiple mole-melanoma syndrome. J Med Genet 1978; 15:352-356.
13. Newton Bishop JA, Bataille V, Pinney E et al. Family studies in melanoma: identification of the atypical mole syndrome (AMS) phenotype. Melanoma Res 1994; 4(4):199-206.
14. Friedman RJ, Rigel DS, Kopf AW. Early detection of malignant melanoma: the role of physician examination and self-examination of the skin. CA Cancer J Clin 1985; 35(3):130-151.
15. Hofmann-Wellenhof R, Blum A, Wolf IH et al. Dermoscopic classification of Clark's nevi (atypical melanocytic nevi). Clin Dermatol 2002; 20(3):255-258.
16. Roesch A, Burgdorf W, Stolz W et al. Dermatoscopy of "dysplastic nevi": a beacon in diagnostic darkness. Eur J Dermatol 2006; 16(5):479-493.
17. Goldstein AM, Fraser MC, Clark WH Jr et al. Age at diagnosis and transmission of invasive melanoma in 23 families with cutaneous malignant melanoma/dysplastic nevi. J Natl Cancer Inst 1994; 86(18):1385-1390.
18. Greene MH, Clark WH Jr, Tucker MA et al. High risk of malignant melanoma in melanoma-prone families with dysplastic nevi. Ann Intern Med 1985; 102(4):458-465.
19. Marks R, Dorevitch AP, Mason G. Do all melanomas come from "moles"? A study of the histological association between melanocytic naevi and melanoma. Australas J Dermatol 1990; 31(2):77-80.
20. Carey WP Jr, Thompson CJ, Synnestvedt M et al. Dysplastic nevi as a melanoma risk factor in patients with familial melanoma. Cancer 1994; 74(12):3118-3125.
21. Hussussian CJ, Struewing JP, Goldstein AM et al. Germline p16 mutations in familial melanoma. Nat Genet 1994; 8(1):15-21.
22. Kamb A, Gruis NA, Weaver-Feldhaus J et al. A cell cycle regulator potentially involved in genesis of many tumor types. Science 1994; 264(5157):436-440.
23. de Snoo FA, Hayward NK. Cutaneous melanoma susceptibility and progression genes. Cancer Lett 2005; 230(2):153-186.
24. Platz A, Egyhazi S, Ringborg U et al. Human cutaneous melanoma; a review of NRAS and BRAF mutation frequencies in relation to histogenetic subclass and body site. Mol Oncol 2008; 1(4):395-405.
25. Pollock PM, Harper UL, Hansen KS et al. High frequency of BRAF mutations in nevi. Nat Genet 2003; 33(1):19-20.
26. Michaloglou C, Vredeveld LC, Soengas MS et al. BRAFE600-associated senescence-like cell cycle arrest of human naevi. Nature 2005; 436(7051):720-724.
27. Gray-Schopfer VC, Cheong SC, Chong H et al. Cellular senescence in naevi and immortalisation in melanoma: a role for p16? Br J Cancer 2006; 95(4):496-505.
28. Florell SR, Meyer LJ, Boucher KM et al. Longitudinal assessment of the nevus phenotype in a melanoma kindred. J Invest Dermatol 2004; 123(3):576-582.
29. Sviderskaya EV, Gray-Schopfer VC, Hill SP et al. p16/cyclin-dependent kinase inhibitor 2A deficiency in human melanocyte senescence, apoptosis and immortalization: possible implications for melanoma progression. J Natl Cancer Inst 2003; 95(10):723-732.
30. Dahl C, Guldberg P. The genome and epigenome of malignant melanoma. APMIS 2007; 115(10):1161-1176.
31. Bennett DC. How to make a melanoma: what do we know of the primary clonal events? Pigment Cell Melanoma Res 2008; 21(1):27-38.
32. Rizos H, Woodruff S, Kefford RF. p14ARF interacts with the SUMO-conjugating enzyme Ubc9 and promotes the sumoylation of its binding partners. Cell Cycle 2005; 4(4):597-603.
33. Rizos H, Scurr LL, Irvine M et al. p14ARF regulates E2F-1 ubiquitination and degradation via a p53-dependent mechanism. Cell Cycle 2007; 6(14):1741-1747.
34. Monzon J, Liu L, Brill H et al. CDKN2A mutations in multiple primary melanomas. N Engl J Med 1998; 338(13):879-887.
35. MacKie RM, Andrew N, Lanyon WG et al. CDKN2A germline mutations in UK patients with familial melanoma and multiple primary melanomas. J Invest Dermatol 1998; 111(2):269-272.
36. Hashemi J, Platz A, Ueno T et al. CDKN2A germ-line mutations in individuals with multiple cutaneous melanomas. Cancer Res 2000; 60(24):6864-6867.

37. Auroy S, Avril MF, Chompret A et al. Sporadic multiple primary melanoma cases: CDKN2A germline mutations with a founder effect. Genes Chromosomes Cancer 2001; 32(3):195-202.
38. Bahuau M, Vidaud D, Jenkins RB et al. Germ-line deletion involving the INK4 locus in familial proneness to melanoma and nervous system tumors. Cancer Res 1998; 58(11):2298-2303.
39. Hewitt C, Lee Wu C, Evans G et al. Germline mutation of ARF in a melanoma kindred. Hum Mol Genet 2002; 11(11):1273-1279.
40. Petronzelli F, Sollima D, Coppola G et al. CDKN2A germline splicing mutation affecting both p16(ink4) and p14(arf) RNA processing in a melanoma/neurofibroma kindred. Genes Chromosomes Cancer 2001; 31(4):398-401.
41. Randerson-Moor JA, Harland M, Williams S et al. A germline deletion of p14(ARF) but not CDKN2A in a melanoma-neural system tumour syndrome family. Hum Mol Genet 2001; 10(1):55-62.
42. Rizos H, Puig S, Badenas C et al. A melanoma-associated germline mutation in exon 1beta inactivates p14ARF. Oncogene 2001; 20(39):5543-5547.
43. Hayward NK. Genetics of melanoma predisposition. Oncogene 2003; 22(20):3053-3062.
44. Goldstein AM, Chan M, Harland M et al. High-risk melanoma susceptibility genes and pancreatic cancer, neural system tumors and uveal melanoma across GenoMEL. Cancer Res 2006; 66(20):9818-9828.
45. Platz A, Hansson J, Mansson-Brahme E et al. Screening of germline mutations in the CDKN2A and CDKN2B genes in Swedish families with hereditary cutaneous melanoma. J Natl Cancer Inst 1997; 89(10):697-702.
46. Gruis NA, Sandkuijl LA, van der Velden PA et al. CDKN2 explains part of the clinical phenotype in Dutch familial atypical multiple-mole melanoma (FAMMM) syndrome families. Melanoma Res 1995; 5(3):169-177.
47. Goldstein AM, Stacey SN, Olafsson JH et al. CDKN2A Mutations and Melanoma Risk in the Icelandic Population. J Med Genet 2008.
48. Ciotti P, Struewing JP, Mantelli M et al. A single genetic origin for the G101W CDKN2A mutation in 20 melanoma-prone families. Am J Hum Genet 2000; 67(2):311-319.
49. Zuo L, Weger J, Yang Q et al. Germline mutations in the p16INK4a binding domain of CDK4 in familial melanoma. Nat Genet 1996; 12(1):97-99.
50. Soufir N, Avril MF, Chompret A et al. Prevalence of p16 and CDK4 germline mutations in 48 melanoma-prone families in France. The French Familial Melanoma Study Group. Hum Mol Genet 1998; 7(2):209-216.
51. Molven A, Grimstvedt MB, Steine SJ et al. A large Norwegian family with inherited malignant melanoma, multiple atypical nevi and CDK4 mutation. Genes Chromosomes Cancer 2005; 44(1):10-18.
52. Pjanova D, Molven A, Akslen LA et al. Identification of a CDK4 R24H mutation-positive melanoma family by analysis of early-onset melanoma patients in Latvia. Melanoma Res 2009; 19(2):119-122.
53. Gillanders E, Juo SH, Holland EA et al. Localization of a novel melanoma susceptibility locus to 1p22. Am J Hum Genet 2003; 73(2):301-313.
54. Walker GJ, Indsto JO, Sood R et al. Deletion mapping suggests that the 1p22 melanoma susceptibility gene is a tumor suppressor localized to a 9-Mb interval. Genes Chromosomes Cancer 2004; 41(1):56-64.
55. Jonsson G, Bendahl PO, Sandberg T et al. Mapping of a novel ocular and cutaneous malignant melanoma susceptibility locus to chromosome 9q21.32. J Natl Cancer Inst 2005; 97(18):1377-1382.
56. Begg CB, Orlow I, Hummer AJ et al. Lifetime risk of melanoma in CDKN2A mutation carriers in a population-based sample. J Natl Cancer Inst 2005; 97(20):1507-1515.
57. Goldstein AM, Fraser MC, Struewing JP et al. Increased risk of pancreatic cancer in melanoma-prone kindreds with p16INK4 mutations. N Engl J Med 1995; 333(15):970-974.
58. Borg A, Sandberg T, Nilsson K et al. High frequency of multiple melanomas and breast and pancreas carcinomas in CDKN2A mutation-positive melanoma families. J Natl Cancer Inst 2000; 92(15):1260-1266.
59. Goldstein AM. Familial melanoma, pancreatic cancer and germline CDKN2A mutations. Hum Mutat 2004; 23(6):630.
60. de Snoo FA, Bishop DT, Bergman W et al. Increased risk of cancer other than melanoma in CDKN2A founder mutation (p16-Leiden)-positive melanoma families. Clin Cancer Res 2008; 14(21):7151-7157.
61. Hille ET, van Duijn E, Gruis NA et al. Excess cancer mortality in six Dutch pedigrees with the familial atypical multiple mole-melanoma syndrome from 1830 to 1994. J Invest Dermatol 1998; 110(5):788-792.
62. Newton Bishop JA, Gruis NA. Genetics: what advice for patients who present with a family history of melanoma? Semin Oncol 2007; 34(6):452-459.

63. Botkin JR, Smith KR, Croyle RT et al. Genetic testing for a BRCA1 mutation: prophylactic surgery and screening behavior in women 2 years post testing. Am J Med Genet A 2003; 118(3):201-209.
64. Hadley DW, Jenkins JF, Dimond E et al. Colon cancer screening practices after genetic counseling and testing for hereditary nonpolyposis colorectal cancer. J Clin Oncol 2004; 22(1):39-44.
65. Leachman SA, Carucci J, Kohlmann W et al. Selection criteria for genetic assessment of patients with familial melanoma. J Am Acad Dermatol 2009; 61(4):677 e671-614.
66. Kefford RF, Newton Bishop JA, Bergman W et al. Counseling and DNA testing for individuals perceived to be genetically predisposed to melanoma: A consensus statement of the Melanoma Genetics Consortium. J Clin Oncol 1999; 17(10):3245-3251.
67. Bishop DT, Demenais F, Goldstein AM et al. Geographical variation in the penetrance of CDKN2A mutations for melanoma. J Natl Cancer Inst 2002; 94(12):894-903.
68. National Institutes of Health Consensus Development Conference Statement on Diagnosis and Treatment of Early Melanoma, 1992. Am J Dermatopathol 1993; 15(1):34-43; discussion 46-51.
69. Ferrini RL, Perlman M, Hill L. American College of Preventive Medicine practice policy statement: skin protection from ultraviolet light exposure. The American College of Preventive Medicine. Am J Prev Med 1998; 14(1):83-86.
70. Krien PM, Moyal D. Sunscreens with broad-spectrum absorption decrease the trans to cis photoisomerization of urocanic acid in the human stratum corneum after multiple UV light exposures. Photochem Photobiol 1994; 60(3):280-287.
71. Kenet RO, Kang S, Kenet BJ et al. Clinical diagnosis of pigmented lesions using digital epiluminescence microscopy. Grading protocol and atlas. Arch Dermatol 1993; 129(2):157-174.
72. Menzies SW, Ingvar C, McCarthy WH. A sensitivity and specificity analysis of the surface microscopy features of invasive melanoma. Melanoma Res 1996; 6(1):55-62.
73. Kelly JW, Yeatman JM, Regalia C et al. A high incidence of melanoma found in patients with multiple dysplastic naevi by photographic surveillance. Med J Aust 1997; 167(4):191-194.
74. McGovern TW, Litaker MS. Clinical predictors of malignant pigmented lesions. A comparison of the Glasgow seven-point checklist and the American Cancer Society's ABCDs of pigmented lesions. J Dermatol Surg Oncol 1992; 18(1):22-26.
75. Shaw HM, McCarthy WH. Small-diameter malignant melanoma: a common diagnosis in New South Wales, Australia. J Am Acad Dermatol 1992; 27(5 Pt 1):679-682.
76. Rhodes AR. Intervention strategy to prevent lethal cutaneous melanoma: use of dermatologic photography to aid surveillance of high-risk persons. J Am Acad Dermatol 1998; 39(2 Pt 1):262-267.
77. Masri GD, Clark WH Jr, Guerry Dt et al. Screening and surveillance of patients at high risk for malignant melanoma result in detection of earlier disease. J Am Acad Dermatol 1990; 22(6 Pt 1):1042-1048.
78. MacKie RM, McHenry P, Hole D. Accelerated detection with prospective surveillance for cutaneous malignant melanoma in high-risk groups. Lancet 1993; 341(8861):1618-1620.
79. Tucker MA, Fraser MC, Goldstein AM et al. A natural history of melanomas and dysplastic nevi: an atlas of lesions in melanoma-prone families. Cancer 2002; 94(12):3192-3209.
80. Hansson J, Bergenmar M, Hofer PA et al. Monitoring of kindreds with hereditary predisposition for cutaneous melanoma and dysplastic nevus syndrome: results of a Swedish preventive program. J Clin Oncol 2007; 25(19):2819-2824.
81. Parker JF, Florell SR, Alexander A et al. Pancreatic carcinoma surveillance in patients with familial melanoma. Arch Dermatol 2003; 139(8):1019-1025.
82. Brand RE, Lerch MM, Rubinstein WS et al. Advances in counselling and surveillance of patients at risk for pancreatic cancer. Gut 2007; 56(10):1460-1469.
83. Lynch HT, Fusaro RM, Lynch JF et al. Pancreatic cancer and the FAMMM syndrome. Fam Cancer 2008; 7(1):103-112.

Chapter 14

Primary Immunodeficiency Syndromes

Mary A. Slatter and Andrew R. Gennery*

Abstract

Several DNA repair pathways have evolved to recognise and repair DNA damaged by exogenous and endogenous agents, in order to maintain genomic integrity. Defects in these pathways can lead to replication errors, loss or rearrangement of genomic material, mutation or cancer and eventual death. The creation of many diverse lymphocyte receptors to identify potential pathogens has evolved by breaking and randomly resorting the gene segments coding for antigen receptors. Subsequent steps utilise the ubiquitous repair proteins. Individuals with defective repair pathways are increasingly recognised with immunodeficiency, many of whom exhibit radiosensitivity. Our understanding of the role of repair proteins in the development of adaptive immunity by VDJ recombination, antibody isotype class switching and affinity maturation by somatic hyper-mutation has made significant progress over the last few years, partly by the identification of new genes involved in human disease. We describe the mechanisms involved in the development of adaptive immunity relating to DNA repair and describe the clinical consequences and treatment developments of primary immunodeficiency resulting from such defects.

Introduction

DNA is constantly exposed to exogenous and endogenous damaging agents. Consequently, a number of DNA repair pathways have evolved to recognise and repair this damage in order to maintain genomic integrity. Defects in these pathways can lead to replication errors, loss or rearrangement of genomic material, mutation or cancer and eventual death. During adaptive immune development, three specific mechanisms, critical for the development of T- and B-lymphocyte receptor rearrangement, immunoglobulin class switch recombination (CSR) and somatic hypermutation (SHM) require repair to DNA damage that has been introduced during these processes. This chapter details the molecular mechanisms, clinical presentation and treatment of human primary immunodeficiency disorders associated with defects in normal DNA damage recognition and repair.

Role of DNA Repair Proteins in Adaptive Immunity

Generation of Lymphocyte Antigen Receptors

An effective immune response necessitates recognition of a wide array of foreign antigens, requiring the generation of $\approx 10^{18}$ genetically diverse cells, each with a unique receptor that recognises a unique antigen/MHC combination. In higher organisms, these genetically diverse receptors are created by breaking, randomly resorting and joining DNA sequences, coding for the antigen capture region of the receptor, by adapting the ubiquitous DNA repair mechanisms that maintain genome stability.

*Corresponding Author: Andrew R. Gennery—Newcastle General Hospital, Westgate Road, Newcastle upon Tyne, NE4 6BE, England. Email: a.r.gennery@ncl.ac.uk

Diseases of DNA Repair, edited by Shamim I. Ahmad. ©2010 Landes Bioscience and Springer Science+Business Media.

VDJ recombination is a site-specific event occurring at 6 loci: the T-lymphocyte (cell) receptor (TCR) α, β, γ and δ chain loci and the B-lymphocyte (cell) receptor (BCR) immunoglobulin heavy (IgH) and κ or λ light chain (IgL) loci. Recombination occurs between component variable (V), junction (J) and for TCRβ, TCRδ and BCR IgH loci, diversity (D) gene segments, with fused VJ or VDJ coding sequence which is subsequently joined to a constant (C) region segment through RNA splicing. Two recombination activating gene proteins, (RAG1/2) initiate this process by introducing site-specific DNA-double strand breaks (dsbs) at recombination signal sequence (RSS) sites either side of the segments to be rearranged, during the G1 phase of the cell cycle.

After the introduction of DNA-dsb at the coding sequence/RSS junction, two types of DNA ends arise; coding sequence ends that reconstitute the Ig and TCR genes are generated as hairpin intermediates, whereas noncoding signal ends containing the motifs targeting site-specific cleavage are generated as blunt double stranded (ds)-DNA ends. DNA-dsbs are repaired using the ubiquitous nonhomologous end joining (NHEJ) repair pathway. The KU70/KU80 heterodimer binds DNA ends, present at RAG1/2-generated coding ends and recruits DNA-PKcs that phosphorylates and activates Artemis endonuclease activity to process the coding sequence hairpin intermediates. Following cleavage, both coding and signal ends are directly ligated by the XRCC4/DNA ligase 4/Cernunnos-XRCC4-like factor (XLF) complex. Further diversity between coding joins is created by asymmetric nicking of the hairpin bends and by deletions, mutations and addition of nontemplated nucleotides to processed coding ends by terminal deoxynucleotidyl transferase (TdT) and DNA polymerase μ[1] (Fig. 1). Rejoining of signal ends does not require DNA-PKcs or Artemis.

Although artemis-mediated cleavage of coding joint hairpins does not require ATM or MRN,[2] ATM does have important roles in stabilising DNA-ends in the RAG postsynaptic cleavage complex, facilitating NHEJ repair of VDJ recombination associated breaks.[3-5] ATM, NBS1, γH2AX and 53BPI are associated with RAG-induced DNA-dsb in developing lymphocytes and localise to DNS-dsb and the chromatin region surrounding the recombining loci.[6-10] ATM may contribute to the efficiency of VDJ recombination by activating cell cycle checkpoint proteins, to enable monitoring of recombination intermediates. In the absence of ATM, lymphocytes with RAG-induced DNA-dsb may enter the S phase of cell cycle, leading to a reduction in productive VDJ recombination and an increased number of abnormal translocations involving Ig and TCR loci (chromosome 7/14 translocations).[11] VDJ recombination is not completely abolished if any of the 7 NHEJ proteins are impaired, as an alternative end-joining pathway exists in which the frequent use of microhomology and excessive deletions are characteristic. RAG proteins play an essential role in the joining phase of V(D)J recombination, but allow a small degree of alternative NHEJ activity.[12]

Receptor recombination occurs in the thymus and bone marrow respectively for T- and B-lymphocytes. Early lymphocyte progenitors move to the site of development and undergo successive stages of lineage commitment, intense cellular proliferation and selection, generating a functional lymphocyte receptor repertoire. Between critical developmental stages including VDJ re-arrangement of the T-lymphocyte β and α chain and B-lymphocyte IgH and IgL chain, the lymphocyte precursors undergo intense proliferation. During this phase, cells experience the normal replicative stress of proliferating cells, thus accumulating abnormal replication intermediates, which are normally resolved by Bloom syndrome helicase and associated proteins.

Immunoglobulin Class Switch Recombination

Maturation of the antibody repertoire is a necessary mechanism for optimized antibody responses with high antigen affinity. Antibody maturation occurs mostly in the germinal centres of the secondary lymphoid organs, following antigen and T-lymphocyte-driven activation; B-lymphocytes undergo intense cell proliferation, dividing every 6-8 hours and accumulating abnormal replication intermediates acquired during the normal replication of proliferating cells and resolved by Bloom syndrome helicase and associated proteins. CSR is a somatic DNA arrangement process, which leads to a switch in the IgH C region of the BCR, expressed from the

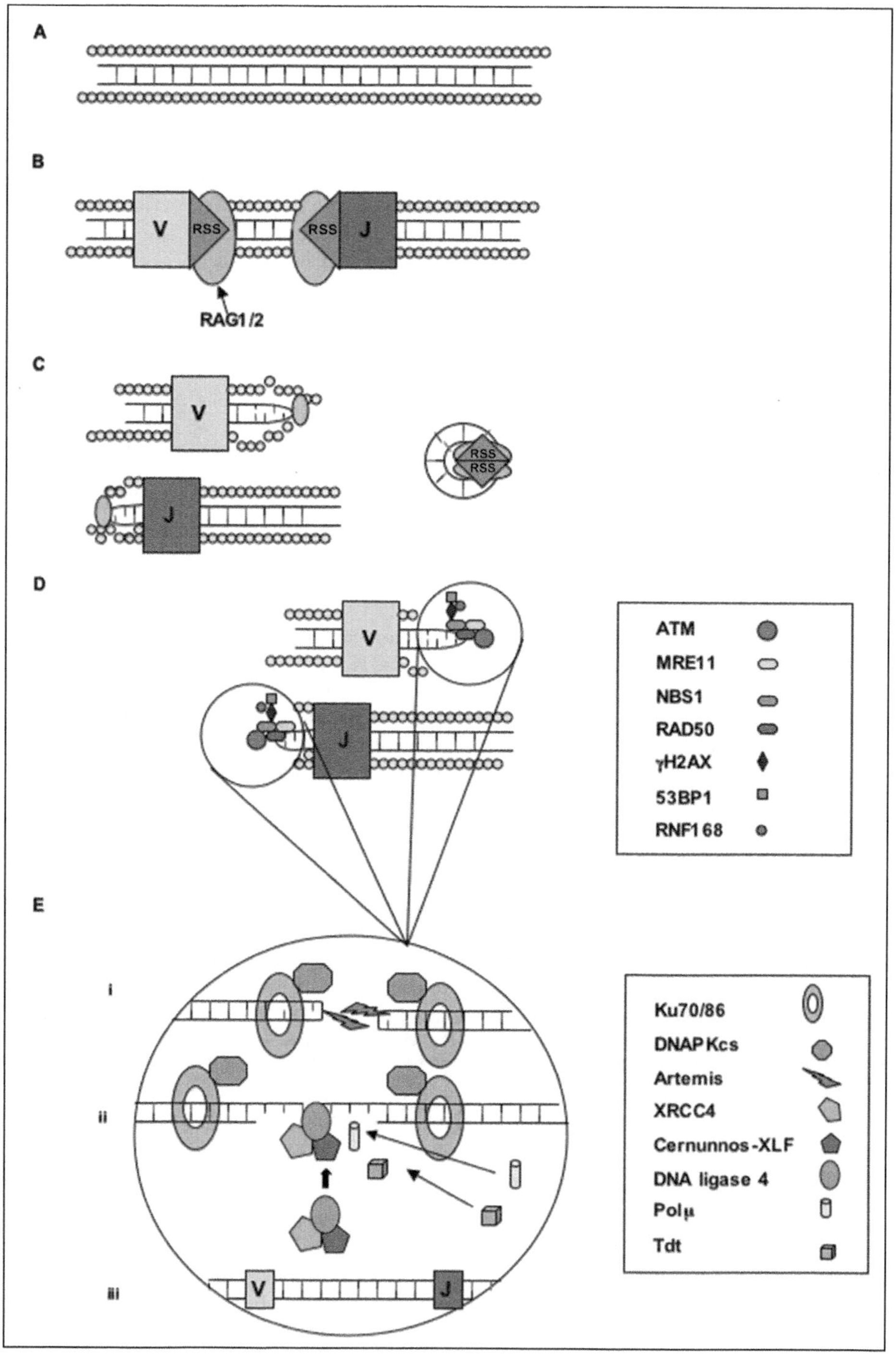

Figure 1. VDJ recombination. Please see legend on following page.

Figure 1, viewed on previous page. VDJ recombination. A) DNA is uncoiled at transcription factories within the cell, where the associated recombination and repair proteins colocalise. B) The lymphoid specific recombinase activating gene 1 and 2 (RAG1/2) proteins recognise and bind the recombination signal sequences (RSS) that flank the VDJ gene segments and introduce site-specific DNA-dsb. C) The phosphorylated blunt signal ends and the covalently sealed hairpin intermediate of the coding end are held together by the RAG complex. D) The MRN complex binds the broken DNA ends and activates ATM which initiates cell cycle arrest and attraction of the repair proteins. H2AX, 53BP1 and RNF168, with other proteins stabilise the damaged chromatin. E) (i) Ku70/Ku80 heterodimer binds the coding ends and recruits DNA-PKcs and Artemis, which is required to open the hairpin intermediates. The covalently sealed hairpin intermediate is randomly nicked by the DNA-PKcs/Artemis complex, which generates a single stranded break with 3′ or 5′ overhangs. (ii) XRCC4, DNA ligase 4 and cernunnos-XLF (C-XLF) co-associate and are recruited to the ends. The signal ends are directly ligated by the XRCC4/DNA-LIG4/C-XLF complex. The opened hairpin intermediate is modified by polymerases, exonucleases and the lymphoid-specific terminal deoxynucleotidyl transferase (TdT), before (iii) being repaired and ligated by the XRCC4/DNA-LIG4/C-XLF complex. (Reproduced with permission of the Paediatric HSCT unit, Newcastle General Hospital.)

region encoded by Cμ to a downstream C region such as that encoded by Cα, Cγ or Cε. This results in the production of antibodies of different isotypes (IgG, IgA and IgE) with the same V(D)J specificity and therefore the same antigen affinity.[13] Activation-induced cytidine deaminase (AID), a B-lymphocyte-specific enzyme critical for CSR, induces DNA-dsb, to initiate CSR.[14,15] AID selectively deaminates cytosine (C) residues to uracil (U) in the switch (S) and V regions.[16] Uracil DNA-glycosylase (UNG) removes U, producing an abasic site, which is cleaved by one of the base excision repair enzymes to create a DNA single strand break (DNA-ssb).[17,18] The mismatch repair (MMR) proteins MSH2-MSH6 recognise U at U:G mismatched bases and create a further DNA-ssb.[19] If U is on the complimentary strand to a previous DNA-ssb, a DNA-dsb results, enabling CSR to occur.[20] ATM is associated with MMR factors, including MSH2, MSH6, MLH1 and PMS2.[21] PMS2, possibly coupled with MHL1 as a heterodimer, may convert AID/UNG—induced DNA-ssb into the DNA-dsb required for CSR by stabilizing the recombination intermediate.[22,23] A significant role for the Bloom syndrome helicase is unlikely during CSR,[24] although it does interact with MLH1 and MSH6 and so could conceivably be involved in the process.[22,25,26] CSR DNA-dsb repair is achieved mainly through NHEJ during the G1 phase of the cell-cycle,[27] although an alternative end-joining mechanism is used when there is impairment of NHEJ.[28] ATM is required for efficient CSR, although exact function remains unclear. A number of factors involved in CSR are phosphorylated by ATM, including NBS1, Mre11, γH2AX and 53BP1, foci of which can be detected at the DNA-dsb switch region during CSR.[29-31] One role of ATM may be to recruit or activate these factors and organize the damaged DNA ends for subsequent repair steps, or arrest cell cycle progression until the repair is complete. ATM may have a more direct role in the end-processing step through phosphorylation of a nuclease that participates in NHEJ, i.e., Mre11 or Artemis. Artemis is downstream in the ATM signaling pathway for repair of a subset of radiation-induced DNA-dsb, but dispensable for ATM-dependent cell-cycle checkpoint arrest[1] and appears dispensable for processing of DNA-dsb for efficient CSR,[33] although may be required for repair of the CSR-related chromosomal breaks at the Ig locus.[34] The altered pattern of CSR junctions in Artemis-deficient patients also suggests that Artemis is required in the predominant NHEJ pathway during CSR.[35] The functional role of Artemis in CSR probably relies on its endonuclease activity on 3′ or 5′ overhangs, rather than hairpin structures. Deficiency in MRE11, NBS1, KU70/80 DNA-PKcs and DNA ligase 4 appears to alter the balance between the predominantly used NHEJ and alternative end-joining mechanism in CSR DNA-dsb repair, suggesting that these proteins are also involved directly or indirectly in CSR.[36]

Little is known about the proteins or mechanism involved in the alternative end-joining mechanism. Poly (ADP-ribose) polymerase 1 (PARP-1) is involved in many cellular responses including base excision repair (BER) and possibly HR.[37] It may be involved in the alternative end-joining pathway, with XRCC1 and DNA ligase III.[38-40] Ligases I and III are required for microhomology-mediated end joining but it is unknown which is involved in alternative end-joining during CSR.[41]

The Bloom syndrome helicase may also play a role in CSR as it has been shown to associate with MHL1.[26] MSH5 is also involved in CSR and may have a specific role in facilitating CSR between Sμ and Sα.[42]

Somatic Hypermutation

CSR and SHM occur together in germinal centres with BCR/CD40 activation, although neither is a prerequisite for the other, because IgM may be mutated in the absence of any such feature in IgG or IgA isotypes.[43] SHM introduces random mutations into the BCR V region resulting in minor conformational changes, enabling positive selection of B-lymphocytes, carrying a BCR with high antigen affinity.[44] SHM is initiated by AID, by RNA editing of V region C residues to U.[45] Repair of DNA-dsb is not an intermediate step during SHM. Instead, the MMR proteins MSH2-MSH6 are required to recognise single strand AID-induced U/G residues and to recruit the exonuclease EXO1 and DNA polymerase η, which causes G:C to T:A transversions.[46] Whilst NHEJ is not employed in SHM, the MRN complex is involved in DNA cleavage at AID-induced abasic sites during SHM[47] and NBS1 has a role in regulating strand-biased repair.[35] Bloom syndrome helicase appears not to be involved in SHM.[48]

Genetic Defects Associated with Primary Immunodeficiency Syndromes

An increasing number of genetic defects in the DNA-dsb and -MMR pathways have been identified in humans, providing insight into the clinical phenotype of genomic instability on the immune system. Immunodeficiency is increasingly recognised as a feature of these syndromes (Table 1).

Genetic Defects Critical for Lymphocyte Development

Recombination Activating Gene 1 and 2

The recombination activating genes (RAG) 1 and 2 are critical for initiating DNA-dsb prior to VDJ recombination of TCR and BCR.[49] Nonsense mutations in *RAG1/2* give rise to a T-B-NK+ severe combined immunodeficiency (SCID) phenotype with absent immunoglobulins. Infants classically present with viral or pneumocystis pneumonitis, persistent viral diarrhoea and growth failure, as with other forms of SCID. Missense mutations give rise to 'leaky' SCID, known as Omenn syndrome,[50] characterised by lymphadenopathy, hepatosplenomegaly, erythroderma, alopecia, agammaglobulinaemia apart from a raised IgE, T-lymphocytosis and absent B-lymphocytes with accompanying respiratory and gastrointestinal inflammation or infection and failure to thrive.[51] T-lymphocytes are activated and show a restricted VB repertoire.[52] Biopsies of affected skin demonstrate a histopathological pattern consistent with graft versus host disease, although T-lymphocytes are autologous.

Two further phenotypes of RAG deficiency have been described:-

i. Normal immunoglobulin levels, specific antibody responses to some infectious agents or vaccine antigens, production of autoantibodies, a predominance of γδ T-lymphocytes and development of Epstein-Barr virus (EBV)-associated lymphoproliferation.[53,54]
ii. T- and B-lymphocytopenia, hypogammaglobulinaemia, recurrent viral infection, EBV-associated lymphoma and extensive granulomatous lesions, associated with compound heterozygous mutations in RAG 1 and 2.[55]

Artemis Deficiency

Artemis is critical for VDJ recombination; null mutations in *Artemis* give rise to a T-B-NK+ SCID phenotype,[56] originally described in Athabascan-speaking native Americans.[57] Bone marrow cells and fibroblasts, from these patients, exhibited increased cellular sensitivity to ionising radiation (IR) (radiosensitive, RS-SCID),[58] but otherwise the clinical presentation is identical to RAG-deficient SCID.

Table 1. Proteins associated with human primary immunodeficiency and DNA-break sensing and repair defects

Gene, Chromosome Protein	Disease	Effect on Lymphocyte Development	Microcephaly	Lymphoid Tumours	Immunodeficiency
Lymphocyte Development					
DCLRE1C 10p Artemis	RS SCID CID	CJ formation CSR	No	EBV-associated B cell lymphoma Chromosome 7:14 Translocations	Agammaglobulinaemia Hypogammaglobulinaemia lymphocytopenia
PRKDC, 8q11 DNA protein kinase	RS SCID	CJ formation ?CSR	Not described	Not described	Agammaglobulinaemia lymphocytopenia
DNA ligase 4, 13q33-34 DNA Ligase 4	IR sensitivity Susceptibility to lymphoid malignancy RS SCID CID	CJ fidelity SJ fidelity CSR	Most patients	EBV-associated lymphoma T cell ALL	Hypogammaglobulinaemia (IgA, IgG), SPAD Lymphocytopenia
NHEJ1, 2q35 Cernonnus-XLF	RS SCID CID	CJ fidelity SJ fidelity ? CSR	Some patients	Not reported	Hypogammaglobulinaemia (IgA, IgG) Hyper IgM Lymphopenia
DNA-dsb Damage Sensing					
ATM, 11q22.3 Ataxia telangiectasia mutated	Ataxia Telangiectasia	SJ fidelity CSR	No	Common Chromosome 7:14 Translocations	Hypogammaglobulinaemia (IgA, IgG), SPAD Lymphocytopenia
hMRE11, 11q21 MRE11	Ataxia Telangiectasia Like disorder	CSR	Some patients	Not reported	SPAD Lymphopenia not reported

continued on next page

Table 1. Continued

Gene, Chromosome Protein	Disease	Effect on Lymphocyte Development	Microcephaly	Lymphoid Tumours	Immunodeficiency
Rad50, 5q31 RAD50	Nijmegen Breakage Like syndrome	SJ fidelity	Yes	Not reported	Nil reported
RNF168, 3q29 RNF168	RIDDLE syndrome	CSR	No	Not reported	Low IgG
CSR and SHM					
AICDA, 12p13 activation-induced cytidine deaminase	Hyper IgM Type 2	CSR SHM	No	No	High IgM Low/absent IgA, IgG
UNG, 12q23-q24.1 uracil-DNA glycosylase	Hyper IgM Type 5	CSR SHM	No	No	High IgM Low/absent IgA, IgG
PMS2, 7p22 PMS2		CSR ?SHM	No	Leukaemias, lymphomas, cerebral tumours colorectal tumours	High IgM Low/absent IgA, IgG
MSH5, 6p22.1-p21.3 MSH5	CVID	CSR	No	Lymphoma in CVID	Low IgM, IgA, IgG, IgA deficiency
Other DNA Repair Disorders and PID					
DNA ligase I, 19q13.2-q13.3 DNA ligase I	DNA ligase I deficiency	?CSR	Elf-like features	Lymphoma	Normal IgM Low IgA, IgG lymphocytopenia
BLM, 15q21.1	Bloom syndrome	?CSR	Proportionate growth deficiency	Lymphoma, carcinoma Increased sister chromatid exchange	Hypogammaglobulinaemia lymphocytopenia

RS SCID, radiosensitive severe combined immunodeficiency; CID, combined immunodeficiency; CSR, class switch recombination; CJ, coding join; SJ, signal join; EBV, Epstein Barr virus; ALL, acute lymphoblastic leukaemia; SPAD, specific pneumococcal polysaccharide antibody deficiency; CVID, common variable immunodeficiency; PID, primary immunodeficiency

A further two clinical phenotypes are described due to hypomorphic mutations in *Artemis*.

i. Infants may present with Omenn syndrome,[59] analogous to the clinical presentation due to hypomorphic RAG mutations;
ii. Patients may present with a progressive combined immunodeficiency (CID) in later infancy, characterised by recurrent sinopulmonary or gastrointestinal infection, T- and B-lymphopenia, hypogammaglobulinaemia and autoimmune cytopenias.[60,61]

Some patients show susceptibility to EBV-associated B-lymphomas.[60] Chromosome 7:14 inversions and translocations have also been described in these patients. No specific dysmorphic features or microcephaly have been noted. Artemis is not essential for viability as many patients with RS-SCID have complete loss of functional alleles.

DNA-PK Deficiency

One Turkish patient, of consanguineous parents, has been described with a homozygous 3-nucleotide deletion and homozygous missense mutation in *DNA-PKcs*.[62] At 5 months of age she presented with classical SCID symptoms of recurrent oral candidiasis and lower respiratory tract infections from the age of 3 months leading to hypoxia. She also had a large oral aphthous ulcer, but no microcephaly or developmental delay. She had a T-B-NK+ SCID lympho-phenotype. The block in B-lymphocyte precursor differentiation was similar to that seen in *Artemis*- and *RAG*-deficient SCID, consistent with a defect in VDJ recombination. Fibroblasts were sensitive to gamma-irradiation, with a DNA-dsb repair defect comparable with that seen in *Artemis*-deficient cells. Coding joints showed long stretches of P nucleotides and an end-joining assay demonstrated increased microhomology use, similar to that seen in *Artemis*-deficient cells. She underwent unconditioned stem cell transplantation from an HLA-identical cousin leading to full reconstitution of all lymphocyte subsets.

DNA Ligase 4 Deficiency

The first patient to be described with DNA ligase 4 defect was clinically and developmentally normal until T-lymphocyte acute lymphoblastic leukaemia developed. Disproportionately severe cytopenia followed treatment with the MRC-UKALL X protocol and standard chemotherapy consolidation therapy was omitted.[63] Cranial irradiation prophylaxis led to an extreme reaction to radiotherapy, including marked and prolonged cytopenia, severe desquamation and death from radiation-induced encephalopathy.[63] Subsequently, a number of patients have been described; six patients had microcephaly, developmental delay, growth failure, lymphopenia, hypogammaglobulinaemia and recurrent infection.[64,65] Bone marrow hypoplasia was a feature in some of these patients. A further 3 patients have been reported with T-B-NK+ RS-SCID, all with microcephaly and growth delay.[66,67] Four patients with microcephaly and a CID phenotype have been reported, of whom two developed an EBV-associated diffuse large cell non-Hodgkin's lymphoma and one developed T-lymphocyte acute lymphoblastic leukaemia.[68-70] (Fig. 2). One patient presented with features consistent with Omenn syndrome.[71] The phenotypic spectrum ranges from normal individuals with extreme sensitivity to ionising radiation, CID with microcephaly and developmental delay (analogous to patients with Nijmegen breakage syndrome), to radiosensitive T-B-NK+ SCID or Omenn syndrome. Other clinical features overlap with those seen in ataxia telangiectasia including photosensitivity and psoriatic-like lesions.[64]

In VDJ recombination assays moderate impairment of VDJ recombination has been observed in LIG-4 deficient fibroblasts: an almost normal frequency of coding and signal joint formation is observed, but fidelity of both coding and signal joint formation is impaired.[64,67] These in vitro findings are less severe than the clinical immunodeficiency, suggesting that DNA ligase 4 is required during lymphocyte development or homeostasis at stages beyond VDJ recombination, possibly to repair DNA damage that may occur during lymphocyte proliferation. Similar observations have been identified in patients with hypomorphic *Artemis* mutations.[61] Patients with LIG 4 defects also show altered resolution of CSR junctions, with greater use of microhomology at $S\mu$-$S\alpha$ junctions.[36]

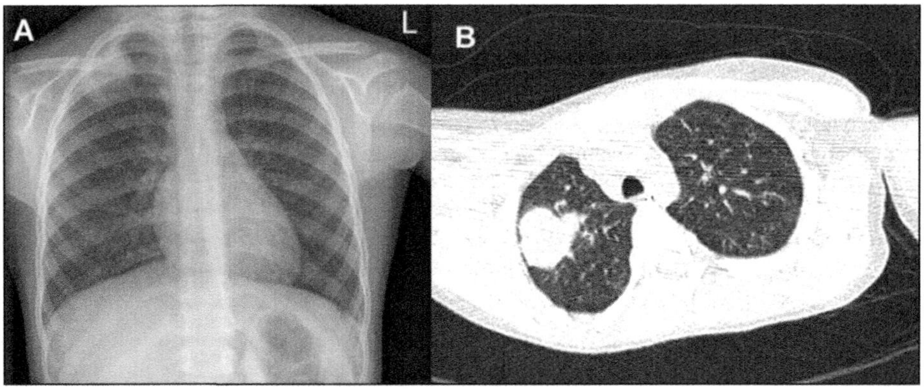

Figure 2. Right apical lung B-lymphocyte lineage lymphoma on (A) chest radiograph and (B) computerised tomography in a patient with DNA ligase 4 deficiency. (Reproduced with permission of the Paediatric HSCT unit, Newcastle General Hospital.)

Cernunnos-XLF Deficiency

Deficiency of Cernunnos-XLF, critical for lymphocyte receptor development, has been described in eight patients;[72,73] Cernunnos-XLF interacts with the XRCC4/DNA ligase 4 complex. The first 2 patients presented with T- and B-lymphocytopenia, with a normal number of NK cells.[74] Subsequently, five patients with CID have been described;[72] all had similar lymphocyte phenotype to the original kindred, but had low IgA and IgG. Two had raised IgM, suggesting a role for Cernunnos-XLF in CSR. Some patients were microcephalic with 'bird-like' dysmorphism, two exhibited autoimmune cytopenia and all suffered from recurrent bacterial and other opportunistic infections. Two had several chromosomal alterations, although chromosome 7:14 translocations were not found. Lymphomas have not been described to date. However, a further patient has been described, who had similar morphological features of microcephaly, small stature and 'bird-like' facies. He suffered recurrent respiratory infections and demonstrated normal IgM, but low IgA and IgG levels and absent response to vaccine protein antigens. His lymphocyte phenotype was characteristic[75] and he developed pancytopenia with tri-lineage marrow dysplasia, enteropathy and underwent successful haematopoietic stem cell transplantation at 10 years of age. The treatment became complicated by EBV-associated posttransplant lymphoproliferative disease.

Other clinical abnormalities described include bone malformations (low implantation of the thumb, hypoplasia of the middle phalanx of the fifth finger), nephroptosis and one patient demonstrated developmental delay, features that overlap with those described in DNA ligase 4 syndrome.

In vitro coding and signal joint formation are reduced in patients compared to controls, with an increase in nucleotide loss at coding joins. The fidelity of signal joins was severely impaired in patient cells, with possible use of microhomology during joining,[72] features previously described in LIG4 patients.[63,66] VDJ deficiency in these patients is less severe than in the *Artemis* deficient radiosensitive SCID and probably accounts for the low numbers of T- and B-lymphocytes found in the patients. As in patients with LIG 4 syndrome, the in vitro findings are less severe than the clinical immunodeficiency, suggesting that Cernunnos-XLF may also be required during lymphocyte development at stages beyond VDJ recombination. There is now some evidence that Cernunnos-XLF is also important for cell replication-induced DNA-dsb repair.[76] Thus, lymphocytes that have successfully rearranged one chain of the antigen receptor despite cernunnos-XLF deficiency may still be eliminated during the intense cell replication that occurs during lymphocyte development.

Primary Immunodeficiency Associated with Disorders in DNA-dsb Damage Sensing Proteins

Ataxia Telangiectasia

Ataxia telangiectasia (A-T) is a rare systemic autosomal recessive disorder caused by mutations in *ATM*, manifested by progressive cerebellar ataxia, oculocutaneous telangiectasia, gonadal sterility, general growth retardation and a high incidence of mainly lymphoid tumours.[77] Patients normally present with cerebellar ataxia before telangiectasia are manifest. Recurrent sino-pulmonary infection can be a presenting feature and may be associated with a raised IgM and low or absent IgG.[78] Sinopulmonary infection combined with recurrent aspiration, can lead to chronic lung disease.[79] The incidence of infections is variable and correlates with the presence of two null mutations in *ATM*.[80] Immunological responses to bacterial antigens are generally reduced, particularly to polysaccharide antigen.[81] Lymphocytic interstitial pneumonitis has been described.[82] Hepatic veno-occlusive disease has rarely been noted.[83] Median survival of patients is 22 years.[84] Thymic output is low; the peripheral T-lymphocyte population is biased towards terminally differentiated effector T-lymphocytes, as reflected by a low ratio of naïve to memory T-lymphocytes and also a skewed TCRBV repertoire of peripheral lymphocytes marked by oligoclonal expansions and a restricted TCRBV chain repertoire.[85] Chromosomal inversions and translocations, particularly chromosome 7:14 translocations are seen in A-T. ATM deficiency does not result in a profound blockage in lymphocyte development, but fidelity rather than completion of VDJ recombination may be affected in the absence of ATM. B-lymphocyte repertoire is restricted and skewed by diffused oligoclonal expansions with normal VDJ joints. B-lymphocytes from AT patients have an intrinsic defect in switching from IgM to other Ig classes, due to a defect in CSR to the most distant immunoglobulin loci, reflecting the requirement of ATM for efficient recombination between immunoglobulin switch regions.[86]

Nijmegen Breakage Syndrome

Nijmegen breakage syndrome (NBS) has characteristic facial appearances with receding forehead, receding mandible and prominent midface ('bird-like' facies).[87] Additional features include epicanthal folds, large ears and sparse hair with microcephaly and mild mental retardation. Patients are susceptible to B-lymphocyte lineage lymphomas and are prone to sino-pulmonary infection. Cellular immunity is consistently depressed in NBS patients with reduced T-lymphocyte proliferation. Lymphopenia is common and whilst CD8+ lymphocytes are generally of normal number, there are reduced proportions of CD3 and CD4 T-lymphocytes.[88] Agammaglobulinaemia is reported in about a third of NBS patients whilst in others the humoral immune deficiency is more variable. Deficiencies of either IgA or IgG4 or in combination are common.[89] About 10% of patients have normal immunoglobulins. The immunodeficiency may result from reduced fidelity of VDJ recombination, as NBS1 is involved with ATM in inducing cell cycle arrest during this process.[90] Frequency of VDJ recombination in NBS patients is normal however with normal IgH rearrangement. The deficiency of serum IgG and IgA with normal or raised IgM is likely due to impaired CSR.[91,92] Photosensitivity has not been reported, but porokeratosis and noncaseating granulomas have both been reported albeit rarely.[93,94]

Chromosomal inversions and translocations, particularly chromosome 7:14 translocations are characteristic of NBS (Fig. 3). NBS and Fanconi anaemia (FA) share certain overlapping features.[96,97] FA is an autosomal recessive chromosomal instability disorder characterised by developmental defects, progressive bone marrow failure and cancer susceptibility; patients develop primarily marrow failure and haematological malignancies but also squamous cell carcinomas. Thirteen genetic groups have been defined.[95] FA as well as NBS cells are sensitive to DNA cross-linking agents such as mitomycin C (MMC) and diepoxybutane. Patients with FANCD2 subtype are sensitive to both MMC and IR. NBS1 and FANCD2 proteins interact and NBS cells show reduced monoubiquitynation of FANCD2. FANCD2 therefore functions at the intersection of 2 signalling pathways, one involving IR activation by ATM, the other

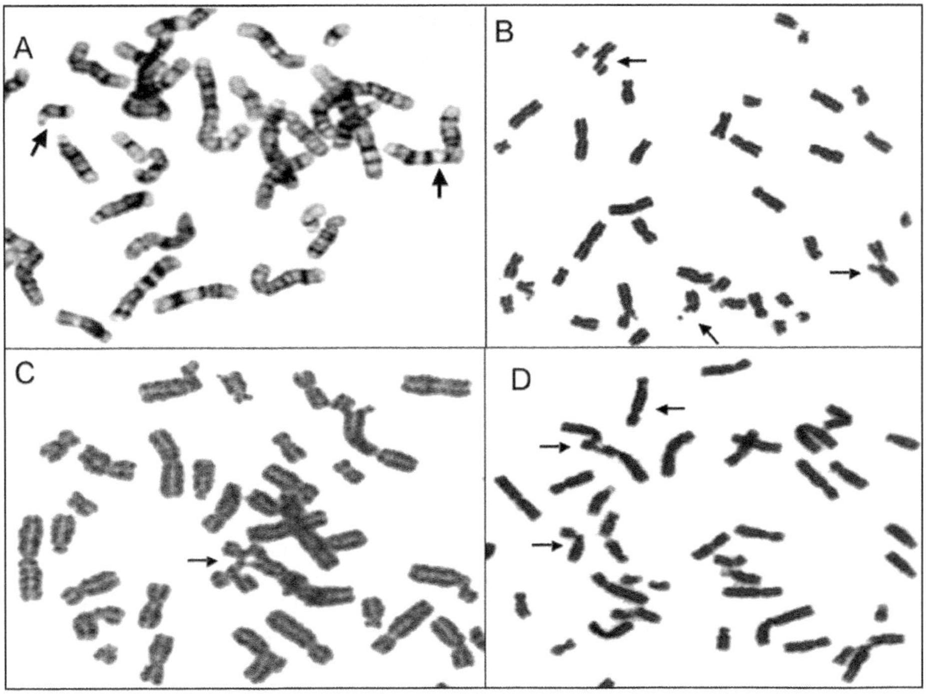

Figure 3. Karyotype from a patient with Nijmegen breakage syndrome showing (A) chromosome t(7; 14) re-arrangement (arrow), (B) chromosomal breakage following exposure to 50 centiGray ionizing radiation, (C) multiradial formation (arrow) after 72-h culture following exposure to mitomycin C at 0.32 Ag/ml for 60 min, (D) chromosome breakage (arrow) following lymphocyte culture with DEB for 72 h. (Reproduced with permission of Paediatric HSCT unit, Newcastle General Hospital.)

involving MMC activation by the FA complex.[96] The enzyme/substrate interaction of ATM and FANCD2 accounts at least in part for the common clinical and cellular phenotypes of A-T and some FA patients. Patients with particular NBS1 mutations have certain features similar to FA,[96,97] although immunodeficiency is more pronounced in NBS, presumably because of the role of NBS1 in VDJ recombination and CSR.

Ataxia telangiectasia-Like Disorder

Ataxia telangiectasia-like disorder (ATLD) is extremely rare with only 17 patients reported worldwide with mutations in *hMRE11*.[98-101] Clinical features are similar to those in patients with A-T with progressive cerebellar ataxia that is of later onset and slower progression of the disease than of those with A-T.[102] Telangiectasia is absent in ATLD and immunoglobulin levels are normal although deficiency in antigen specific antibodies has been reported, particularly to pneumococcal polysaccharide antigen.[103] Some patients are microcephalic, although intelligence is normal. The gene defect in *hMRE11* encodes for a protein that associates in the MRN complex and patients have features overlapping with both A-T and NBS. The absence of reported associated recurrent pulmonary infection due to immunodeficiency may reflect clinical variability as seen in A-T, or the type of mutation present, the majority of which are homozygous missense mutations. Given the role of *hMRE11* in CSR, it would appear logical that a degree of hypogammaglobulinaemia may be seen in some patients. Defective CSR has been reported with reduced switching from Sμ-Sα and an increased usage of microhomology at switch junctions.[103]

RAD50 Deficiency

One patient has been described with features suggestive of NBS, in whom compound heterozygous mutations in RAD50 (one of the components of the MRN complex) were found. One mutation created a premature stop codon, the other led to an abnormally large polypeptide.[104] The patient had intrauterine growth retardation with microcephaly. Growth remained poor and she developed 'bird-like' facies. Delay occurred in speech development and subsequent follow-up has demonstrated moderate psychomotor retardation, with mild spasticity and a nonprogressive ataxic gait. She developed multiple cutaneous pigmented naevi and hypopigmented areas. She did not suffer from excessive infections and immunoglobulin levels were normal. Lymphocyte numbers and proliferations to mitogens were normal. Chromosomal instability with 7:14 translocations were noted and there was radiosensitivity.[105] By 23 years of age she had not developed evidence of myelodysplasia or lymphoid malignancy. No report is available on the use of microhomology at V-J junctions. One report implicates RAD50 in fidelity of end-joining of VDJ signal join substrates.[106] The phenotype of RAD50 deficiency resembles more closely that of NBS than A-T. Whilst immunodeficiency has not been reported in this patient, given the function of RAD50 in the MRN complex in TCR and BCR formation and in CSR, it is likely that immunodeficiency will be a feature of the expanded phenotype in a larger group of patients.

Radiosensitivity, Immunodeficiency, Dysmorphic Features and Learning Difficulties (RIDDLE) Syndrome

To date only one caucasian patient has been described with Radiosensitivity, Immunodeficiency, Dysmorphic features and Learning difficulties (RIDDLE syndrome).[107] He presented with mild facial dysmorphism, short stature, learning difficulties and mild motor abnormalities. There were no oculocutaneous telangiectasia. At the age of one year he had low serum IgG levels, with normal IgM and IgA. By 3 years of age his IgG remained low but IgA and IgM were at normal levels. T- and B-lymphocyte numbers were normal. From the age of 3 years he was treated with replacement immunoglobulin and has remained well. His father has subsequently developed B-cell chronic lymphocytic leukaemia. Biallelic mutations in RNF168, a ubiquitin ligase, have subsequently been reported.[108] Analysis of B-lymphocytes revealed increased levels of microhomology across the Sμ-Sα and Sα-Sγ3 switch regions with a reduced frequency of mutations and insertions, findings similar, although less severe, to those found in LIG4 deficiency and suggestive of abnormal CSR, although SHM was normal. Fibroblasts exhibited moderately increased sensitivity to ionising radiation and failed to localize 53BP1 to damaged chromatin. It is postulated that RNF168 may be playing a role in organising chromatin to facilitate long-range NHEJ, which is essential for CSR, but not VDJ recombination; thus explaining normal cellular immunity in the patient.[109-111]

Other Uncharacterised Disorders

A number of genetically undefined disorders, with phenotypic and cellular features characteristic of NBS but with no mutations in candidate genes, have been described,[112-114] suggesting defects in further DNA repair genes in human primary immunodeficiency may exist.

Human Primary Immunodeficiency Due to Genetic Defects in Class Switch Recombination and Somatic Hypermutation

Autosomal Recessive Hyper-IgM Syndromes

Three autosomal recessive hyper-IgM syndromes have been described due to defects in DNA break and repair mechanisms, which lead to decreased or abolished isotype switching and impaired somatic hypermutation (see also chapter by Kracker, Gardes and Durandy in this volume).

AID Deficiency

Activation-induced cytidine deaminase (AID) deficiency, an autosomal recessive disease due to mutations in *AICDA*, usually presents in early childhood with severe, recurrent infections, of which recurrent sino-pulmonary infections are the most common. Gastrointestinal infections are also common.[115,116] Despite this early presentation, many patients are not diagnosed and treated until the second or third decade of life.[117] Massive lymphadenopathy, with giant germinal centres on histological examination are characteristic. Immunological features include raised IgM and low or absent IgA and IgG. There is an increased incidence of organ-specific autoimmune disease in these patients, particularly diabetes mellitus, polyarthritis, autoimmune hepatitis and Crohn's disease.[117] Although the precise mechanism of these features remains unresolved, it could include:

i. Lack of inhibitory signalling through the FcγIIB receptor to the B-lymphocyte, resulting in immune complex-mediated tissue damage;
ii. Decrease in peripheral tolerance due to unregulated B-lymphocyte proliferation;
iii. Compromised elimination of external antigen due to a lack of CSR and SHM, driving B-lymphocyte activation.[118]

Patients with impaired CSR but normal SHM have mutations in the carboxy-terminal region of AID protein and present with milder disease[119] and a small subset of patients may present with autosomal dominant disease.[120]

UNG Deficiency

Three patients have been described with a defect in uracil-DNA glycosylase (*UNG*). Clinical presentation was similar to those with AID deficiency including recurrent respiratory tract infections from early childhood and lymphoid hyperplasia.[121] Raised IgM and profoundly decreased IgA and IgG serum levels were found, with depressed antigen-specific antibody responses. A skewed pattern of SHM was found with almost all mutations being transitions (G>A, C>T).

PMS2 Deficiency

PMS2 forms a heterodimer with MLH1 to form hMutα which plays important roles in mismatch repair. Defects have been described in 3 individuals.[23] In addition to raised serum IgM and decreased IgA and IgG with recurrent infections, café au lait spots and malignancy including leukaemias, lymphomas, cerebral tumours and colorectal tumours are characteristic.[122,123] Increased levels of microhomology were found across the Sμ-Sα switch junctions. SHM may have been mildly reduced.

MSH5 Deficiency

MSH5 has been implicated in IgA deficiency and common variable immunodeficiency.[42] Increased microhomology at Sμ-Sα switch junctions was found in patients carrying disease-associated *MSH5* alleles, with fewer mutations at Sμ-Sα switch junctions than in controls. The precise mechanism by which defects in MSH5 contribute to the abnormalities observed in CSR has not been worked out although a regulatory role is proposed.

Undefined Defects of Class Switch Recombination and Somatic Hypermutation

It is likely that further, as yet undescribed, gene defects give rise to a clinical picture of hyper IgM syndrome. Hyper IgM Type 4 has been described in 15 patients with characteristic features of recurrent respiratory and gastrointestinal tract infection, lymphoid hyperplasia and autoimmune features.[124] CSR was defective, but SHM normal. *AIDCA* and *UNG* mutations were excluded in all patients.

A further clinical entity, consisting of increased radiosensitivity but normal checkpoint arrest and NHEJ, increased levels of microhomology across Sμ-Sα switch junctions and a skewed SHM toward transitions at G/C residues, has been described in 5 patients.[125] All had recurrent respiratory infections; lymphoid hyperplasia and autoimmunity were also described. Raised IgM and decreased IgA and IgG levels were also noted. Genetic defect(s) in these cases have yet to be identified. A further group of patients has been identified, in whom recurrent bacterial infection, autoimmunity and lymphadenopathy are observed, although the lymphadenopathy was less marked than in AID-deficient patients. A lack of class-switched B-lymphocytes in these patients was observed, although SHM was normal. There was no sensitivity to IR.[126]

Human Primary Immunodeficiency Due to Genetic Defects in Other DNA Repair Genes

DNA Ligase 1

To date, one patient has been described with two compound missense mutations in DNA ligase I.[127,128] Clinical features, overlapping with those of Bloom syndrome and ataxia telangiectasia, included intrauterine growth retardation, general growth retardation, developmental delay but normal cognitive development, some dysmorphism with elf-like features, photosensitivity and immunodeficiency. Her immunodeficiency manifested as recurrent middle ear and chest infections from age 2 years, with evolving IgA deficiency, relative hypogammaglobulinaemia of IgG and normal IgM. She had an evolving lymphocytopenia with poor proliferative response to mitogens. During her teenage years her respiratory status deteriorated and secondary sexual characteristics did not develop. At the age of 17 years, patches of venous dilatation appeared on her skin mainly on the limbs and there was some bulbar conjunctival telangiectasia. Thyroid, pituitary and adrenal functions were normal. She developed hepatosplenomegaly and a liver biopsy showed lymphocyte infiltration of the portal tract suggesting lymphoma. Her splenomegaly became massive and she developed neutropenia and increasing lymphopenia. She had a severe cutaneous herpes zoster infection and died from pneumonia at the age of 19 years. No information on end-joining of VDJ substrates is available, or on the use of microhomology. An increase in the number of DNA-ssb and -dsb in newly replicated DNA molecules was seen in an immortalised fibroblast line, possibly because of the failure of dealing with damage at replication forks.[129] The cause of the immunodeficiency can only be hypothesised, given the paucity of data. In that (i) there is no current evidence linking DNA ligase 1 with VDJ recombination,[130] (ii) DNA ligase I complexes with Nbs1 and (iii) DNA ligase I and Nbs1 colocalize at replication factories to repair DNA-dsb by homologous recombination at stalled replication forks,[131] defects in DNA ligase I may be associated with failure to repair DNA damage during lymphocyte proliferation, rather than failure to complete NHEJ in TCR and BCR formation. The finding of low IgA and IgG, but normal IgM is tantalising and further work needs to be done to investigate what role, if any, DNA ligase I has in CSR.

Bloom Syndrome

Bloom syndrome is an autosomal recessive disorder characterized by proportionate pre and postnatal growth deficiency, sun-sensitivity, telangiectatic, hypo- and hyperpigmented skin, predisposition to malignancy and chromosomal instability. There is an increased incidence of diabetes mellitus. Immunodeficiency, whilst common, is variable and generally not severe,[132,133] although life-threatening infection may occur.[134] Low concentrations of one or more immunoglobulin isotypes are found most frequently.[132,133,135] Impaired T-lymphocyte proliferation, diminished CD4+ T-lymphocyte numbers and impaired CD4+ T-lymphocyte function have been observed in Bloom syndrome patients.[132,136] There is a characteristic increase in sister chromatid exchange on cytogenetic analysis (Fig. 4). The Bloom syndrome helicase appears not to play a role in VDJ recombination[137,138] and has only a minor role to play in CSR, if any,[139] although microhomology—mediated end joining was observed at Sμ-Sγ3 switch regions, possibly implicating Blm in the resolution phase of CSR.[139]

Treatment

Treatment for DNA-repair deficient primary immunodeficiencies depends on the clinical presentation and is generally the same as for similar disorders due to other underlying molecular defects. The degree of immunodeficiency varies from mild to severe B- and T-lymphocytopenia. Patients with B-lymphocyte deficiency can be treated with prophylactic antibiotics and immunoglobulin replacement therapy. However patients with severe T-lymphocyte deficiency require more aggressive therapy and haematopoietic stem cell transplantation may be recommended. In particular children, with recurrent infection and failure to thrive and those at risk of developing malignancy, may benefit. Because of the risk of generalised chromosome damage with radiotherapy, this should be omitted from conditioning regimens. The use of low intensity chemotherapy conditioning appears to favour a successful outcome.[69,75,140,141] It is important to note that modified chemotherapy regimens, for the

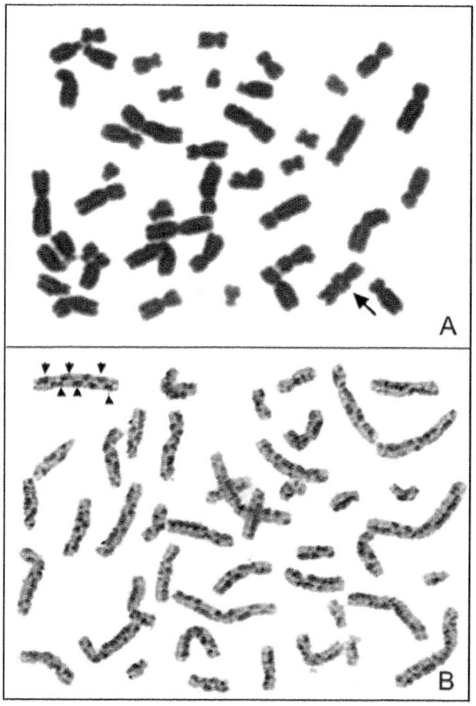

Figure 4. Karyotype from a patient with Bloom syndrome showing (A) a symmetrical quadrid-radial (arrow) and (B) increased numbers of sister chromatid exchanges (arrows). (Reproduced with permission of Paediatric HSCT unit, Newcastle General Hospital.)

treatment of lymphoid malignancy, are needed due to the high level of toxicity using conventional regimens.[142] The incidence of secondary malignancies is more frequent in these patients so careful follow-ups are required. Gene therapy may be valuable treatment for some conditions, although clinical trials are yet to be perfected.[143] Another novel approach to treatment under investigation is the use of antisense oligonucleotides to correct splicing, frameshift and missense mutations leading to production of missing or unstable protein to convert to partially or fully functional type.[144]

Conclusion

Much has been learnt about the molecular mechanisms of lymphocyte receptor formation, immunoglobulin isotype switching and the affinity maturation from the careful studies of human primary immunodeficiencies. Conversely, an understanding of the mechanisms has aided discovery of novel genetic immune defects in patients. Many questions still remain unanswered. Further studies on patients with these extremely rare diseases such as Ligase I deficiency, RAD50 deficiency, RIDDLE syndrome and PMS2 deficiency will expand the clinical phenotype and give greater understanding of the role these proteins play in immune development. Key proteins of the DNA-dsb repair pathways, such as XRCC4, known to cause immunodeficiency in animal models, have yet to be linked to human disease. It is likely that further studies of milder antibody deficiencies will reveal further defects in MMR proteins responsible for SHM and antibody affinity maturation. An appreciation of the role that these defects play in immunodeficiency and in the wider biological processes of other cells is likely to lead to better, more targeted treatments for these patients, for instance, modified chemotherapy conditioning regimens for patients with DNS-dsb repair defects undergoing haematopoietic stem cell transplantation, or use of antisense oligonucleotides to correct nonsense mutations by read-through compounds such as aminoglycosides.

References

1. Lieber MR, Ma Y, Pannicke U et al. The mechanism of vertebrate nonhomologous DNA end joining and its role in V(D)J recombination. DNA Repair (Amst) 2004; 3:817-826.
2. Riballo E, Kuhne M, Rief N et al. A pathway of double-strand break rejoining dependent upon ATM, Artemis and proteins locating to gamma-H2AX foci. Mol Cell 2004; 16:715-724.
3. Bredemeyer AL, Sharma GG, Huang CY et al. ATM stabilizes DNA double-strand-break complexes during V(D)J recombination. Nature 2006; 442:466-470.
4. Huang CY, Sharma GG, Walker LM et al. Defects in coding joint formation in vivo in developing ATM-deficient B- and T-lymphocytes. J Exp Med 2007; 204:1371-1381.
5. Helmink BA, Bredemeyer AL, Lee BS et al. MRN complex function in the repair of chromosomal Rag-mediated DNA double-strand breaks. J Exp Med 2009; 206:669-679.
6. Perkins EJ, Nair A, Cowley DO et al. Sensing of intermediates in V(D)J recombination by ATM. Genes Dev 2002; 16:159-164.
7. Chen HT, Bhandoola A, Difilippantonio MJ et al. Response to RAG-mediated VDJ cleavage by NBS1 and γ-H2AX. Science 2000; 290:1962-1965.
8. Celeste A, Fernandez-Capetillo O, Kruhlak MJ et al. Histone H2AX phosphorylation is dispensable for the initial recognition of DNA breaks. Nat Cell Biol 2003; 5:675-679.
9. Stracker TH, Theunissen JW, Morales M et al. The Mre11 complex and the metabolism of chromosome breaks: the importance of communicating and holding things together. DNA Repair (Amst) 2004; 3:845-854.
10. Difilippantonio S, Celeste A, Fernandez-Capetillo O et al. Role of Nbs1 in the activation of the Atm kinase revealed in humanized mouse models. Nat Cell Biol 2005; 7:675-685.
11. Kobayashi Y, Tycko B, Soreng AL et al. Transrearrangements between antigen receptor genes in normal human lymphoid tissues and in ataxia telangiectasia. J Immunol 1991; 147:3201-3209.
12. Corneo B, Wendland RL, Deriano L et al. Rag mutations reveal robust alternative end joining. Nature 2007; 449:483-486.
13. Iwasato T, Shimizu A, Honjo T et al. Circular DNA is excised by immunoglobulin class switch recombination. Cell 1990; 62:143-149.
14. Muramatsu M, Kinoshita K, Fagarasan S et al. Class switch recombination and hypermutation require activation-induced cytidine deaminase (AID), a potential RNA editing enzyme. Cell 2000; 102:553-563.
15. Revy P, Muto T, Levy Y et al. Activation-Induced cytidine Deaminase (AID) deficiency causes the autosomal recessive form of Hyper-IgM syndrome (HIGM2). Cell 2000; 102:565-575.
16. Bransteitter R, Pham P, Scharff MD et al. Activation-induced cytidine deaminase deaminates deoxycytidine on single-stranded DNA but requires the action of RNase. Proc Natl Acad Sci USA 2003; 100:4102-4107.
17. Rada C, Williams GT, Nilsen H et al. Immunoglobulin isotype switching is inhibited and somatic hypermutation perturbed in UNG-deficient mice. Curr Biol 2002; 12:1748-1755.
18. Guikema JE, Linehan EK, Tsuchimoto D et al. APE1- and APE2-dependent DNA breaks in immunoglobulin class switch recombination. J Exp Med 2007; 204:3017-3026.
19. Xue K, Rada C, Neuberger MS. The in vivo pattern of AID targeting to immunoglobulin switch regions deduced from mutation spectra in msh2−/− ung−/− mice. J Exp Med 2006; 203:2085-2094.
20. Wilson TM, Vaisman A, Martomo SA et al. MSH2-MSH6 stimulates DNA polymerase eta, suggesting a role for A:T mutations in antibody genes. J Exp Med 2005; 201:637-645.
21. Matsuoka S, Ballif BA, Smogorzewska A et al. ATM and ATR substrate analysis reveals extensive protein networks responsive to DNA damage. Science 2007; 316:1160-1166.
22. Schrader CE, Vardo J, Stavnezer J. Role for mismatch repair proteins Msh2, Mlh1 and Pms2 in immunoglobulin class switching shown by sequence analysis of recombination junctions. J Exp Med 2002; 195:367-373.
23. Péron S, Metin A, Gardès P et al. Human PMS2 deficiency is associated with impaired immunoglobulin class switch recombination. J Exp Med 2008; 205:2465-2472.
24. Babbe H, Chester N, Leder P et al. The Bloom's syndrome helicase is critical for development and function of the alphabeta T-cell lineage. Mol Cell Biol 2007; 27:1947-1959.
25. Pedrazzi G, Bachrati CZ, Selak N et al. The Bloom's syndrome helicase interacts directly with the human DNA mismatch repair protein hMSH6. Biol Chem 2003; 384:1155-1164.
26. Pedrazzi G, Perrera C, Blaser H et al. Direct association of Bloom's syndrome gene product with the human mismatch repair protein MLH1. Nucleic Acids Res 2001; 29:4378-4386.
27. Schrader CE, Guikema JE, Linehan EK et al. Activation-induced cytidine deaminase-dependent DNA breaks in class switch recombination occur during G1 phase of the cell cycle and depend upon mismatch repair. J Immunol 2007; 179:6064-6071.
28. Yan CT, Boboila C, Souza EK et al. IgH class switching and translocations use a robust nonclassical end-joining pathway. Nature 2007; 449:478-482.

29. Berkovich E, Monnat RJ Jr, Kastan MB. Roles of ATM and NBS1 in chromatin structure modulation and DNA double-strand break repair. Nat Cell Biol 2007; 9:683-690.
30. Burma S, Chen BP, Murphy M et al. ATM phosphorylates histone H2AX in response to DNA double-strand breaks. J Biol Chem 2001; 276:42462-42467.
31. Kobayashi J, Tauchi H, Chen B et al. Histone H2AX participates the DNA damage-induced ATM activation through interaction with NBS1. Biochem Biophys Res Commun 2009; 380:752-757.
32. Ward IM, Reina-San-Martin B, Olaru A et al. 53BP1 is required for class switch recombination. J Cell Biol 2004; 165:459-464.
33. Rooney S, Alt FW, Sekiguchi J et al. Artemis-independent functions of DNA-dependent protein kinase in Ig heavy chain class switch recombination and development. Proc Natl Acad Sci USA 2005; 102:2471-2475.
34. Franco S, Murphy MM, Li G et al. DNA-PKcs and Artemis function in the end-joining phase of immunoglobulin heavy chain class switch recombination. J Exp Med 2008; 205:557-564.
35. Du L, van der Burg M, Popov SW et al. Involvement of Artemis in nonhomologous end-joining during immunoglobulin class switch recombination. J Exp Med 2008; 205:3031-3040.
36. Pan-Hammarstrom Q, Jones AM, Lahdesmaki A et al. Impact of DNA ligase IV on nonhomologous end joining pathways during class switch recombination in human cells. J Exp Med 2005; 201:189-194.
37. Helleday T, Bryant HE, Schultz N. Poly (ADP-ribose) polymerase (PARP-1) in homologous recombination and as a target for cancer therapy. Cell Cycle 2005; 4:1176-1178.
38. Audebert M, Salles B, Calsou P. Involvement of poly (ADP-ribose) polymerase-1 and XRCC1/DNA ligase III in an alternative route for DNA double-strand breaks rejoining. J Biol Chem 2004; 279:55117-55126.
39. Wang H, Rosidi B, Perrault R et al. DNA ligase III as a candidate component of backup pathways of nonhomologous end joining. Cancer Res 2005; 65:4020-30.
40. Robert I, Dantzer F, Reina-San-Martin B. Parp1 facilitates alternative NHEJ, whereas Parp2 suppresses IgH/c-myc translocations during immunoglobulin class switch recombination. J Exp Med 2009; 206:1047-1056.
41. Liang L, Deng L, Nguyen SC et al. Human DNA ligases I and III, but not ligase IV, are required for microhomology-mediated end joining of DNA double-strand breaks. Nucleic Acids Res 2008; 36:3297-3310.
42. Sekine H, Ferreira RC, Pan-Hammarström Q et al. Role for Msh5 in the regulation of Ig class switch recombination. Proc Natl Acad Sci USA 2007; 104:7193-7198.
43. Kaartinen M, Griffiths GM, Markham AF et al. mRNA sequences define an unusually restricted IgG response to 2-phenyloxazolone and its early diversification. Nature 1983; 304:320-324.
44. Storb U. Progress in understanding the mechanism and consequences of somatic hypermutation. Immunol Rev 1998; 162:5-11.
45. Shivarov V, Shinkura R, Doi T et al. Molecular mechanism for generation of antibody memory. Philos Trans R Soc Lond B Biol Sci 2009; 364:569-575.
46. Schanz S, Castor D, Fischer F et al. Interference of mismatch and base excision repair during the processing of adjacent U/G mispairs may play a key role in somatic hypermutation. Proc Natl Acad Sci USA 2009; 106:5593-5598.
47. Larson ED, Cummings WJ, Bednarski DW et al. MRE11/RAD50 cleaves DNA in the AID/UNG-dependent pathway of immunoglobulin gene diversification. Mol Cell 2005; 20:367-375.
48. Sack SZ, Liu Y, German J et al. Somatic hypermutation of immunoglobulin genes is independent of the Bloom's syndrome DNA helicase. Clin Exp Immunol 1998; 112:248-254.
49. Schwarz K, Glaus GH, Ludwig L et al. RAG mutations in human B-cell-negative SCID. Science 1996; 274:97-99.
50. Villa A, Santagata S, Bozzi F et al. Partial V(D)J recombination activity leads to Omenn syndrome. Cell 1998; 93:885-896.
51. Omenn GS. Familial reticuloendotheliosis with eosinophilia. N Engl J Med 1965; 273:427-432.
52. Rieux-Laucat F, Bahadoran P, Brousse N et al. Highly restricted human T-cell repertoire beta (TCRB) chain diversity in peripheral blood and tissue-infiltrating lymphocytes in Omenn's syndrom (severe combined immunodeficiency with hypereosinophilia). J Clin Invest 1998; 102:312-321.
53. Ehl S, Schwarz K, Enders A et al. A variant of SCID with specific immune responses and predominance of gamma delta T-cells. J Clin Invest 2005; 115:3140-3148.
54. de Villartay JP, Lim A, Al-Mousa H et al. A novel immunodeficiency associated with hypomorphic RAG1 mutations and CMV infection. J Clin Invest 2005; 115:3291-3299.
55. Schuetz C, Huck K, Gudowius S et al. An immunodeficiency disease with RAG mutations and granulomas. N Engl J Med 2008; 358:2030-2038.
56. Moshous D, Callebaut I, de Chasseval R et al. Artemis, a novel DNA double-strand break repair/V(D)J recombination protein, is mutated in human severe combined immune deficiency. Cell 2001; 105:177-186.

57. Jones JF, Ritenbaugh CK, Spence MA et al. Severe combined immunodeficiency among the Navajo. I. Characterization of phenotypes, epidemiology and population genetics. Hum Biol 1991; 63:669-682.
58. Cavazzana-Calvo M, Le Deist F, de Saint Basile G et al. Increased radiosensitivity of granulocyte macrophage colony-forming units and skin fibroblasts in human autosomal recessive severe combined immunodeficiency. J Clin Invest 1993; 91:1214-1218.
59. Ege M, Ma Y, Manfras B et al. Omenn syndrome due to ARTEMIS mutations. Blood 2005; 105:4179-4186.
60. Moshous D, Pannetier C, de Chasseval R et al. Partial T- and B-lymphocyte immunodeficiency and predisposition to lymphoma in patients with hypomorphic mutations in Artemis. J Clin Invest 2003; 111:381-387.
61. Evans PM, Woodbine L, Riballo E et al. Radiation-induced delayed cell death in a hypomorphic Artemis cell line. Hum Mol Genet 2006; 15:1303-1311.
62. van der Burg M, Ijspeert H, Verkaik NS et al. A DNA-PKcs mutation in a radiosensitive T-B- SCID patient inhibits Artemis activation and nonhomologous end-joining. J Clin Invest 2009; 119:91-98.
63. Riballo E, Critchlow SE, Teo SH et al. Identification of a defect in DNA ligase IV in a radiosensitive leukaemia patient. Curr Biol 1999; 9:699-702.
64. O'Driscoll M, Cerosaletti KM, Girard PM et al. DNA ligase IV mutations identified in patients exhibiting developmental delay and immunodeficiency. Mol Cell 2001; 8:1175-1185.
65. Unal S, Cerosaletti K, Uckan-Cetinkaya D et al. A novel mutation in a family with DNA ligase IV deficiency syndrome. Pediatr Blood Cancer 2009; 53:482-484.
66. van der Burg M, van Veelen LR, Verkaik NS et al. A new type of radiosensitive T-B-NK+ severe combined immunodeficiency caused by a LIG4 mutation. J Clin Invest 2006; 116:137-145.
67. Buck D, Moshous D, de Chasseval R et al. Severe combined immunodeficiency and microcephaly in siblings with hypomorphic mutations in DNA ligase IV. Eur J Immunol 2006; 36:224-235.
68. Ben-Omran TI, Cerosaletti K, Concannon P et al. A patient with mutations in DNA Ligase IV: clinical features and overlap with Nijmegen breakage syndrome. Am J Med Genet A 2005; 137:283-287.
69. Enders A, Fisch P, Schwarz K et al. A severe form of human combined immunodeficiency due to mutations in DNA ligase IV. J Immunol 2006; 176:5060-5068.
70. Toita N, Hatano N, Ono S et al. Epstein-Barr virus-associated B-cell lymphoma in a patient with DNA ligase IV (LIG4) syndrome. Am J Med Genet A 2007; 143:742-745.
71. Grunebaum E, Bates A, Roifman CM. Omenn syndrome is associated with mutations in DNA ligase IV. J Allergy Clin Immunol 2008; 122:1219-1220.
72. Buck D, Malivert L, de Chasseval R et al. Cernunnos, a novel nonhomologous end-joining factor, is mutated in human immunodeficiency with microcephaly. Cell 2006; 124:287-299.
73. Ahnesorg P, Smith P, Jackson SP. XLF interacts with the XRCC4-DNA ligase IV complex to promote DNA nonhomologous end-joining. Cell 2006; 124:301-313.
74. Dai Y, Kysela B, Hanakahi LA et al. Nonhomologous end joining and V(D)J recombination require an additional factor. Proc Natl Acad Sci USA 2003; 100:2462-2467.
75. Faraci M, Lanino E, Micalizzi C et al. Unrelated hematopoietic stem cell transplantation for Cernunnos-XLF deficiency. Pediatr Transplant 2008. [Epub ahead of print].
76. Schwartz M, Oren YS, Bester AC et al. Impaired replication stress response in cells from immunodeficiency patients carrying Cernunnos/XLF mutations. PLoS One 2009; 4:e4516.
77. Ahmad SI, ed. Molecular mechanisms of ataxia telangiectasis. medical intelligence unit, Landes Bioscience Publication, Texas, 2009.
78. Noordzij JG, Wulffraat NM, Haraldsson A et al. Ataxia-telangiectasia patients presenting with hyper-IgM syndrome. Arch Dis Child 2009; 94:448-449.
79. Lefton-Greif MA, Crawford TO, Winkelstein JA et al. Oropharyngeal dysphagia and aspiration in patients with ataxia-telangiectasia. J Pediatr 2000; 136:225-231.
80. Staples ER, McDermott EM, Reiman A et al. Immunodeficiency in ataxia telangiectasia is correlated strongly with the presence of two null mutations in the ataxia telangiectasia mutated gene. Clin Exp Immunol 2008; 153:214-220.
81. Sanal O, Ersoy F, Yel L et al. Impaired IgG antibody production to pneumococcal polysaccharides in patients with ataxia-telangiectasia. J Clin Immunol 1999; 19:326-334.
82. Tangsinmankong N, Wayne AS, Howenstine M et al. Lymphocytic interstitial pneumonitis, elevated IgM concentration and hepatosplenomegaly in ataxia-telangiectasia. J Pediatr 2001; 138:939-941.
83. Srisirirojanakorn N, Finegold MJ, Gopalakrishna GS et al. Hepatic veno-occlusive disease in ataxia-telangiectasia. J Pediatr 1999; 134:786-788.
84. Crawford TO, Skolasky RL, Fernandez R et al. Survival probability in ataxia telangiectasia. Arch Dis Child 2006; 91:610-611.
85. Giovannetti A, Mazzetta F, Caprini E et al. Skewed T-cell receptor repertoire, decreased thymic output and predominance of terminally differentiated T-cells in ataxia telangiectasia. Blood 2002; 100:4082-4089.

86. Reina-San-Martin B, Chen HT, Nussenzweig A et al. ATM is required for efficient recombination between immunoglobulin switch regions. J Exp Med 2004; 200:1103-1110.
87. Weemaes CM, Hustinx TW, Scheres JM et al. A new chromosomal instability disorder: the Nijmegen breakage syndrome. Acta Paediatr Scand 1981; 70:557-564.
88. Digweed M, Sperling K. Nijmegen breakage syndrome: clinical manifestation of defective response to DNA double-strand breaks. DNA Repair (Amst) 2004; 3:1207-1217.
89. Gregorek H, Chrzanowska KH, Michalkiewicz J et al. Heterogeneity of humoral immune abnormalities in children with Nijmegen breakage syndrome: an 8-year follow-up study in a single centre. Clin Exp Immunol 2002; 130:319-324.
90. Xu Y. ATM in lymphoid development and tumorigenesis. Adv Immunol 1999; 72:179-189.
91. Kracker S, Bergmann Y, Demuth I et al. Nibrin functions in Ig class-switch recombination. Proc Natl Acad Sci USA 2005; 102:1584-1589.
92. Reina-San-Martin B, Nussenzweig MC, Nussenzweig A et al. Genomic instability, endoreduplication and diminished Ig class-switch recombination in B-cells lacking Nbs1. Proc Natl Acad Sci USA 2005; 102:1590-1595.
93. Wolf EK, Shwayder TA. Nijmegen breakage syndrome associated with porokeratosis. Pediatr Dermatol 2009; 26:106-8.
94. Yoo J, Wolgamot G, Torgerson TR et al. Cutaneous noncaseating granulomas associated with Nijmegen breakage syndrome. Arch Dermatol 2008; 144:418-419.
95. Green AM, Kupfer GM. Fanconi anemia. Hematol Oncol Clin North Am 2009; 23:193-214.
96. Ahmad SI, Kirk SH, eds. Molecular Mechanisms of Fanconi anemia. Austin: Landes Bioscience, 2006.
97. Gennery AR, Slatter MA, Bhattacharya A et al. The clinical and biological overlap between Nijmegen Breakage Syndrome and Fanconi anemia. Clin Immunol 2004; 113:214-219.
98. Stewart GS, Maser RS, Stankovic T et al. The DNA double-strand break repair gene hMRE11 is mutated in individuals with an ataxia-telangiectasia-like disorder. Cell 1999; 99:577-587.
99. Delia D, Piane M, Buscemi G et al. MRE11 mutations and impaired ATM-dependent responses in an Italian family with ataxia-telangiectasia-like disorder. Hum Mol Genet 2004; 13:2155-2163.
100. Fernet M, Gribaa M, Salih MA et al. Identification and functional consequences of a novel MRE11 mutation affecting 10 Saudi Arabian patients with the ataxia telangiectasia-like disorder. Hum Mol Genet 2005; 14:307-318.
101. Khan AO, Oystreck DT, Koenig M et al. Ophthalmic features of ataxia telangiectasia-like disorder. J AAPOS 2008; 12:186-189.
102. Taylor AM, Groom A, Byrd PJ. Ataxia-telangiectasia-like disorder (ATLD)-its clinical presentation and molecular basis. DNA Repair (Amst) 2004; 3:1219-1225.
103. Lahdesmaki A, Taylor AM, Chrzanowska KH et al. Delineation of the role of the Mre11 complex in class switch recombination. J Biol Chem 2004; 279:16479-16487.
104. Waltes R, Kalb R, Gatei M et al. Human RAD50 deficiency in a Nijmegen breakage syndrome-like disorder. Am J Hum Genet 2009; 84:605-616.
105. Barbi G, Scheres JM, Schindler D et al. Chromosome instability and X-ray hypersensitivity in a microcephalic and growth-retarded child. Am J Med Genet 1991; 40:44-45.
106. Donahue SL, Tabah AA, Schmitz K et al. Defective signal joint recombination in fanconi anemia fibroblasts reveals a role for Rad50 in V(D)J recombination. J Mol Biol 2007; 370:449-458.
107. Stewart GS, Stankovic T, Byrd PJ et al. RIDDLE immunodeficiency syndrome is linked to defects in 53BP1-mediated DNA damage signaling. Proc Natl Acad Sci USA 2007; 104:16910-16915.
108. Stewart GS, Panier S, Townsend K et al. The RIDDLE syndrome protein mediates a ubiquitin-dependent signaling cascade at sites of DNA damage. Cell 2009; 136:420-434.
109. Difilippantonio S, Gapud E, Wong N et al. 53BP1 facilitates long-range DNA end-joining during V(D)J recombination. Nature 2008; 456:529-533.
110. Manis JP, Morales JC, Xia Z et al. 53BP1 links DNA damage-response pathways to immunoglobulin heavy chain class-switch recombination. Nat Immunol 2004; 5:481-487.
111. Ward IM, Reina-San-Martin B, Olaru A et al. 53BP1 is required for class switch recombination. J Cell Biol 2004; 165:459-464.
112. Berardinelli F, di Masi A, Salvatore M et al. A case report of a patient with microcephaly, facial dysmorphism, chromosomal radiosensitivity and telomere length alterations closely resembling "Nijmegen breakage syndrome" phenotype. Eur J Med Genet 2007; 50:176-187.
113. Maraschio P, Spadoni E, Tanzarella C et al. Genetic heterogeneity for a Nijmegen breakage-like syndrome. Clin Genet 2003; 63:283-290.
114. Hiel JA, Weemaes CM, van Engelen BG et al. Nijmegen breakage syndrome in a Dutch patient not resulting from a defect in NBS1. J Med Genet 2001; 38:E19.
115. Revy P, Muto T, Levy Y et al. Activation-Induced cytidine Deaminase (AID) deficiency causes the autosomal recessive form of Hyper-IgM syndrome (HIGM2). Cell 2000; 102:565-575.

116. Quartier P, Bustamante J, Sanal O et al. Clinical, immunologic and genetic analysis of 29 patients with autosomal recessive hyper-IgM syndrome due to Activation-Induced Cytidine Deaminase deficiency. Clin Immunol 2004; 110:22-29.
117. Minegishi Y, Lavoie A, Cunningham-Rundles C et al. Mutations in activation-induced cytidine deaminase in patients with hyper IgM syndrome. Clin Immunol 2000; 97:203-210.
118. Hase K, Takahashi D, Ebisawa M et al. Activation-induced cytidine deaminase deficiency causes organ-specific autoimmune disease. PLoS One 2008; 3:e3033.
119. Ta VT, Nagaoka H, Catalan N et al. AID mutant analyses indicate requirement for class-switch-specific cofactors. Nat Immunol 2003; 4:843-848.
120. Imai K, Zhu Y, Revy P et al. Analysis of class switch recombination and somatic hypermutation in patients affected with autosomal dominant hyper-IgM syndrome type 2. Clin Immunol 2005; 115:277-285.
121. Imai K, Slupphaug G, Lee WI et al. Human uracil-DNA glycosylase deficiency associated with profoundly impaired immunoglobulin class-switch recombination. Nat Immunol 2003; 4:1023-1028.
122. De Vos M, Hayward BE, Charlton R et al. PMS2 mutations in childhood cancer. J Natl Cancer Inst 2006; 98:358-361.
123. Kratz CP, Niemeyer CM, Jüttner E et al. Childhood T-cell non-Hodgkin's lymphoma, colorectal carcinoma and brain tumor in association with café-au-lait spots caused by a novel homozygous PMS2 mutation. Leukemia 2008; 22:1078-1080.
124. Imai K, Catalan N, Plebani A et al. Hyper-IgM syndrome type 4 with a B-lymphocyte-intrinsic selective deficiency in Ig class-switch recombination. J Clin Invest 2003; 112:136-142.
125. Péron S, Pan-Hammarström Q, Imai K et al. A primary immunodeficiency characterized by defective immunoglobulin class switch recombination and impaired DNA repair. J Exp Med 2007; 204:1207-1216.
126. Durandy A. Immunoglobulin class switch recombination: study through human natural mutants. Philos Trans R Soc Lond B Biol Sci 2009; 364:577-582.
127. Webster AD, Barnes DE, Arlett CF et al. Growth retardation and immunodeficiency in a patient with mutations in the DNA ligase I gene. Lancet 1992; 339:1508-1509.
128. Barnes DE, Tomkinson AE, Lehmann AR et al. Mutations in the DNA ligase I gene of an individual with immunodeficiencies and cellular hypersensitivity to DNA-damaging agents. Cell 1992; 69:495-503.
129. Soza S, Leva V, Vago R et al. DNA ligase I deficiency leads to replication-dependent DNA damage and impacts cell morphology without blocking cell cycle progression. Mol Cell Biol 2009; 29:2032-2041.
130. Petrini JH, Donovan JW, Dimare C et al. Normal V(D)J coding junction formation in DNA ligase I deficiency syndromes. J Immunol 1994; 152:176-178.
131. Vago R, Leva V, Biamonti G et al. DNA ligase I and Nbs1 proteins associate in a complex and colocalize at replication factories. Cell Cycle 2009; 8(16). [Epub ahead of print].
132. Hütteroth TH, Litwin SD, German J. Abnormal immune responses of Bloom's syndrome lymphocytes in vitro. J Clin Invest 1975; 56:1-7.
133. Van Kerckhove CW, Ceuppens JL, Vanderschueren-Lodeweyckx M et al. Bloom's syndrome. Clinical features and immunologic abnormalities of four patients. Am J Dis Child 1988; 142:1089-1093.
134. German J. Bloom's syndrome. Dermatol Clin 1995; 13:7-18.
135. Kondo N, Ozawa T, Kato Y et al. Reduced secreted mu mRNA synthesis in selective IgM deficiency of Bloom's syndrome. Clin Exp Immunol 1992; 88:35-40.
136. Taniguchi N, Mukai M, Nagaoki T et al. Impaired B-cell differentiation and T-cell regulatory function in four patients with Bloom's syndrome. Clin Immunol Immunopathol 1982; 22:247-258.
137. Hsieh CL, Arlett CF, Lieber MR. V(D)J recombination in ataxia telangiectasia, Bloom's syndrome and a DNA ligase I-associated immunodeficiency disorder. J Biol Chem 1993; 268:20105-20109.
138. Babbe H, Chester N, Leder P et al. The Bloom's syndrome helicase is critical for development and function of the alphabeta T-cell lineage. Mol Cell Biol 2007; 27:1947-1959.
139. Babbe H, McMenamin J, Hobeika E et al. Genomic instability resulting from Blm deficiency compromises development, maintenance and function of the B-cell lineage. J Immunol 2009; 182:347-360.
140. Gruhn B, Seidel J, Zintl F et al. Successful bone marrow transplantation in a patient with DNA ligase IV deficiency and bone marrow failure. Orphanet J Rare Dis 2007; 2:5.
141. Albert MH, Gennery AR, Greil J et al. Successful Stem cell transplantation for Nijmegen breakage syndrome. Bone Marrow Transplant 2009 (in press).
142. Dembowska-Baginska B, Perek D, Brozyna A et al. Non-Hodgkin lymphoma (NHL) in children with Nijmegen Breakage syndrome (NBS). Pediatr Blood Cancer 2009; 52:186-190.
143. Benjelloun F, Garrigue A, Demerens-de Chappedelaine C et al. Stable and functional lymphoid reconstitution in artemis-deficient mice following lentiviral artemis gene transfer into hematopoietic stem cells. Mol Ther 2008; 16:1490-1499.
144. Lai CH, Chun HH, Nahas SA et al. Correction of ATM gene function by aminoglycoside-induced read-through of premature termination codons. Proc Natl Acad Sci USA 2004; 101:15676-15681.

CHAPTER 15

Inherited Defects of Immunoglobulin Class Switch Recombination

Sven Kracker, Pauline Gardës and Anne Durandy*

Abstract

The investigation of an inherited primary immunodeficiency, the immunoglobulin class switch recombination deficiency, has allowed the delineation of complex molecular events that underlie antibody maturation in humans. The Activation-induced cytidine deaminase (AID)-deficiency, characterized by a defect in Class Switch Recombination (CSR) and somatic hypermutation, has revealed the master role of this molecule in the induction of DNA damage, the first step required for these two processes. The description that mutations in the gene encoding the Uracil-DNA glycosylase (UNG) lead to defective CSR has been essential for defining the DNA-editing activity of AID. Analysis of post meiotic segregation 2 (PMS2)-deficient patients gave evidence for the role of this mismatch repair enzyme in the generation of the DNA breaks that are required for CSR. Novel findings are awaited from the study of yet-genetically undefined CSR-deficiencies, probably leading to the identification of AID cofactor(s) and/or proteins involved in CSR-induced DNA repair.

Introduction

Maturation of the antibody repertoire results in the production of efficient antibodies of various isotypes, via a two-step process. The first step occurs in the fetal liver or in the bone marrow in an antigen and T-cell-independent way. It is achieved by genomic rearrangement between Variable (V), Diversity (D) and Joining (J) elements of the immunoglobulin (Ig) heavy chain gene and between V and J elements of the Ig light chain gene, resulting in the membrane expression by mature B-cells of a B-Cell Receptor (BCR) of the IgM and IgD isotypes.[1] The second step is shaped in the secondary lymphoid organs and is driven by antigen and T-cell interaction. It involves two processes: immunoglobulin class switch recombination (CSR) leading to the production of antibodies of various isotypes and Somatic Hypermutation (SHM), which results in the production of antibodies with high affinity for antigen.

CSR is accomplished by a DNA recombination event between two switch (S) regions, located upstream of the C_μ (S_μ) and a target C_x region of another isotype (S_x) with deletion of the intervening DNA. By means of CSR, the C_μ heavy chain of Ig is replaced by a different C_x heavy chain, thus allowing the production of immunoglobulins of various isotypes with the same antigen affinity since the V region is left unchanged.[2] Conversely, the SHM process modifies the antigen affinity (but not the isotype) of an antibody by introducing mutations in the V region at a very high frequency (1×10^{-3} bases per generation).[3] Mutations are generally missense mutations but deletions or insertions can also occur. This step is followed by the positive selection and proliferation of B-cells harboring a BCR with a higher affinity for antigen through a close interaction with antigen-loaded follicular dendritic cells[4] and follicular B helper T-cells.[5] Understanding the

*Corresponding Author: Anne Durandy—INSERM, U768; Université Paris Descartes, Faculté de Médecine René Descartes, Paris, F-75005 France. Email: anne.durandy@inserm.fr

Diseases of DNA Repair, edited by Shamim I. Ahmad. ©2010 Landes Bioscience and Springer Science+Business Media.

molecular mechanisms and signals involved in CSR and SHM have recently benefited from the characterization of a group of inherited disorders, the Ig CSR-deficiencies. Patients with this condition exhibit normal or high serum IgM levels, but markedly reduced serum levels of the other isotypes. Most of these patients also display impaired SHM generation. Besides the abnormalities in the T/B cooperation or the CD40 activation pathway,[6-8] that give evidence for the master role for CD40L/CD40 interaction in both CSR and SHM, other CSR deficiencies are caused by an intrinsic B-cell defect. The recent elucidation of their molecular basis has provided considerable insight into the complex mechanisms that govern the antibody maturation in humans.

CSR Deficiency Caused by Activation-Induced Cytidine Deaminase (AID)-Defect

This condition is transmitted as an autosomal recessive (AR) disease, although a rare autosomal dominant (AD) transmission has been reported to be associated with some peculiar mutations (see below). The Activation-Induced cytidine Deaminase (AID) is a B-cell specific molecule only expressed in B-cells activated to CSR and SHM.[9] Ectopic expression of AID in fibroblasts is able to induce CSR and SHM on a proper substrate, indicating that AID is the only B-cell specific component necessary for CSR and SHM.[10,11] AID belongs to the cytidine deaminase family able to deaminate deoxycytidine (dC) into deoxyuracil (dU) residues. Besides its cytidine deaminase domain, AID possesses an APOBEC-like domain, a leucine-rich domain prone to protein-protein interaction, a nuclear localization signal (NLS) and a nuclear export signal (NES) located respectively in the N and C ter parts, suggesting that AID, which is mostly expressed in the cytoplasm of activated B-cells, can shuttle into and off the nucleus.[12-14] Moreover, AID presence is not maintained in the nucleus because of its rapid degradation by the proteasome.[15] Although it has been firstly postulated that, according to its sequence similarity with APOBEC-1, AID was a RNA-editing enzyme,[16] several pieces of argument are in favor of a direct activity of AID on DNA.[17-22] Deamination by AID of dC to dU on single-stranded DNA during DNA S and V region transcription induces the DNA damage, necessary for both CSR and SHM. This DNA lesion step is followed by removal of dU by the Uracil-N glycosylase, an event that represents a physiologic trigger for the base excision repair pathway leading to the generation of DNA breaks.[22-24] Since double strand DNA breaks are required at least for CSR, AID should act on both DNA strands. Indeed, it has been shown that AID can also act on DNA template strand in the transcription bubbles,[21] or on double stranded DNA when DNA is super coiled.[25] Recent data suggest that the protein kinase A (PKA) in the nucleus phosphorylates AID on single strand DNA of S regions and interaction between AID and PKA recruits the replication protein A (RPA), a single-stranded DNA-binding protein.[26,27] AID thus introduces the DNA damage by introducing uracil residues on DNA.

A deficiency in AID leads to both defective CSR and SHM, as shown by the phenotype of AID-deficient patients.[28] In most cases, the disease is symptomatic from childhood and patients suffer mainly from recurrent bacterial infections of the upper and lower respiratory and digestive tract. Lymphoid hyperplasia is a prominent feature of the disease and is due to massive enlargement of germinal centres, that are filled of actively proliferating B-cells that co-express CD38, sIgM and sIgD, all markers of germinal centre founder cells. Such lymphoid hyperplasia, also affecting Peyer's patches as consequence of intestinal microbial infections, is also found in AID knock-out mice.[16,29] Autoimmunity (haemolytic anaemia, thrombocytopenia, hepatitis) affects about 20% of the patients, with the presence of auto-antibodies of IgM isotype.[30]

The proportion of memory ($CD27^+$) B-lymphocytes is normal, despite the complete lack of SHM. B-cells are unable to undergo in vitro CSR upon activation by CD40-agonists and appropriate cytokines and the CSR-defect has been shown to be located downstream from DNA transcription of S regions but upstream from DSB occurrence in switch (S) regions.[31]

AICDA mutations are scattered all along the gene (Fig. 1), resulting in an absence of AID expression as shown by immunohistochemistry in germinal centres and, dependent on the *AICDA* mutations, in a decrease or an absence of AID protein in EBV B-cell lines as shown by Western Blot analysis (A. Durandy, unpublished data). Examination of a cohort of 72 AID-deficient patients has

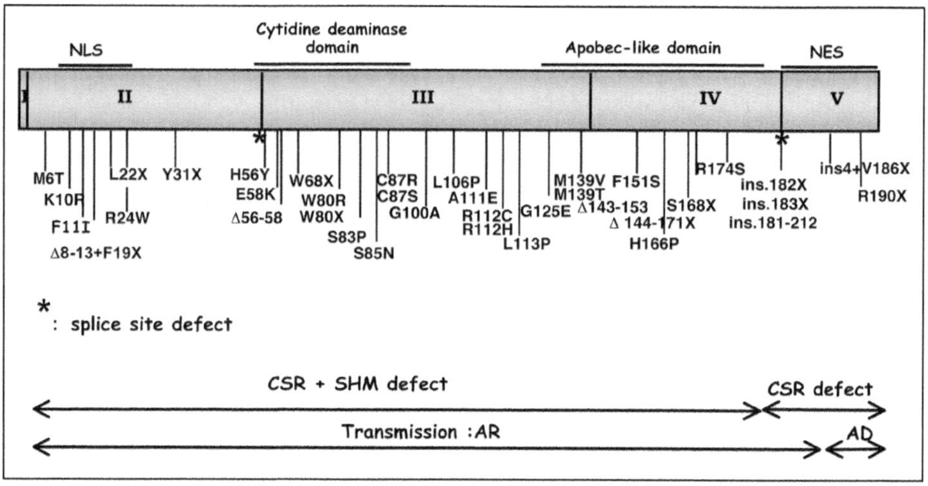

Figure 1. AICDA gene and mutations leading to CSR deficiencies. NLS: nuclear localization signal; NES: nuclear export signal; AR: autosomal recessive, AD: autosomal dominant.

revealed unique features associated with specific mutations of *AICDA*. In particular, mutations in the C-terminal part of the gene result in the selective impairment of CSR, leaving SHM intact[32] (Fig. 1). Since these mutants exert a normal cytidine activity after transfection either in *E. coli* or in fibroblasts and mutations in the Sμ regions are normally found in murine B-cells lacking the C ter part of AID,[33] it is postulated that the C ter part of AID should interact with a protein uniquely involved in CSR-induced DNA repair. In addition, while AID deficiency is usually inherited as an AR trait, nonsense mutations in the C-terminus part, that result in deletion of the NES, are associated with an AD inheritance[34] (Fig. 1). Two, non mutually exclusive, hypotheses can account for this observation: (i) the deletion which affects the NES domain of the protein lead to nuclear accumulation of the mutant allele, overriding the normal allele function;[12] (ii) the mutant allele exerts a dominant negative effect indicating that AID multimerization and/or a complex formation with other not yet defined partners is required for efficient CSR.

CSR Deficiency Caused by Uracil-N Glycosylase (UNG)-Defect

This description of a rare AR CSR defect, caused by mutations in the *UNG* gene strongly reinforced the hypothesis of a DNA-editing activity of AID.[24] UNG has two isoforms, with 2 different promoters, UNG1 which is ubiquitously expressed in mitochondria and UNG2 expressed transiently in the nucleus of proliferating cells, including B-cells activated for CSR.[24] Indeed, UNG2 interacts with RPA on single strand DNA and mediates the uracil deglycosylation and removal[23,35] leading to an abasic site which is eventually cleaved by the apurinic/apyrimidinic endonucleases, APEX1 and APEX2 as shown in mice.[36] The SHM process involves different error-prone repair pathways leading to mutations in the V region: the mismatch repair (MMR) and Polη for unprocessed U:G residues, the translesion synthesis polymerases for abasic sites, although the participation of polymeraseζ is controversially discussed and error-prone polymerases for nick repair.[37-41] For CSR, cleavage of abasic site by the APEX and occurrence of DSB are required downstream from UNG activity.

Up till now, three cases of UNG deficiency have been reported in humans.[24] The phenotype is identical to that of AID-deficiency, with an in vitro CSR defect located downstream from transcription but upstream of the DSB in S regions. Strikingly, in contrast to what is observed in AID-deficiency, the frequency of SHM was found normal, with a markedly skewed nucleotide substitution towards transitions on G and C residues. This abnormal pattern was likely due to the

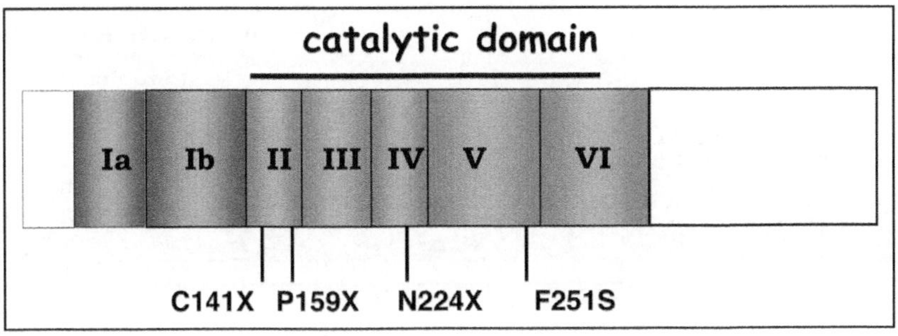

Figure 2. UNG gene and mutations leading to CSR deficiencies.

replication pathway occurring on the U:G residues, the few mutations arising on A:T residues being related to the activity of the MMR, MSH2-MSH6 and Polη.[23,42]

In the three cases described above, both alleles were mutated, affecting both the active domain of UNG1 and UNG2 (Fig. 2). Two patients carried frame shift mutations that led to premature termination, whereas one patient was homozygous for a missense mutation (F251S). Interestingly, this mutation caused mislocalization of the mutant UNG2 protein to the mitochondria, rather than to the nucleus.[43] Since UNG, a base excision repair, constitutes a major anti-mutagenic strategy, patients might be at risk for cancers, as observed in UNG-deficient mice.[44]

The description in humans of an Ig-CSR deficiency caused by UNG defect is reminiscent to what is observed in mice;[23] however, the human CSR defect is much more pronounced, suggesting the lack of any compensatory mechanism. Although the description of a CSR deficiency in UNG-deficient humans and mice strongly suggests that UNG activity is required for CSR, it has been proposed that UNG is only required as a docking protein.[45]

CSR Deficiency Caused by Post-Meiotic Segregation 2 (PMS2) Defect

Recent analysis of patients affected by PMS2-deficiency led to the first description of the role of this mismatch repair (MMR) enzyme in human CSR, in agreement with the previous description of PMS2-deficient mice.[46-48] PMS2 belongs to the MMR pathway that recognizes and repairs mismatched nucleotides on DNA. There are two main MMR components: the MutS homologue (MSH1–6) and the MutL homologue (PMS2/MLH1/PMS1). The MSH2—MSH6 complex appears to recognize AID-induced mismatches which escape the UNG processing and to recruit the PMS2/MLH1 complex. PMS2 which possesses an endonuclease domain could therefore introduce a nick nearby of the mismatch repair leading to exonuclease EXO1 activity.[49,50] Therefore, PMS2 could be involved in the generation of DNA double strand breaks of S regions.

Four patients affected by homozygous nonsense mutations in PMS2 could thoroughly be studied (Fig. 3).[51] As other patients carrying bi-allelic mutations in MMR genes, the main symptom of the disease was the occurrence of cancers from the first years of age.[52,53] Presence of café-au-lait skin spots was also evocative of a MMR defect. The two youngest patients present both with decreased serum IgA levels, in one associated with low IgG levels and severe susceptibility to bacterial infections. The two others, tested as young adults, only display a profound defect in IgG2 and IgG4 isotypes. However, no switched B-cells could be observed ex vivo and in vitro CSR towards IgA and IgE was defective, both observations arguing in favor of a B-cell intrinsic CSR defect. Accumulation of long lived plasma cells with age could compensate this defective CSR, as already shown in MMR or UNG-deficient mice.[42,54] As in AID and UNG-deficiencies, the CSR defect was located downstream from transcription and upstream of the occurrence of DSB in S regions. SHM frequency was slightly decreased in CD19+CD27+ B-cell population, but the nucleotide substitution pattern was found normal.

Our data provide clear evidence that PMS2 plays an important role in Ig-CSR in humans. Because of a very severe Ig-CSR defect in UNG-deficient patients, we suspect that PMS2 acts

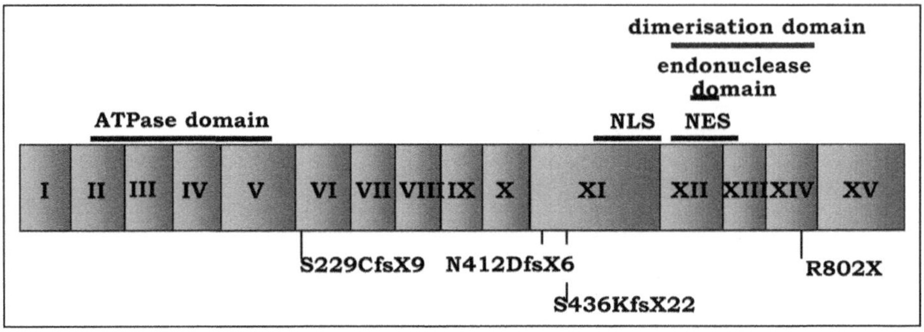

Figure 3. PMS2 gene and mutations leading to CSR deficiencies.

downstream from UNG in the same pathway, rather than being an alternative mechanism for base excision repair.

CSR Deficiency Caused by An Unknown DNA Repair Factor Defect

This CSR deficiency, although not yet molecularly defined, appears as likely transmitted as an AR disease since the M:F ratio is around 1 and consanguinity is observed in 20% cases. A CSR defect located downstream from the DNA cleavage step, characterizes this condition and suggests a DNA repair defect of switching B-cells, a hypothesis reinforced by the observation of tumor occurrence (non EBV-induced B-cell lymphomas) in 4 out of 45 patients. Moreover, a strong decrease of CD27+B-cells, a defect in switch junctions repair with a preferential usage of microhomology and especially a significant increased radiosensitivity of fibroblasts and EBV B-cell lines argue in favor of this hypothesis.[55]

DNA repair of S regions is achieved through a complex mechanism: it has been shown that upon CSR-activation and in an AID-dependent manner, the conformation of the Ig locus is modified, leading the Sμ-Sx regions to be in close proximity.[56] The maintenance of this synapsis requires a multimolecular complex involving the phosphorylated γ-H2AX- and the 53BP1 protein, the complex MRE11/RAD50/NBS1 and ATM,[57-61] leading to i) cell cycle arrest and ii) DNA repair through the Non Homologous End Joining pathway (NHEJ).[62-64] Since the involvement of all these molecules has been excluded in these patients, it is likely that (an) other(s) molecule(s), up to now undefined and deficient in these patients, play a role in CSR-induced DNA repair of S regions.

CSR-Deficiency in Molecularly Defined Syndromes Affecting the DNA Repair Machinery

a. As ATM and the MRE11/RAD50/NBS1 (MRN) complex are involved in Sμ-Sx synapsis maintenance and DSB DNA repair, a CSR deficiency was not unexpected in these syndromes caused by a defect in one of these molecules. ATM-(ataxia-telangiectasia: AT), MRE11-(AT-like disease: ATLD) and NBS1-(Nijmegen breakage syndrome)-deficiencies are associated with variable defect in CSR, a symptom sometimes preceding the neurological abnormalities in AT.[65] Study of recombined switch junctions in Ig gene locus indicates that the DNA repair does not occur properly during CSR.[66] Normal SHM generation in AT and AT-like disorder confirms that neither ATM nor MRE11 are essential for DNA repair of V regions.[67] A slightly abnormal pattern of SHM in NBS1 patients has been reported, but the role of this molecule in DNA repair of V regions remains obscure.[68]

b. Hypomorphic mutations in genes encoding the molecules of the NHEJ pathway are also associated to variable CSR deficiency, as shown by the serum Ig levels and especially by the defective DNA repair of switch junctions observed in patients. Indeed in ligase-IV or Artemis-deficient B-cells, switch junctions are repaired by using microhomology, strongly suggesting that, in the absence of a functional classical NHEJ pathway, an alternative end joining pathway can promote CSR.[69-71]

Conclusion

The ongoing delineation of inherited CSR deficiencies is shedding new light on the complex molecular mechanisms that govern CSR and SHM. Briefly, according to data from mice models and from these human natural mutants, the different events required for CSR and SHM have been precised:

a. **Transcription of target DNA sequences:** Transcription of specific RGYW sequences (R is A or G, Y is C or T and W is A or T) in S and V regions is required for both processes. During CSR, activation of B-cells by cytokines activates the specific I promoter located 5' to the target Cμ or C_x gene and induce the synthesis and splicing of so-called germ line or sterile transcripts.[72] This transcription step leads to the formation of RNA/DNA hybrids on the template DNA strand, leading to DNA accessibility for DNA damage.[18,19,73] Transcription is also required for SHM,[74] even if its induction has not been so far elucidated.

b. **DNA damage:** DNA lesion is induced by AID which by cytidine deamination introduces uracil residues on single strand DNA. AID is phosphorylated by PKA on S region DNA and recruits the single strand DNA binding protein RPA. UNG, which also binds RPA, deglycosylates and removes the uracil residues from DNA, leading to an abasic site.

c. **DNA repair:** During the SHM process, the AID-induced U:G mismatched residues, the UNG-induced abasic site and DNA nick are processed and repaired by the MMR pathway and error-prone polymerases, leading to introduction of mutations in the V region. CSR require the generation of blunt double strand DNA breaks which occur by a mechanism not yet completely elucidated. According to mice models, the APEX 1 and 2 cleave the abasic sites induced by UNG, leading to staggered DSB. Downstream from UNG activity at least in humans, the U:G residues remaining on DNA are repaired by the MMR system, likely involving PMS2 endonuclease activity and its recruitment of the exonuclease EXO1. The staggered DSB are therefore processed into blunt DSB. Thereafter, DNA repair can occur, requiring (i) the generation and the maintenance of a synapsis between Sμ-Sx regions, (ii) cell cycle arrest (iii) NHEJ repair. However, study of patients affected by not yet characterized CSR deficiencies strongly suggests that other important factors play a role in generation or repair of AID-induced DNA damage.

References

1. Tonegawa S. Somatic generation of antibody diversity. Nature 1983; 302:575-581.
2. Kinoshita K, Honjo T. Unique and unprecedented recombination mechanisms in class switching. Curr Opin Immunol 2000; 12:195-198.
3. Storb U, Peters A, Klotz E et al. Cis-acting sequences that affect somatic hypermutation of Ig genes. Immunol Rev 1998; 162:153-160.
4. Rajewsky K. Clonal selection and learning in the antibody system. Nature 1996; 381:751-758.
5. Vinuesa CG, Tangye SG, Moser B et al. Follicular B helper T-cells in antibody responses and autoimmunity. Nat Rev Immunol 2005; 5:853-865.
6. Korthauer U, Graf D, Mages HW et al. Defective expression of T-cell CD40 ligand causes X-linked immunodeficiency with hyper-IgM. Nature 1993; 361:539-541.
7. Ferrari S, Giliani S, Insalaco A et al. Mutations of CD40 gene cause an autosomal recessive form of immunodeficiency with hyper IgM. Proc Natl Acad Sci USA 2001; 98:12614-12619.
8. Doffinger R, Smahi A, Bessia C et al. X-linked anhidrotic ectodermal dysplasia with immunodeficiency is caused by impaired NF-kappaB signaling. Nat Genet 2001; 27:277-285.
9. Muramatsu M, Sankaranand VS, Anant S et al. Specific expression of activation-induced cytidine deaminase (AID), a novel member of the RNA-editing deaminase family in germinal center B-cells. J Biol Chem 1999; 274:18470-18476.
10. Okazaki IM, Kinoshita K, Muramatsu M et al. The AID enzyme induces class switch recombination in fibroblasts. Nature 2002; 416:340-345.
11. Yoshikawa K, Okazaki IM, Eto T et al. AID enzyme-induced hypermutation in an actively transcribed gene in fibroblasts. Science 2002; 296:2033-2036.
12. Ito S, Nagaoka H, Shinkura R et al. Activation-induced cytidine deaminase shuttles between nucleus and cytoplasm like apolipoprotein B mRNA editing catalytic polypeptide 1. Proc Natl Acad Sci USA 2004; 101:1975-1980.

13. McBride KM, Barreto V, Ramiro AR et al. Somatic hypermutation is limited by CRM1-dependent nuclear export of activation-induced deaminase. J Exp Med 2004; 199:1235-1244.
14. Brar S, Watson M, Diaz M. Activation-induced cytosine deaminase, AID, is actively exported out of the nucleus but retained by the induction of DNA breaks. J Biol Chem 2004.
15. Aoufouchi S, Faili A, Zober C et al. Proteasomal degradation restricts the nuclear lifespan of AID. J Exp Med 2008; 205:1357-1368.
16. Muramatsu M, Kinoshita K, Fagarasan S et al. Class switch recombination and hypermutation require activation-induced cytidine deaminase (AID), a potential RNA editing enzyme. Cell 2000; 102:553-563.
17. Petersen-Mahrt SK, Harris RS, Neuberger MS. AID mutates E. coli suggesting a DNA deamination mechanism for antibody diversification. Nature 2002; 418:99-103.
18. Chaudhuri J, Tian M, Khuong C et al. Transcription-targeted DNA deamination by the AID antibody diversification enzyme. Nature 2003; 422:726-730.
19. Ramiro AR, Stavropoulos P, Jankovic M et al. Transcription enhances AID-mediated cytidine deamination by exposing single-stranded DNA on the nontemplate strand. Nat Immunol 2003; 4:452-456.
20. Dickerson SK, Market E, Besmer E et al. AID Mediates Hypermutation by Deaminating Single Stranded DNA. J Exp Med 2003; 197:1291-1296.
21. Bransteitter R, Pham P, Scharff MD et al. Activation-induced cytidine deaminase deaminates deoxycytidine on single-stranded DNA but requires the action of RNase. Proc Natl Acad Sci USA 2003; 100:4102-4107.
22. Petersen-Mahrt SK, Harris RS, Neuberger MS. AID mutates E. coli suggesting a DNA deamination mechanism for antibody diversification. Nature 2002; 418:99-104.
23. Rada C, Williams GT, Nilsen H et al. Immunoglobulin isotype switching is inhibited and somatic hypermutation perturbed in UNG-deficient mice. Curr Biol 2002; 12:1748-1755.
24. Imai K, Slupphaug G, Lee WI et al. Human uracil-DNA glycosylase deficiency associated with profoundly impaired immunoglobulin class-switch recombination. Nat Immunol 2003; 4:1023-1028.
25. Shen HM, Ratnam S, Storb U. Targeting of the activation-induced Cytosine deaminase is strongly influenced by the sequence and structure of the targeted DNA. Mol Cell Biol 2005; 25:10815-10821.
26. Basu U, Chaudhuri J, Alpert C et al. The AID antibody diversification enzyme is regulated by protein kinase A phosphorylation. Nature 2005; 438:508-511.
27. Vuong BQ, Lee M, Kabir S et al. Specific recruitment of protein kinase A to the immunoglobulin locus regulates class-switch recombination. Nat Immunol 2009; 10:420-426.
28. Revy P, Muto T, Levy Y et al. Activation-induced cytidine deaminase (AID) deficiency causes the autosomal recessive form of the Hyper-IgM syndrome (HIGM2). Cell 2000; 102:565-575.
29. Fagarasan S, Muramatsu M, Suzuki K et al. Critical roles of activation-induced cytidine deaminase in the homeostasis of gut flora. Science 2002; 298:1424-1427.
30. Quartier P, Bustamante J, Sanal O et al. Clinical, immunologic and genetic analysis of 29 patients with autosomal recessive hyper-IgM syndrome due to Activation-Induced Cytidine Deaminase deficiency. Clin Immunol 2004; 110:22-29.
31. Catalan N, Selz F, Imai K et al. The block in immunoglobulin class switch recombination caused by activation-induced cytidine deaminase deficiency occurs prior to the generation of DNA double strand breaks in switch mu region. J Immunol 2003; 171:2504-2509.
32. Ta VT, Nagaoka H, Catalan N et al. AID mutant analyses indicate requirement for class-switch-specific cofactors. Nat Immunol 2003; 4:843-848.
33. Barreto V, Reina-San-Martin B, Ramiro AR et al. C-terminal deletion of AID uncouples class switch recombination from somatic hypermutation and gene conversion. Mol Cell 2003; 12:501-508.
34. Imai K, Zhu Y, Revy P et al. Analysis of class switch recombination and somatic hypermutation in patients affected with autosomal dominant hyper-IgM syndrome type 2. Clin Immunol 2005; 115:277-285.
35. Di Noia J, Neuberger MS. Altering the pathway of immunoglobulin hypermutation by inhibiting uracil-DNA glycosylase. Nature 2002; 419:43-48.
36. Guikema JE, Linehan EK, Tsuchimoto D et al. APE1- and APE2-dependent DNA breaks in immunoglobulin class switch recombination. J Exp Med 2007; 204:3017-3026.
37. Zeng X, Winter DB, Kasmer C et al. DNA polymerase eta is an A-T mutator in somatic hypermutation of immunoglobulin variable genes. Nat Immunol 2001; 2:537-541.
38. Zan H, Komori A, Li Z et al. The translesion DNA polymerase zeta plays a major role in Ig and bcl-6 somatic hypermutation. Immunity 2001; 14:643-653.
39. Schenten D, Kracker S, Esposito G et al. Pol zeta ablation in B-cells impairs the germinal center reaction, class switch recombination, DNA break repair and genome stability. J Exp Med 2009; 206:477-490.
40. Zan H, Shima N, Xu Z et al. The translesion DNA polymerase theta plays a dominant role in immunoglobulin gene somatic hypermutation. EMBO J 2005; 24:3757-3769.

41. Masuda K, Ouchida R, Takeuchi A et al. DNA polymerase theta contributes to the generation of C/G mutations during somatic hypermutation of Ig genes. Proc Natl Acad Sci USA 2005; 102:13986-13991.
42. Rada C, Di Noia JM, Neuberger MS. Mismatch recognition and uracil excision provide complementary paths to both Ig switching and the A/T-focused phase of somatic mutation. Mol Cell 2004; 16:163-171.
43. Kavli B, Andersen S, Otterlei M et al. B-cells from hyper-IgM patients carrying UNG mutations lack ability to remove uracil from ssDNA and have elevated genomic uracil. J Exp Med 2005; 201:2011-2021.
44. Nilsen H, Stamp G, Andersen S et al. Gene-targeted mice lacking the Ung uracil-DNA glycosylase develop B-cell lymphomas. Oncogene 2003; 22:5381-5386.
45. Begum NA, Stanlie A, Doi T et al. Further evidence for involvement of a noncanonical function of uracil DNA glycosylase in class switch recombination. Proc Natl Acad Sci USA 2009; 106:2752-2757.
46. Ehrenstein MR, Rada C, Jones AM et al. Switch junction sequences in PMS2-deficient mice reveal a microhomology-mediated mechanism of Ig class switch recombination. Proc Natl Acad Sci USA 2001; 98:14553-14558.
47. Schrader CE, Edelmann W, Kucherlapati R et al. Reduced isotype switching in splenic B-cells from mice deficient in mismatch repair enzymes. J Exp Med 1999; 190:323-330.
48. Schrader CE, Vardo J, Stavnezer J. Role for mismatch repair proteins Msh2, Mlh1 and Pms2 in immunoglobulin class switching shown by sequence analysis of recombination junctions. J Exp Med 2002; 195:367-373.
49. Kadyrov FA, Dzantiev L, Constantin N et al. Endonucleolytic function of MutLalpha in human mismatch repair. Cell 2006; 126:297-308.
50. Schrader CE, Guikema JE, Linehan EK et al. Activation-induced cytidine deaminase-dependent DNA breaks in class switch recombination occur during G1 phase of the cell cycle and depend upon mismatch repair. J Immunol 2007; 179:6064-6071.
51. Peron S, Metin A, Gardes P et al. Human PMS2 deficiency is associated with impaired immunoglobulin class switch recombination. J Exp Med 2008; 205:2465-2472.
52. De Vos M, Hayward BE, Charlton R et al. PMS2 mutations in childhood cancer. J Natl Cancer Inst 2006; 98:358-361.
53. Kratz CP, Niemeyer CM, Juttner E et al. Childhood T-cell non-Hodgkin's lymphoma, colorectal carcinoma and brain tumor in association with cafe-au-lait spots caused by a novel homozygous PMS2 mutation. Leukemia 2007.
54. Shen HM, Tanaka A, Bozek G et al. Somatic hypermutation and class switch recombination in Msh6(-/-) Ung(-/-) double-knockout mice. J Immunol 2006; 177:5386-5392.
55. Peron S, Pan-Hammarstrom Q, Imai K et al. A primary immunodeficiency characterized by defective immunoglobulin class switch recombination and impaired DNA repair. J Exp Med 2007; 204:1207-1216.
56. Wuerffel R, Wang L, Grigera F et al. S-S synapsis during class switch recombination is promoted by distantly located transcriptional elements and activation-induced deaminase. Immunity 2007; 27:711-722.
57. Manis JP, Morales JC, Xia Z et al. 53BP1 links DNA damage-response pathways to immunoglobulin heavy chain class-switch recombination. Nat Immunol 2004; 5:481-487.
58. Celeste A, Petersen S, Romanienko PJ et al. Genomic instability in mice lacking histone H2AX. Science 2002; 296:922-927.
59. Petersen S, Casellas R, Reina-San-Martin B et al. AID is required to initiate Nbs1/gamma-H2AX focus formation and mutations at sites of class switching. Nature 2001; 414:660-665.
60. Reina-San-Martin B, Chen HT, Nussenzweig A et al. ATM is required for efficient recombination between immunoglobulin switch regions. J Exp Med 2004; 200:1103-1110.
61. Kracker S, Bergmann Y, Demuth I et al. Nibrin functions in Ig class-switch recombination. Proc Natl Acad Sci USA 2005; 102:1584-1589.
62. Manis JP, Gu Y, Lansford R et al. Ku70 is required for late B-cell development and immunoglobulin heavy chain class switching. J Exp Med 1998; 187:2081-2089.
63. Casellas R, Nussenzweig A, Wuerffel R et al. Ku80 is required for immunoglobulin isotype switching. EMBO J 1998; 17:2404-2411.
64. Manis JP, Dudley D, Kaylor L et al. IgH class switch recombination to IgG1 in DNA-PKcs-deficient B-cells. Immunity 2002; 16:607-617.
65. Etzioni A, Ben-Barak A, Peron S et al. Ataxia-telangiectasia in twins presenting as autosomal recessive hyper-immunoglobulin M syndrome. Isr Med Assoc J 2007; 9:406-407.
66. Lahdesmaki A, Taylor AM, Chrzanowska KH et al. Delineation of the role of the Mre11 complex in class switch recombination. J Biol Chem 2004; 279:16479-16487.

67. Pan-Hammarstrom Q, Dai S, Zhao Y et al. ATM is not required in somatic hypermutation of VH, but is involved in the introduction of mutations in the switch mu region. J Immunol 2003; 170:3707-3716.
68. Du L, Dunn-Walters DK, Chrzanowska KH et al. A regulatory role for NBS1 in strand-specific mutagenesis during somatic hypermutation. PLoS ONE 2008; 3:e2482.
69. Yan CT, Boboila C, Souza EK et al. IgH class switching and translocations use a robust nonclassical end-joining pathway. Nature 2007; 449:478-482.
70. Pan-Hammarstrom Q, Jones AM, Lahdesmaki A et al. Impact of DNA ligase IV on nonhomologous end joining pathways during class switch recombination in human cells. J Exp Med 2005; 201:189-194.
71. Soulas-Sprauel P, Le Guyader G, Rivera-Munoz P et al. Role for DNA repair factor XRCC4 in immunoglobulin class switch recombination. J Exp Med 2007; 204:1717-1727.
72. Hein K, Lorenz MG, Siebenkotten G et al. Processing of switch transcripts is required for targeting of antibody class switch recombination. J Exp Med 1998; 188:2369-2374.
73. Yu K, Chedin F, Hsieh CL et al. R-loops at immunoglobulin class switch regions in the chromosomes of stimulated B-cells. Nat Immunol 2003; 4:442-451.
74. Fukita Y, Jacobs H, Rajewsky K. Somatic hypermutation in the heavy chain locus correlates with transcription. Immunity 1998; 9:105-114.

Chapter 16

Ligase IV Syndrome

Dimitry A. Chistiakov*

Abstract

Ligase IV (LIG4) syndrome belongs to the group of hereditary disorders associated with impaired DNA damage response mechanisms. Clinically and morphologically, patients affected with this syndrome are characterized by microcephaly, unusual facial features, growth retardation, developmental delay, skin anomalies and are typically pancytopenic. The disease leads to acute radiosensitivity, immunodeficiency and bone marrow abnormalities. LIG4 syndrome arises from hypomorphic mutations in the *LIG4* gene encoding DNA ligase IV; a component of the nonhomologous end-joining machinery, which represents a major mechanism of repair of double strand DNA breaks in mammals. The hypomorphic mutations do not completely abolish but significantly reduce enzyme function. This results in impaired V(D)J recombination, the essential rejoining process in T- and B-cell development, in whose ligase IV plays the key role. As a consequence, patients with LIG4 syndrome frequently develop multiple immune abnormalities, clinically overlapping with severe combined immunodeficiency syndrome.

Ligase IV Syndrome: Clinical Features

Ligase IV (LIG4) syndrome (MIM 606593) belongs to the group of hereditary disorders associated with impaired DNA damage response mechanisms. This autosomal recessive disease is extremely rare. It is difficult to estimate the incidence of this disorder since LIG4 syndrome has been recognized only in few cases. Clinically and morphologically, patients affected with this syndrome are characterized by microcephaly, unusual facial features, growth retardation, developmental delay, skin anomalies and are typically pancytopenic.[1] The disease leads to acute radiosensitivity, immunodeficiency and bone marrow abnormalities.[2] LIG4 syndrome shares similar clinical phenotypes with other rare DNA damage response diseases such as Seckel syndrome, Nijmegen breakage syndrome (NBS) and Fanconi anemia (Table 1). Cell lines from the patients with LIG syndrome show pronounced radiosensitivity;[3] however, unlike NBS cell lines, they show normal cell cycle checkpoint responses but impaired DSB rejoining. An unexpected V(D)J recombination phenotype was observed involving a small decrease in rejoining frequency coupled with elevated imprecision at signal junctions.[3]

DNA Ligase IV Is a Component of Nonhomologous End-Joining Pathway of DNA Repair

LIG4 syndrome arises from hypomorphic mutations in the *LIG4* gene encoding DNA ligase IV (MIM 601837, EC 6.5.1.1), a component of the nonhomologous end-joining (NHEJ) machinery, which represents a major mechanism of repair of double strand DNA breaks (DSBs) in

*Dimitry A. Chistiakov—Department of Molecular Diagnostics, National Research Center GosNIIGenetika, 1st Dorozhny Proezd 1, 117545 Moscow, Russia.
Email: dimitry.chistiakov@lycos.com

Diseases of DNA Repair, edited by Shamim I. Ahmad. ©2010 Landes Bioscience and Springer Science+Business Media.

Table 1. Clinical overlaps between LIG4 syndrome, Nijmegen breakage syndrome, Fanconi anemia and Seckel syndrome

Clinical Phenotype	Nijmegen Breakage Syndrome*	*Fanconi anemia**	Seckel Syndrome*	LIG4 Syndrome**
Microcephaly	+	+	+	+
Sloping forehead	+	–	+	+
Palpebral fissures	Upslanting	–	Downslanting	Upslanting
Epicanthic folds	+	–	–	+
Micrognathia	+	–	+	–
External ear abnormalities	+	+	+	–
Long/large nose	+	–	+	+
Long philtrum	+	–	–	–
Polydactyly	+	+	–	–
Clinodactyly	+	–	+	+
Sindactyly	+	–	–	–
Skin abnormalities	+	+	+	–
Mental retardation/developmental delay	+	+	+	+
Malignancy	+	+	+	+
Recurrent infections	+	+	–	+
Pancytopenia	–	+	–	+
Genital abnormalities	+	+	+	+
Radiosensitivity	+	+	+	+
Gene	NBS1	FANC	ATR	LIG4

+Features present. -Features absent. *Features obtained from The London Dysmorphology Database.[4] **Data taken from Ben-Omran et al.[5]

mammals.[6] DSBs are generally induced by ionizing radiation and during V(D)J recombination, the essential rejoining process that serves to rearrange the variable, diversity and joining segments in T- and B-cell development.[7]

NHEJ repair is a multistep process that requires the involvement of several proteins (Fig. 1). Briefly, the heterodimeric Ku 70/80 protein binds to the DNA ends in a site where a DSB occurs. DNA-bound Ku serves to recruit the complex between Artemis and DNA-dependent protein kinase catalytic subunit (DNA-PKcs) and then activate its kinase activity as well as to attract binding the DNA ligase IV/XRCC4 complex to the DSB site. DNA-PKcs undergoes autophosphorylation and also phosphorylates Artemis, which regulates its binding to DNA and is required for its function in DSB rejoining.[8] The phosphorylation permits the Artemis/DNA-PKcs complex to function as an endonuclease allowing it to cleave both 5′ and 3′ overhangs of any length.[9] The DNA ligase IV/XRCC4 complex exists endogenously in a pre-adenylated form. Recruitment of the DNA ligase IV/XRCC4 complex to DNA causes inward translocation of Ku allowing the AMP moiety of the ligase-adenylate complex to transiently attach to the DNA end promoting ligation.[10]

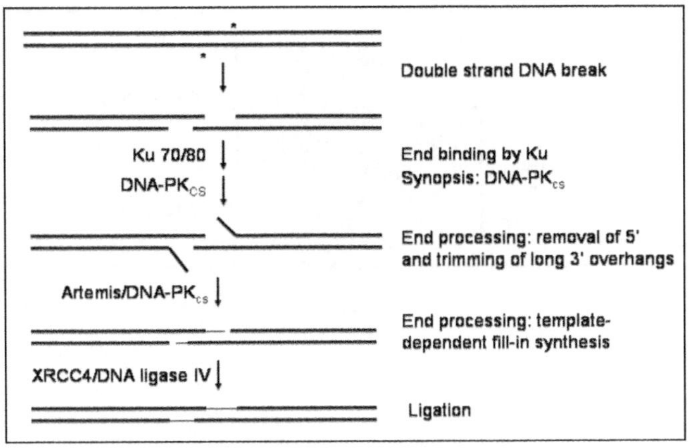

Figure 1. The mechanism of the vertebrate nonhomologous DNA end joining (NHEJ) pathway. A double-stranded DNA break (DSB) formation induces Ku70/80 heterodimer binding to DNA ends in the DSB site. End binding by Ku allows holding the DNA ends in proximity. This step can be referred to as synapsis. Though Ku alone may not be able to achieve synapsis, there is some evidence that DNA-PKcs is capable of noncovalently holding on to each of the two DNA ends. For incompatible DNA ends generally generated by ionizing radiation, there is often need for nucleolytic processing. If so, Artemis:DNA-PKcs are likely to carry out most and perhaps all, such processing. For only a subset of DNA ends, end alignment can occur at chance sites of terminal microhomology of 1-4 nucleotides. End processing not only includes nucleolytic removal of sections of the DNA termini, but also the filling in of gaps by polymerases. Ligation is the final step and it requires a ligatable nick on each strand. Ligation in NHEJ is performed by the XRCC4:DNA ligase IV complex. Rejoining of a simple pair of DNA ends (blunt or compatible), may not require not require a nuclease or a polymerase.

DNA Ligase IV: Structure and Function

The *LIG4* gene encompassing 10.9 kb and consisting of 2 exons and one intron was mapped at chromosome 13q33-q34.[11] A cDNA encoding a polypeptide of 911 amino acids, with a predicted mass of 96 kDa, was first isolated from HeLa cells.[11]

DNA ligase IV has a complex structure formed by several domains (Fig. 2A). The enzyme contains a conserved ligase domain at its N-terminus and a tandem BRCT domain at its C-terminus.[12] Interaction with XRCC4 requires the region between the BRCT domains and likely part of the BRCT domain.[13] The first step of ligation involves the formation of a covalent AMP-enzyme intermediate with AMP being attached to the enzyme via a highly conserved lysine residue. The second step involves the formation of a DNA-adenylate complex followed finally by rejoining. All ATP-dependent DNA ligases have a modular structure of two domains, an adenylation domain (AdD) and an oligo-binding domain (OBD).[14] The larger eukaryotic ligases such as ligase I and ligase IV also possess an additional N-terminal DNA-binding domain (DBD) that is required for efficient ligation and enables these ligases to encircle DNA.[15] Six conserved motifs, designated motifs I, III, IIIa, IV, V and VI, have been identified among covalent nucleotide transferases, of which five are found in the AdD. Motifs I, III, IIIa, IV and V are essential for ATP binding and the autoadenylation reaction. Motif I, encompassing the conserved lysine residue, forms the active site loop of the enzyme and constitutes part of the ATP binding pocket. Based on the crystal structures of a number of DNA ligase complexes, it has been proposed that ligases undergo profound conformational changes upon ATP binding and/or DNA binding.[16] This is exemplified by the rotation of the OBD. Motif VI lies within the OBD, distant from the active site on AdD.[17] However, upon ATP-binding this face of OBD

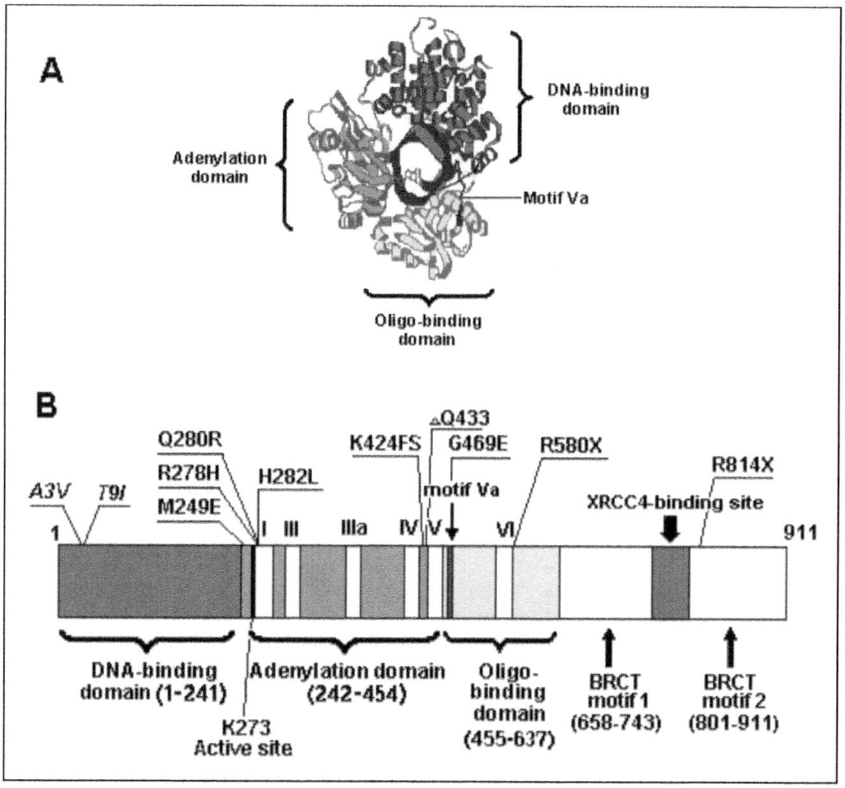

Figure 2. A) Structure of DNA ligase IV. The enzyme comprises three domains: the DNA-binding domain (designated in red), adenylation domain (green) and oligo-binding domain (yellow). The adenylation and oligo-binding domains constitute the catalytic domain. The highly conserved motif Va (colored as dark blue) forms a loop-like structure on the surface of the oligo-binding domain and makes a direct contact to the DNA helix (grey). B) Localization of mutations found in the LIG4 syndrome patients. The green regions (designated I-VI) represent conserved motifs identified in DNA ligases. Two C-terminal BRCT motifs and XRCC4-binding site (grey) are required for interaction with XRCC4. Two N-terminal polymorphisms (A3V and T9I) are designated in italic. Adopted from Marchetti et al.[19] A color version of this figure is available online at www.landesbioscience.com/curie.

moves towards the active site and residues, including those from motif VI, participate in the adenylation reaction.[16] Subsequently, the OBD moves away from the active site and swivels around placing motif VI far from the AdD, orientating the DNA-binding surface of OBD towards the now adenylated AdD.[18] This switching is essential for the catalytic cycle as it most likely prevents the formation of nonproductive complexes between non-adenylated ligase and unnicked/unbroken DNA.

In a recent study, Marchetti et al[19] identified a highly conserved motif in ligase IV called motif Va. The motif Va, which consists of 9 amino acids (codons 468-476), forms a loop-like structure on the surface of OBD and makes direct contact with the DNA. Experiments with genetically engineered mutants showed the involvement of two glycine residues at positions 468 and 469 in adenylate complex formation in vitro, whereas the remaining residues in the motif Va do not play critical roles in DNA ligase IV-adenylate complex formation but significantly impair the double-stranded ligation activity.[19]

Animal Models with Genetic Defects in *Lig4*

Lig4-deficient mice die on late embryonic stage from massive neuronal apoptosis, arrested lymphogenesis and multiple cellular defects.[20] This could happen due to p53-directed apoptosis caused by unrepaired DNA damage since mice with double deficiency for *Lig4/p53* is able to escape from the embryonic neuronal apoptosis but not from defects in lymphocyte development. Finally, the *Lig4/p53* double null mouse routinely dies from pro-B lymphomas.[20] *Rad54/Lig4* double-mutant mice showed serious defects in cell proliferation due to the inability to repair spontaneous DSBs and maintain chromosome stability.[21] *Lig4* inactivation, along with other DSB repair factors and p53, caused embryonic medulloblastoma development accompanied with chromosomal rearrangements and amplification of genomic regions containing medulloblastoma-specific protooncogenes.[22] DT40 chicken B-lymphocyte cells knocked out for the *Lig4* gene showed marked sensitivity to X-rays and DNA-damaging agents.[23] Taken together, these findings suggest for a key role of DNA-ligase IV in NHEJ, which is vitally essential for cell growth and proliferation.

Since DNA ligase IV is essential, mutations identified in LIG4 syndrome are hypomorphic, that is they confer residual activity but impair NHEJ repair. In support of this, Sharpless et al[24] created a haploinsufficient mouse that lacked one copy of *Lig4* on the basis of a tumor-prone *Ink4a*$^{-}$/*Arf*$^{-}$ strain double deficient for tumor suppressors, p16^{INK4a} and p19^{ARF}. These mice exhibited dramatically increased radiosensitivity and genome instability resulted in various chromosome aberrations and development of soft-tissue sarcomas. Therefore, lack of a single copy of *Lig4* does not arrest but sufficiently alters NHEJ-mediated DNA repair of spontaneously generated DSBs that promotes loss of chromosome stability and tumorigenesis.

X-ray-hypersensitive mutant SX10 of mouse FM3A cells showed marked radiosensitivity due to the defective NHEJ DNA repair.[25] The high vulnerability of this cell line to ionizing radiation was found to arise from the truncating mutation in *Lig4* W471X that leads to loss of the C-terminal XRCC4-binding domain in the mutant enzyme. The functional properties of the W471 mutant have not been evaluated but this mutation is likely to be a null allele and results in nonfunctional unstable product. The mutation is dominant since it greatly impairs DNA repair in SX10 cells despite the presence of the normal functionally active copy of *Lig4*.

Recently, a murine model of LIG4 syndrome has been developed.[26] The mouse strain has the hypomorphic Y288C mutation in the catalytic domain of Lig4 and exhibits clinical features similar to those for human disease, i.e., immunodeficiency, growth retardation, progressive loss of haematopoietic stem cells and bone marrow failure during ageing. The mutant protein is stably complexed with Xrcc4 but has an approximately two-fold less adenylation and DSB ligation activity. The mutation also affects *Lig4* expression decreasing its level by five-fold. The impaired activity of the mutated enzyme led to stable increase in numbers of unrepaired DSBs that in turn decreased proliferative potential, increased turnover of the haematopoietic stem cell population, decreased self-renewal and hence promoted age-dependent decline in multipotent cells, which becomes manifested as loss of bone marrow cellularity and erythropoiesis.[26]

Ligase IV Mutations Causing LIG4 Syndrome in Humans

Mutation R278H

The R278H mutation has been first described by Riballo et al[27] in cell line 180BR derived from a patient with lymphoblastic leukaemia who developed acute tissue reactions in response to radiotherapy. The R278H mutation lies within the highly conserved motif KxDGxR. The motif includes the lysine residue, which forms a covalent bond with AMP in the ligase–AMP complex and the R278H mutation in turn impairs the enzyme–adenylate complex formation that results in 10-fold reduction of the ligase activity (Table 2). The 180BR cells carrying this defect showed markedly reduced ability of DNA ligase IV to form an enzyme–adenylate complex but its ability to interact with XRCC4 remained unimpaired. Using in vitro expression assay for cell-free extracts from the 180BR cells, Riballo et al[27] showed that the mutated enzyme retained residual adenylation activity at ATP concentrations exceeding 50-fold those necessary for the wild-type ligase. Thus, the 180BR mutation severely inhibits, but does not abolish, the ability of DNA ligase IV to form an enzyme—adenylate intermediate and then reduces the rejoining activity.[28] As a consequence, this cell line showed marked cellular radiosensitivity and defective DSB repair but surprisingly did not appear to be seriously defective in the ability to carry out V(D)J recombination.[29]

Mutation G469E

Another mutation, G469E, residing in the conserved motif Va,[3] also greatly affected the DSB ligation reaction, reducing the adenylation activity of the enzyme by 100-fold (Fig. 2B).[19] The G468E mutant protein showed little functional activity either for adenylation or ligation activities, most likely due to a significant impact on conformation. G469, which is only slightly less buried in the OBD than G468, also appears to be important for protein conformation. The G469E substitution generates a protein with significantly reduced conformational stability that impacts upon activity.

Mutation M249V

This mutation has recently been described in a Japanese girl with a progressing combined immunodeficiency accompanied with non-Hodgkin's diffuse large B-cell lymphoma.[30] The patient had several clinical symptoms of LIG4 syndrome, i.e., microcephaly, growth delay, immune defects (decreased serum IgM and IgG, reduced numbers of T- and B-cells), radiosensitivity and skin abnormalities. She was double heterozygous for M249V and a 5-bp deletion at nucleotide position1544 that caused a frameshift at the amino acid position 424. The functional effect of the M249V mutation has not been evaluated yet. The methionine residue at amino acid 249 is located near an ATP-binding site and is conserved among other DNA ligases. Like R278H, the M249V mutation is expected to alter ATP-binding properties of the enzyme but should not completely abolish the activity. The residual activity of the mutant protein seems to be insufficient for proper V(D)J recombination during T-cell and B-cell development that resulted in serious immunodeficiency and B-cell malignancy developed in this patient.

Frameshift Mutation (K424FS)

A 5-bp deletion at nucleotide position 1270-1274, leading to a frameshift at K424 (K424FS), causes a premature stop codon 20 amino acids downstream. The heterozygous mutation was found in three patients,[30,31] all exhibiting acute SCID phenotype. This putative truncated protein would lack the C terminus half of the protein including the two BRCT motifs and the XRCC4-interacting domain. The protein expression level of the K424FS mutant in skin fibroblasts, derived from one of those patients, was profoundly reduced and not detectable with C-terminus-specific antibodies.[31] The frameshift mutation results in a nonfunctional product and therefore represents a null allele.

Mutation Q280R

The heterozygous Q280R substitution was found along with the frameshift mutation K424FS in two siblings, both affected with overt immunodeficiency.[31] The Q280R amino acid change lies in the vicinity to the K273 active site, which is part of the conserved ligase motif (273-278) (Fig. 2B). This mutation is hypomorphic since it does not alter ligase expression but seriously disturbs function of the enzyme. Fibroblasts from patients having Q280R mutation showed normal ligation activity in vitro and a 30-fold decreased fidelity in NHEJ-mediated V(D)J recombination compared to the wild-type fibroblasts (Table 2).[31] Due to location in the ATP-binding site, Q280R is likely to affect ligase function through impairment of the adenylation reaction.

Truncating Mutations (R580X and R814X)

O' Driscoll et al[3] found two truncating mutations (R580X and R814X), both were in heterozygous state. Ben-Omran et al[5] reported a patient homozygous for the R814X mutation. The patient, who had clinical features of LIG4 syndrome, developed acute T-cell lymphoma and severe neutropenia persisted until his death from sepsis. Skin fibroblasts derived from patients carrying these mutations

Table 2. Summary of mutation effects on ligase IV function

Mutation	Effect	Localization of Mutated Protein	Adenylation Activity	DSB Ligation Activity	Nick Ligation
R580X	Truncating mutation: causes loss of the C-terminal domain responsible for XRCC4 binding	Cytoplasm	-	-	-
R814X	Truncating mutation: causes loss of the C-terminal BRCT domain, impairs XRCC4 binding	Nucleus (<10% expressed)	5% (CHO cells)	10-15%	-
G469E	Located in the conserved motif Va; impairs catalytic function	Nucleus	Not detectable	Not detectable	Borderline for detection
R278H	Located in the conserved motif close to the active center; impairs ATP binding	Nucleus	~5%	5-10%	5-10%
A3V+T9I	Located on the N-terminus; moderately impairs catalytic function	Nucleus	~50%	~30-50%	-
A3V+T9I+ R258H	Severely impairs catalytic function	Nucleus	Not detectable	Not detectable	~1%
M249E	Changes a conserved Met close to the ATP-binding site; likely impairs ATP binding	N/A	N/A	N/A	N/A
c.1270-1274del (K424FS)	Causes frameshift in codon 424 resulting in loss of the XRCC4-binding domain	N/A	N/A	N/A	N/A
Q280R	Located in the conserved motif close to the active center; likely impairs ATP binding	N/A	N/A	Normal but 30-fold decreased fidelity	N/A
delQ433	A 3-bp deletion located between the conserved motifs IV and V and resulting in loss of glutamine-433; likely affects protein stability	Not detectable	N/A	N/A	N/A
H282L	Located in the conserved motif close to the active center; likely impairs ATP binding	N/A	N/A	N/A	N/A

N/A, not available.

were highly vulnerable to X-rays and UV exposure due to the impaired DSB repair. Peripheral blood lymphocytes isolated from these patients showed higher chromosome instability resulting in 10-fold increase in spontaneous chromosome breakage and 2.5-fold decrease in fidelity of V(D)J recombination. Defects in V(D)J recombination caused severe immunodeficiency, pancytopenia and myelodisplasia in carriers of these mutations. R580X results in loss of the C terminus encompassing the two BRCT domains involved in XRCC binding (Fig. 2B). This truncation has deleterious effects on the mutated protein. Due to the loss of the bipartite nucleus localization signal sequence at position 632-638, which targets DNA ligase to the nucleus, the R580X mutant protein was detectable only in cytoplasm and not in the nucleus where the enzyme is normally present.[32] In addition, the R580X mutant is rapidly degraded due to the high instability. Therefore, the R580X mutation is likely to represent a null allele since the protein is not stably expressed, does not interact with XRCC4 and does not enter the nucleus.

R814X lies in the intervening region between two BRCT domains, which has been reported to be important for interaction with XRCC4 (Fig. 2B) and leads to the synthesis of the protein lacking one of two BRCT motifs.[33] However, the presence of a single BRCT motif in the R814X mutant is enough to interact with XRCC4, but with significantly lower efficiency, since only ~10% of XRCC4 was co-immunoprecipitated with the R814X mutant enzyme.[33] Due to the impaired binding to XRCC4, the R814X ligase also had markedly reduced adenylation and DSB rejoining activities which did not exceed 10-15% of the wild-type enzyme activity (Table 2). The R814X mutant is expressed at level not exceeding 10% of the wild-type enzyme, most likely due to the nonsense-mediated decay.[3] Heterozygous carriages of the R814X and G489 mutations could result in severe clinical outcome that caused rapid progression from hypoplasia to bone marrow failure and further required bone marrow transplantation.[34]

N-Terminal Polymorphisms (A3V and T9I)

O' Driscoll et al[3] described a patient carrying three homozygous sequence alterations (A3V, T9I and R278H) in *LIG4*. The patient had typical clinical features of LIG4 syndrome, i.e., a 'bird-like' face, microcephaly, global developmental delay, pancytopenia and severe skin abnormalities. Skin fibroblasts (cell line 411BR) taken from this patient exhibited high radiosensitivity and impaired profile of the NHEJ-mediated DSB repair. In 411BR cells, the mutant protein was normally expressed and could interact with XRCC4, but had impaired activity. Complexes between the mutated enzyme containing the N-terminal substitutions, A3V and T9I, with or without the R278H mutation and XRCC4 showed the same ability to interact with XRCC4. Complexes with the N-terminal substitutions alone were also as efficient for nick ligation as the wild-type complex whereas no activity above background was detected with complexes harboring the combined A3V, T9I and R278H alterations.[3] These observations suggest that the impaired activity of the ligase containing all the three mutations primarily arises from R278H, while both N-terminal changes have no or moderate effect.

Further studies revealed that the N-terminal amino acid substitutions are actually two linked single nucleotide polymorphisms [SNPs; rs1806389 (A3V) and rs1805388 (T9I)], with frequency of a minor allele of 7% (A3V) and 19% (T9I) in the general population.[35] These SNPs were shown to constitute a haplotype modulating risk of multiple myeloma, a malignant disorder of plasma cells characterized in 60% of cases, by aberrant class switch recombination.[36] NHEJ and DNA ligation by ligase IV is important in all of these mechanisms of antigen receptor rearrangement. Although the A3V and T9I markers did not influence the ligase IV stability,[3] they were predicted to be functional by increasing the hydrophobicity of the region.[35] Girard et al[32] showed that both polymorphisms do have a mild effect decreasing by 50% of the adenylation activity of the enzyme with a double mutation (A3V + T9I) (Table 2). Furthermore, Girard et al,[3] using a more precise approach to measure the DNA ligation activity of LIG4/XRCC4, found that a triple mutant (A3V + T9I + R278H) possesses a very low but still detectable residual activity in a nick ligation assay, which is less than 1% of the wild-type LIG4/XRCC4 activity. Therefore, in the triple mutant protein, R278H has a major impact abolishing the adenylation and ligation activities of the enzyme by ~90-95%, whereas both N-terminal markers only slightly impair the enzyme function.

LIG4 Mutations Associated with Severe Combined Immunodeficiency

Severe combined immunodeficiency (SCID) with sensitivity to ionizing radiation (RS-SCID) is typically caused by defects in the Artemis gene.[1] Van der Burg et al[37] recognized a patient with a new type of SCID carrying mutation in the ligase IV gene. The patient had no typical symptom of LIG4 syndrome such as microcephaly, neurological abnormalities or growth and developmental delay, but developed a clinical SCID picture characterized with extremely low levels of blood B-cells and radiosensitivity. The patient showed strongly impaired NHEJ resulted in delayed and aberrant V(D)J recombination occurred in B-cell precursors that caused severe abnormalities in B-cell development and maturation.

The *LIG4* mutation in the LIG4-deficient RS-SCID patient concerned a homozygous deletion of 3 nucleotides resulting in the deletion of glutamine at position 433. Q433 is located between two conserved stretches in the catalytic domain of LIG4 (Fig. 2B). The deletion resulted in undetectable levels of LIG4 protein, most probably due to instability of the mutated ligase. In addition to the deleterious effects on protein levels, the delQ433 mutation also greatly reduced efficiency of DSB repair affecting the ligation phase of V(D)J recombination, which resulted in an incomplete but severe block in precursor B-cell differentiation at the transition of the preB-I to the preB-II stage and strongly reduced numbers of B-cells in peripheral blood.[37]

Grunebaum et al[38] reported a patient with Omenn syndrome, a fatal form of SCID, treated with bone marrow transplantation. She had serious abnormalities in the development of T-cell lineages showing low numbers of T-lymphocytes, highly skewed T-cell repertoire and markedly reduced T-cell-mediated immune responses. The patient carried three heterozygous mutations in LIG4: T9I SNP, H282L substitution near the ATP-binding site and a frameshift K424FS mutation that causes a downstream premature stop after codon 442.

The H282L and K424FS mutations were previously found at compound heterozygous state in two siblings developed overt immunodeficiency that caused ~100-fold and 10-fold reduction in number of peripheral B- and T-lymphocytes, respectively, decrease in the variability of circulating immunoglobulins and impaired antimicrobial antibody responses.[39] The precise functional effects of these two mutations have been not studied yet. Like R258H and Q280R, H282L could affect ATP binding by the ligase and influence the adenylation reaction. Since K424FS fully abolishes both expression and function of the enzyme, H282L seems to be hypomorphic and remains residual activity in the mutated protein. However, the repairing activity and fidelity of the H282L mutant is highly impaired during V(D)H recombination since Enders et al[39] observed multiple aberrations in V_H-J_H joint segments.

Conclusion

Since LIG4 null-mutant mice are embryonic lethal[20] and biallelic null mutations have not been described to date in LIG4-deficient patients, viability of the DNA ligase IV deficiency syndrome appears to require at least one allele with a hypomorphic mutation. Mutations R278H, Q280R, H282L, M249E located in the vicinity of the active site are typical hypomorphic because they do not affect ligase expression and remain residual but impaired activity of the enzyme at level of 5-10% of that for the wild-type ligase. Carriers heterozygous for those mutations usually develop moderate defects in V(D)J recombination, mild immune abnormalities and malignancy. In contrast, mutations resided in OBD, i.e., in the C-terminal subdomain of the catalytic domain and in XRCC4-binding domain more dramatically inhibit the ligase function and also greatly decrease its expression. A truncating mutation R580X and a frameshift mutation K424FS resulting in loss of the C-terminal XRCC4-binding domain have deleterious effect on both expression and function of Lig4 and represent a null allele. Patients heterozygous for K424FS are reported to have acute defects in the development of B- and T-lymphocytes and show overt SCID phenotype.[38,39]

LIG4 syndrome defined by hypomorphic mutations in *LIG4* shares common phenotypic features with SCID caused by alterations in *Artemis*, NHEJ DNA repair protein. Similarly to LIG4 syndrome, patients with hypomorphic mutations in *Artemis* exhibit impaired V(D)J recombination resulted in moderate defects in B- and T-cell development, mild immunodeficiency, reduced numbers of lymphocytes, chromosome instability and malignancy (lymphoma).[40,41] However, truncating mutations and deletions leading to deficiency for Artemis strongly increase SCID severity causing the development of RS-SCID, an overt form of combined immunodeficiency characterized by complete depletion of T- and B-cells and acute radiosensitivity.[42,43]

In conclusion, LIG4 syndrome and other related autosomal recessive disorders arisen from hypomorphic mutations represent a unique pathologic condition characterized by moderate but not deadly abnormalities. Like in any autosomal recessive condition, severity of LIG4 syndrome phenotype is determined by features of a mutational alteration. Missense (hypomorphic) mutations remain a residual function and hence lead to mild dysfunction while nonsense (truncating) mutations generally fully abolish function and greatly increase disease severity.

References

1. O'Driscoll M, Gennery AR, Seidel J et al. An overview of three new disorders associated with genetic instability: LIG4 syndrome, RS-SCID and ATR-Seckel syndrome. DNA Repair (Amsterdam) 2004; 3:1227-1235.
2. Revy P, Buck D, le Deist F et al. The repair of DNA damages/modifications during the maturation of the immune system lessons from human primary immunodeficiency disorders and animal models. Adv Immunol 2005; 87:237-295.
3. O'Driscoll M, Cerosaletti KM, Girard P et al. DNA ligase IV mutations identified in patients exhibiting developmental delay and immunodeficiency. Mol Cell 2001; 8:1175-1185.
4. Winter RM, Baraitser M. The London Dysmorphology Database. Great Britain: Oxford University Press, 2003.
5. Ben-Omran TI, Cerosaletti K, Conacannon P et al. A patient with mutations in DNA ligase IV: clinical features and overlap with Nijmegen Breakage Syndrome. Am J Med Genet 2005; 137A:283-287.
6. Helleday T, Lo J, van Gent DC et al. DNA double-strand break repair: from mechanistic understanding to cancer treatment. DNA Repair (Amsterdam) 2007; 6:923-935.
7. Lieber MR, Ma Y, Pannicke U et al. The mechanism of vertebrate nonhomologous DNA end joining and its role in V(D)J recombination. DNA Repair (Amsterdam) 2004; 3:817-826.
8. Chan DW, Chen BP, Prithivirajsingh S et al. Autophosphorylation of the DNA-dependent protein kinase catalytic subunit is required for rejoining of DNA double-strand breaks. Genes Dev 2002; 16:2333-2338.
9. Ma Y, Pannicke U, Schwarz K et al. Hairpin opening and overhang processing by an Artemis:DNA-PKcs complex in V(D)J recombination and in nonhomologous end joining. Cell 2002; 108:781-794.
10. Kysela B, Doherty AJ, Chovanec M et al. Ku stimulation of DNA ligase IV-dependent ligation requires inward movement along the DNA molecule. J Biol Chem 2003; 278:22466-22474.
11. Wei Y-F, Robins P, Carter K et al. Molecular cloning and expression of human cDNAs encoding a novel DNA ligase IV and DNA ligase III, an enzyme active in DNA repair and recombination. Mol Cell Biol 1995; 15:3206-3216.
12. Robins P, Lindahl T. DNA ligase IV from HeLa cell nuclei. J Biol Chem 1996; 271:24257-24261.
13. Sibanda BL, Critchlow SE, Begun J et al. Crystal structure of an Xrcc4-DNA ligase IV complex. Nat Struct Biol 2001; 8:1015-1019.
14. Doherty AJ, Suh SW. Structural and mechanistic conservation in DNA ligases. Nucleic Acids Res 2000; 28:4051-4058.
15. Pascal JM, O'Brien PJ, Tomkinson AE et al. Human DNA ligase I completely encircles and partially unwinds nicked DNA. Nature 2004; 432:473-478.
16. Lee YE, Chang C, Song HK et al. Crystal structure of NAD(+)-dependent DNA ligase: modular architecture and functional implications. EMBO J 2000; 19:1119-1129.
17. Subramanya HS, Doherty AJ, Ashford SR et al. Crystal structure of an ATP- dependent DNA ligase from bacteriophage T7. Cell 1996; 85:607-615.
18. Hakansson K, Doherty AJ, Shuman S et al. X-ray crystallography reveals a large conformational change during guanyl transfer by mRNA capping enzymes. Cell 1997; 89:545-553.
19. Marchetti C, Walker SA, Odreman F et al. Identification of a novel motif in DNA ligases exemplified by DNA ligase IV. DNA Repair (Amsterdam) 2006; 5:788-798.
20. Frank KM, Sekiguchi JM, Seidl KJ et al. Late embryonic lethality and impaired V(D)J recombination in mice lacking DNA ligase IV. Nature 1998; 396:173-177.
21. Mills KD, Ferguson DO, Essers J et al. Rad54 and DNA ligase IV cooperate to maintain mammalian chromatid stability. Genes Dev 2004; 18:1283-1292.

22. Frappart PO, Lee Y, Russell HR et al. Recurrent genomic alterations characterize medulloblastoma arising from DNA double-strand break repair deficiency. Proc Natl Acad Sci USA (in press). Doi: 10.1073/pnas.0806882106.
23. Adachi N, Ishino T, Ishii Y et al. DNA ligase IV-deficient cells are more resistant to ionizing radiation in the absence of Ku70: Implications for DNA double-strand break repair. Proc Natl Acad Sci USA 2001; 98:12109-13.
24. Sharpless NE, Ferguson DO, O'Hagan RC et al. Impaired nonhomologous end-joining provokes soft tissue sarcomas harboring chromosomal translocations, amplifications and deletions. Mol Cell 2001; 8:1187-1196.
25. Sado K, Ayusawa D, Enomoto A et al. Identification of a mutated DNA ligase IV gene in the X-ray-hypersensitive mutant SX10 of mouse FM3A cells. J Biol Chem 2001; 276:9742-9748.
26. Nijnik A, Woodbine L, Marchetti C et al. DNA repair is limiting for haematopoietic stem cells during ageing. Nature 2007; 447:686-690.
27. Riballo E, Critchlow SE, Teo S-H et al. Identification of a defect in DNA ligase IV in a radiosensitive leukaemia patient. Curr Biol 1999; 9:699-702.
28. Riballo E, Doherty AJ, Dai Y et al. Cellular and biochemical impact of a mutation in DNA ligase IV conferring clinical radiosensitivity. J Biol Chem 2001 276; 33:31124-31132.
29. Badie C, Goodhardt M, Waugh A et al. A DNA double-strand break defective fibroblast cell line, (180BR), derived from a radiosensitive patient represents a new mutant phenotype. Cancer Res 1997; 57:4600-4607.
30. Toita N, Hatano N, Ono S et al. Epstein—Barr virus-associated B-cell lymphoma in a patient with DNA ligase IV (LIG4) syndrome. Am J Med Genet A 2007; 143A:742-745.
31. Buck D, Moshous D, de Chasseval R et al. Severe combined immunodeficiency and microcephaly in siblings with hypomorphic mutations in DNA ligase IV. Eur J Immunol 2006; 36:224-235.
32. Girard P-M, Kysela B, Harer CJ et al. Analysis of DNA ligase IV mutations found in LIG4 syndrome patients: the impact of two linked polymorphisms. Hum Mol Genet 2004; 13:2369-2376.
33. Grawunder U, Zimmer D, Lieber MR. DNA ligase IV binds to XRCC4 via a motif located between rather than within its BRCT domains. Curr Biol 1998; 8:873-876.
34. Gruhn B, Seidel J, Zintl F et al. Successful bone marrow transplantation in a patient with DNA ligase IV deficiency and bone marrow failure. Orphanet J Rare Dis 2007; 2:5.
35. Roddam PL, Rollinson S, O'Driscoll M et al. Genetic variants of NHEJ DNA ligase IV can affect the risk of developing multiple myeloma, a tumour characterised by aberrant class-switch recombination. J Med Genet 2002; 39:900-905.
36. Fenton JAL, Pratt G, Rawstron AC et al. Isotype class switching and the pathogenesis of multiple myeloma. Hematolog Oncol 2002; 19:1-12.
37. Van der Burg M, van Veelen LR et al. A new type of radiosensitive T-B-NK+ severe combined immunodeficiency caused by a LIG4 mutation. J Clin Invest 2006; 116:137-145.
38. Grunebaum E, Bates A, Roifman CM. Omenn syndrome is associated with mutations in DNA ligase IV. Allergy Clinical Immunology 2008; 122:1219-1220.
39. Enders A, Fisch P, Schwarz K et al. A severe form of human combined immunodeficiency due to mutations in DNA ligase IV. J Immunol 2006; 176:5060-5068.
40. Moshous D, Pannetier C, Chasseval RR et al. Partial T- and B-lymphocyte immunodeficiency and predisposition to lymphoma in patients with hypomorphic mutations in Artemis. J Clin Invest 2003; 111:381-387.
41. Ege M, Ma Y, Manfras B et al. Omenn syndrome due to ARTEMIS mutations. Blood 2005; 105:4179-4186.
42. Neitzel H, Jeggo PA, Concannon P. DNA ligase IV mutations identified in patients exhibiting developmental delay and immunodeficiency. Mol Cell 2001; 8:1175-1185.
43. Noordzij JG, Verkaik NS, van der Burg M et al. Radiosensitive SCID patients with Artemis gene mutations show a complete B-cell differentiation arrest at the preB-cell receptor checkpoint in bone marrow. Blood 2003; 101:1446-1452.

Chapter 17

Muir-Torre Syndrome

Pedro Mercader*

Abstract

Muir-Torre syndrome (MTS) is an autosomal dominant genodermatosis that consists of unusual cutaneous sebaceous neoplasm, with or without kerathoacantomas and one or more low-grade visceral malignancies, with or without colonic polyps, in the absence of other predisposing factors. This chapter presents a review of the principal clinical and genetic findings in this syndrome and discusses its relation with Lynch syndrome.

Introduction

Muir-Torre syndrome (MTS) is an autosomal dominant genodermatosis that consists of unusual cutaneous sebaceous neoplasm, with or without kerathoacantomas and one or more low-grade visceral malignancies, with or without colonic polyps, in the absence of other predisposing factors.[1] Diagnosis of MTS can be established in the absence of sebaceous tumours if the patient has multiple keratoacanthomas with multiple visceral malignancies and family history of MTS.

The first report of this syndrome was made by Muir et al in a 32- year-old man with cancer of the larynx, four separate synchronous primary carcinomas of the colon and two synchronous primary cancers of the duodenum.[2] These were associated with intestinal polyps and multiple skin tumors in the face at histological examination five of these lesions were diagnosed as keratoacanthomas and one of them as sebaceous adenoma respectively. In the same year Torre independently presented to the New York Dermatologic Society, a similar case of a 57-year-old man with more than 100 skin sebaceous tumors on the face, trunk and scalp together with a primary cancer of the ampulla of Vater and colonic cancer.[3] Later Baker et al reported a similar case and suggested that these 3 cases represented an example of a rare syndrome; this case was the first where a family history of cancer was well documented.[4] Since 1967 several cases have been reported and in the latest review 205 cases were found with 399 internal malignancies.[5]

The etiology of MTS has been controversial, although some authors believed that this syndrome was not related to other hereditary cancer syndromes,[1,6,7] successive findings have demonstrated that this opinion is wrong. Lynch et al found sebaceous neoplasm in patients from four families, affected by hereditary nonpolyposis colorectal cancer and concluded that MTS was a clinical variant of this syndrome, also called Lynch syndrome.[8] Four years later they described a patient with MTS who was a descendant of the Warthin's family G, which is thought to be the first case of Lynch syndrome.[9,10] This has been confirmed in several reports[11,17] that have found germline mutations in DNA mismatch repair genes hMSH2 and hMLH1, in patients with MTS. These genes are the same that are mutated in Lynch syndrome.

The main clinical characteristic in MTS patients is the presence of cutaneous lesions which, at least in some cases, allow us to make the diagnosis before internal cancers develop.[5] The histology of theses cutaneous lesions is the key feature in the diagnosis and has been explained below.

*Pedro Mercader—Dermatology Service, Hospital General Universitario Morales Meseguer, Murcia, Spain. Email: pedromercader@hotmail.com

Diseases of DNA Repair, edited by Shamim I. Ahmad. ©2010 Landes Bioscience and Springer Science+Business Media.

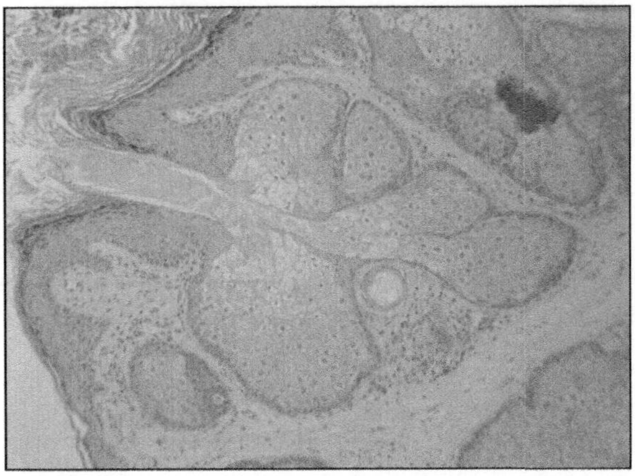

Figure 1. Sebaceous hyperplasia showing mature sebocytes and sebum. This lesion presents well defined sebaceous lobules connected to an infundibulum (Hematoxylin-eosin, x100).

Histopathology

The histopathological diagnosis of sebaceous neoplasm can be complicated. In a recent report on 20 patients, initially diagnosed as having sebaceous neoplasm, 8 had other clear cell neoplasm or sebaceous hyperplasias.[18] When a histopathological diagnosis of sebaceous tumors is made the pathologist must keep in mind that the clinical management of these patients will require the performance of additional tests to rule out internal malignancies, with the cost and emotional stress for patients facing the possibility that they might have an occult tumor.

The diagnosis of sebaceous tumors requires one or more of these findings[19]:

1. Cells with morphological characteristics, similar to those of sebocytes, from normal sebaceous glands. These cells have a pale foamy cytoplasm due to small lipidic vacuoles and the nuclei are scalloped owing to their compression by droplets of lipids (Fig. 1).
2. Ducts that become keratinized in a similar manner to sebaceous ducts. A sebaceous duct is lined by a thin cornifying squamous epithelium with barely detectable granular zone that becomes more noticeable as the wall of the duct widens near the infundibulum of the folliculum-sebaceous unit. The keratinocytes of the sebaceous duct are arranged compactly and its luminal border is marked by distinctive crenulations.
3. Sebum. This is a mixture of components that include sebaceous secretion, desquamated queratinocytes, yeasts and bacteria (Fig. 1).

The differential diagnosis includes others clear cell tumors, like clear-cell basal cell carcinoma, clear-cell squamous cell carcinoma, Paget disease, pagetoid Bowen disease, clear-cell hydradenoma, clear-cell syringoma, clear-cell syringoid carcinoma and skin metastases from renal cell carcinoma. The differential diagnosis also includes other sebaceous lesions that are not true tumors and have no relation with MTS, like sebaceous hyperplasia or nevus sebaceous of Jadassohn.

There is no uniform nomenclature in the classification of sebaceous tumors; some terminologies are obscured and the definitions may change between authors and this can lead to a wrong diagnosis; for example in a report,[20] it was found that 42% of patients with sebaceous tumors have internal cancer, but in this article potentially confusing terminologies such as basal cell carcinoma with sebaceous differentiation and adenomatous lesions with features of sebaceous hyperplasia, epithelioma and keratin pearls were used.

The commonly used terminologies are the following:

Sebaceous Adenoma

Balzer and Menetrier[21] were the firsts to use this term, but the first histological features of this tumor was described by Lever.[22] This tumor is a well circumscribed neoplasm, mostly superficial and

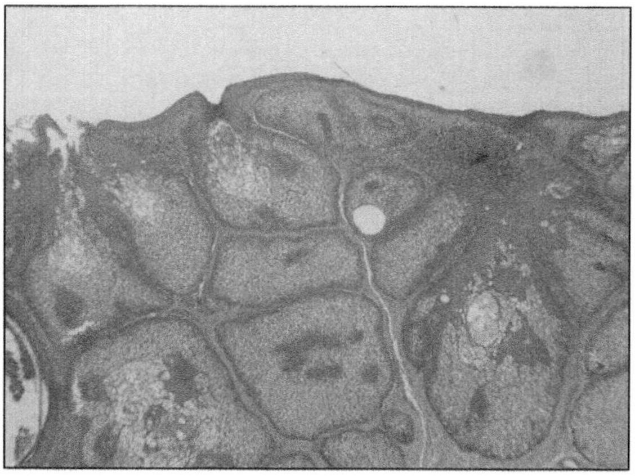

Figure 2. Sebaceous adenoma. Superficial lesion with mature sebocytes nevertheless sebaceous lobules are irregular. (Hematoxylin-eosin, x40).

compound in its entirety of well differentiated sebocytes; also cells with basophilic cytoplasm and a big nucleus can be seen. These cells are similar to the basaloid cells of the basal cell carcinoma and are termed germinative cells (Figs. 2 and 3). One way to differentiate a sebaceous adenoma from a sebaceous hyperplasia is the presence of these cells and intermediate cells between sebocites and germinative cells. Nussen and Ackerman[23] proposed that this tumor is a type of well differentiated sebaceous carcinoma, although this is not accepted by other authors.

Sebaceous Epithelioma

Lever introduced the term sebaceous epithelioma; this neoplasm was shown to have a medium grade of differentiation between sebaceous adenoma and the cystic variant of basal cell carcinoma.[24] This terminology is confusing and some authors have used it as a synonym of basal cell carcinoma with sebaceous differentiation, whereas others have used it to refer benign neoplasm with sebaceous differentiation. At present this terminology is refrained from being used in most reports.[25]

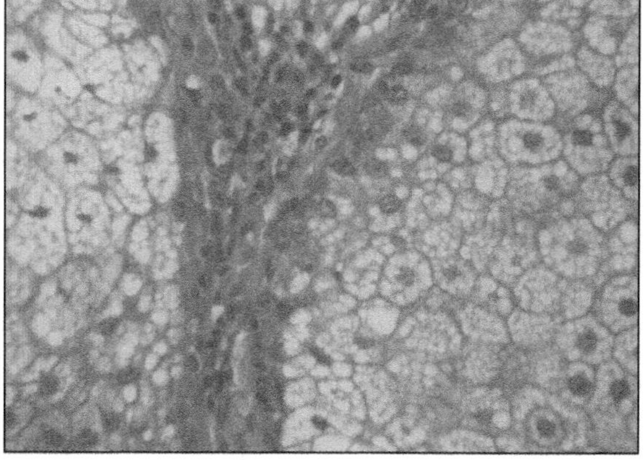

Figure 3. Sebaceous adenoma. Mature sebocytes predominate but germinative cells and intermediate cells are present. (Hematoxylin-eosin, x400).

Sebaceoma

Troy and Ackerman introduced this term to differentiate between sebaceous epithelioma (a benign neoplasm with sebaceous differentiation) and basal cell carcinoma with sebaceous differentiation.[26] This neoplasm can be differentiated form sebaceous adenoma because the germinative cells predominate over sebaceous cells and are located deeply in the dermis. It is a well circumscribed neoplasm without signs of malign neoplasm and can be differentiated from sebaceous carcinoma. Sebaceoma and sebaceous adenoma are widely accepted terminology in the literature.

Basal Cell Carcinoma with Sebaceous Differentiation

This terminology has been used as synonym of sebaceous epithelioma. The true basal cell carcinoma with sebaceous differentiation must have architectural characteristics of basal cell carcinoma with at least some areas of sebaceous differentiation. Most cases of basal cell carcinoma with sebaceous differentiation described in the literature, are in fact clear-cell basal cell carcinomas or sebaceous glands trapped as a result of tumour growth.[27]

Sebomatricoma

Sanchez-Yus et al reviewed the literature about the confusing terminology concerning sebaceous tumours.[28] They concluded that there was no significant difference between sebaceous adenoma and sebaceoma; both were benign neoplasm with variable amount of sebocytes and germinative cells and with different grades of differentiation. They proposed a new terminology that comprised both and named it sebomatricoma. This has been supported by others Spanish authors, but has not been widely accepted.[18,29]

Sebaceous Carcinoma

The first description of sebaceous carcinoma is attributed to Allaire in his doctoral thesis.[30] This is a rare neoplasm classified into two groups: an eyelid aggressive variant and a non-eyelid benign form. More recent observations indicate that this distinction is inappropriate because there are non-eyelid tumours associated with metastasis and appreciable mortality.[31] Nevertheless a population-based study has shown significant clinical differences between eyelid and non-eyelid form of sebaceous carcinoma.[32] For example the non-eyelid form is more frequent in men and associated with a significant risk of other primary cancers than the eyelid variant. These data furthermore suggest heterogeneity in the aetiology of sebaceous carcinoma and perhaps we should maintain the distinction between both variants. Sebaceous carcinoma is distinguished from other sebaceous neoplasm because it has pleomorphism, mitotic activity with abnormal mitotic figures and architectural characteristics of malign neoplasm (Fig. 4).

Keratoacanthoma

This is a common subtype of squamous cell carcinoma. Keratoacanthoma is characterized by well differentiated cells with eosinophilic cytoplasm and tendency towards keratination. A central plug and symmetrical architecture are also present. In patients with MTS there are multiple lesions, sometimes with sebaceous differentiation. For diagnosis of MTS with multiple keratoacathomas, a family history of MTS syndrome must be available.

Clinical Features

MTS is a rare disease and the clinical and epidemiological aspects of this disease have mainly been reported in reviews and small series in the scientific literatures.[1,5,7,33,34] All these studies are retrospective and there are no population based studies on MTS. The data obtained from these studies have limitations. A review showed 205 cases of MTS, in whom 399 internal malignancies has been diagnosed.[5] The median age for the appearance of the skin lesions was 53 years, with a range from 23 to 81 years and the median age for the detection of the initial internal neoplasm was 50 years, with a range from 23 to 81 years.[7] Most cases of MTS involve white patients, with no epidemiologic data on the disease in Asian and African population.[35] The low incidence of sebaceous carcinomas in African population has been confirmed in the

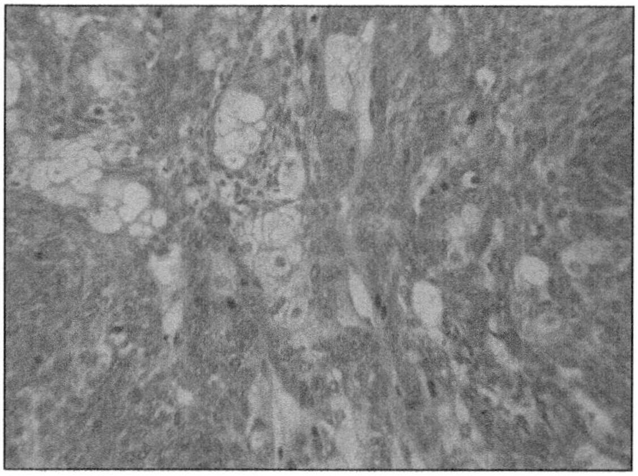

Figure 4. Sebaceous carcinoma. Sebocytes are present between germinative cells that present nuclear pleomorphism and mitoses. (Hematoxylin-eosin, x200).

studies by Dores et al.[32] MTS is present in both sexes, with a small predominance in men (male to female ratio of 3:2).

The benign sebaceous tumors (sebaceoma, sebaceous adenoma or sebomatricoma) appear as small yellowish nodules or papules with smooth surface (Fig. 5). Clinically these lesions have been diagnosed as basal cell carcinomas or sebaceous hyperplasias. In other cases they have been diagnosed as xanthomas or juvenile xanthogranulomas, because of the yellowish colour and in rare cases show marked hyperkeratosis and the clinical characteristics of cutaneous horns.[36] Benign sebaceous tumours appear as solitary lesions, yet in certain MTS patients multiple lesions may be present which are typical and sometimes may appear hundreds in a patient. The most frequent locations of these tumours are face and scalp but may be found in any part of the body.[20] In rare cases some tumours are located in oral mucosa and parotid gland.[37,38]

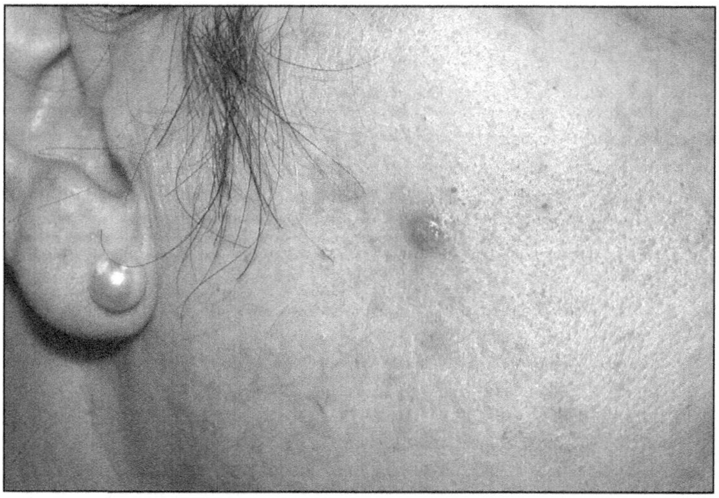

Figure 5. A small yellowish papule in the face of a 52 years old patient diagnosed of endometrium carcinoma and with a known family history of colorectal cancer. Biopsy of the lesion was diagnostic of sebaceous adenoma.

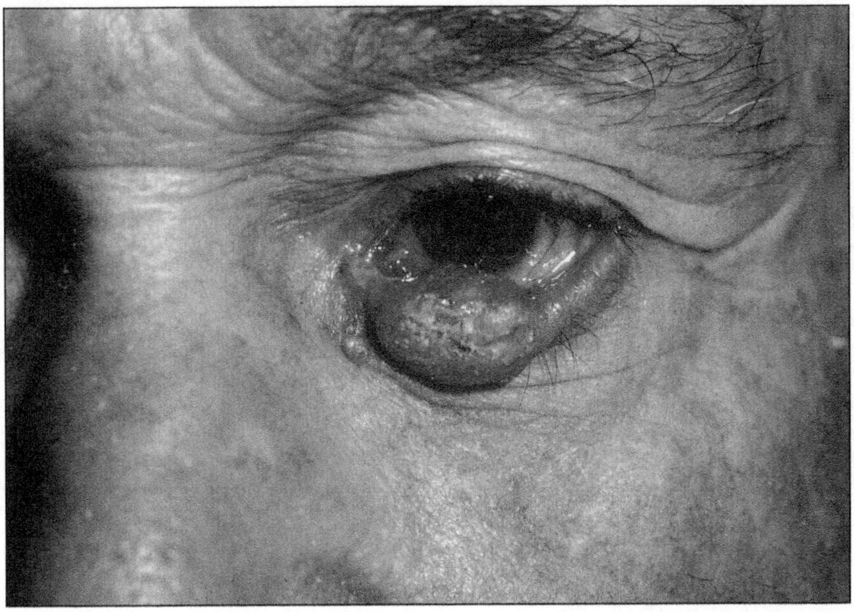

Figure 6. Nonulcerated nodule in the lower eyelid of a 75 years old patient. This lesion was growing for two years. Biopsy of this lesion was diagnostic of sebaceous carcinoma.

The clinical diagnosis of sebaceous carcinoma is also difficult. In the eyelids, sebaceous carcinoma affects upper one more frequently than the lower and a recent report has shown no difference in incidence between men and women.[32,39] Sebaceous carcinoma in the eyelid is presented as a growing nodule or tumour and are not ulcerated, but sometimes may simulate other diseases of the eyelid (Fig. 6). Clinical differential diagnosis includes benign diseases as chalazion,[40] chronic blefaroconjunctivitis,[41] dacryocystitis,[42] or unilateral papillary conjunctivitis.[43] The inflammatory aspect of sebaceous carcinoma looks to be more frequent in those tumours that extend to conjunctiva, the neoplastic cells cause an inflammatory response,[44] that can mimic inflammatory diseases. Non-eyelid sebaceous carcinoma usually affects face and neck of elderly people and may also affect any part of the skin,[44] and other organs. There are cases of sebaceous carcinoma involving oral mucosa,[45] salivary glands,[46,47] breast,[48] and also in dermoid cysts in the ovary.[49] The non-eyelid sebaceous carcinoma predominates in men.[32] In the skin, sebaceous carcinoma appears as a solitary nodule or plaque sometimes ulcerated. The clinical diagnosis is equivocal and in most cases has been diagnosed as squamous cell carcinomas or basal-cell carcinomas.[44] Sebaceous carcinoma are associated with MTS and also with other causes such as HIV,[50] immunodepression after organ transplant[51] and after radiotherapy for retinoblastoma[52] and other benign diseases.[53,54] It is unclear whether immunosuppression induces the expression of a latent MTS phenotype or selects the emergence of a mutator phenotype and thus predisposes individuals to the development of sebaceous carcinomas.[51] Sebaceous carcinoma in MTS are less aggressive than their counterparts not associated with the syndrome. For example, they do not tend to display pagetoid spreading and seem less likely to metastasize.[1]

Fifty six percent of the skin lesions in Muir-Torre syndrome develop after diagnosis of the first internal malignant disease; 6% appear concomitantly and 22% occur as the first tumour of the syndrome.[5] For this reason and also due the low frequency of this tumours and its relationship with internal cancers, it has been recommended that the presence of a unique sebaceous tumour is enough to search for internal malignancies.[55] It is difficult to specify the incidence of MTS in patients with sebaceous neoplasm due the infrequent occurrence of this tumour, the reports vary from 42% to 13'9%.[20,56]

In patients, diagnosed to have sebaceous skin tumours, a complete evaluation of their personal and family history of cancer must be made in order to search for clinical criteria of Lynch syndrome. These criteria are summarized in the Bethesda guidelines and the Amsterdam criteria. Nevertheless some patients with sebaceous tumours may have genetic mutation(s) but not fulfil Bethesda guidelines.[57] For this reason it is important to use screening techniques of immunohistochemistry and microsatellite instability of the cutaneous tumours for selection of those patients and families that may follow screening programs for internal cancers. Internal malignancies, associated with MTS, are same as found in patients with Lynch syndrome variant II and therefore the screening protocols should be the same. Nearly half of patients with MTS had two or more visceral malignant diseases and 10% had more than four.[35] Gastrointestinal cancers are the most common internal malignancies (61%), followed by genitourinary (22%). Other tumours may appear but are less frequent.[5] The most common tumour is colorectal cancer; in these patients the site of origin was at or proximal to the spleenic flexure, in contrast to general population where colorectal cancer are distal. These patients have an early age onset of their visceral cancers but an improved survival when compared stage for stage with their sporadic counterparts. Other features described in MTS include hyperlipidemia and psoriasis-like lesions.[58,59]

Genetics of Muir Syndrome

MTS is a subtype of Lynch syndrome and the genetic features are the same, caused by autosomal dominant germline mutations in the mismatch repair genes, MSH-2 and MLH-1.[11-17] Mutation in MSH2 protein is observed more frequently than in MLH1 protein in patients with MTS.[57] Vast majority of these mutations are truncating mutations that lead to a nonfunctional mismatch repair protein and are distributed over the entire gene with no evidence of a mutational hot spot.[60] The DNA mismatch repair system is critical for the maintenance of genomic stability. It increases the fidelity of single base mismatches and insertion/deletion loops that may arise during DNA replication. When this system fails a pronounced genetic instability occurs which could be detected by microsatellite marker analysis. Microsatellites are repetitive mononucleotide, dinucleotide, trinucleotide, or tetranucleotide sequences that are distributed over the whole genome. They are especially prone to replication errors because of their structure and when detected are microsatellite instability.

Recent studies show that some patients with MTS have mutation in the MSH6 gene,[61-64] the potential role of this gene, in the pathogenesis of MTS, may be more important than previously thought and when discovered alone, the typical microsatellite instability found in patients with MTS may be absent.[64] Other mutation, linked to MTS phenotype was studied by Ponti et al; they found a patient with sebaceous adenomas, colon cancer, multiple colonic polyps and a papillary carcinoma of thyroid that was due to a mutation in the MYH gene, which is related to the attenuated familial adenomatous polyposis. Since no other cases had been found as yet, the significance of this finding is uncertain.[65]

Conclusion

In this moment Muir-Torre syndrome should be considered a phenotypic variant of Lynch syndrome. The genetic findings are similar in both syndromes and have been described patients with Muir-Torre syndrome and a family history of Lynch syndrome. The clinical hallmark of this syndrome is the finding of sebaceous tumors, when present a search for internal tumors should be made.

Acknowledgement

I would like to thank to Dr. Victor Alegre from the Consorcio Hospital General Universitario de Valencia, for provide me with the clinical and histopathological pictures of this chapter.

References

1. Schwart RA, Torre DP. The Muir-Torre syndrome: A 25-year retrospect. J Am Acad Dermatol 1995; 33:90-104.
2. Muir EG, Bell AJY, Barlow KA. Multiple primary carcinomata of the colon, duodenum and larynx associated with keratho-acanthomata of the face. Br J Surg 1967; 54:191-195.
3. Torre D. Society transactions: New York Dermatological Society, Oct 24, 1967 (multiple sebaceous tumors). Arch Dermatol 1968; 98:549-551.
4. Baker PM, Tjon A, Joe SS. Multiple sebaceous gland tumours with multiple tumours of internal organs. A new syndrome? Dermatologica 1971; 142:50-57.
5. Akhtar S, Oza KK, Khan SA et al. Muir-Torre syndrome: case report of a patient with concurrent jejunal and ureteral cancer and a review of the literature. J Am Acad Dermatol 1999; 41:681-686.
6. Anderson DE. An inherited form of large bowel cancer: Muir's syndrome. Cancer 1980; 45:1103-1107.
7. Cohen PR, Kohn SR, Kurzrock R. Association or sebaceous gland tumors and internal malignancy: the Muir-Torre syndrome. Am J Med 1991; 90:606-613.
8. Lynch HT, Lynch PM, Pester J et al. The cancer family syndrome: rare cutaneous phenotypic linkage of Torre's syndrome. Arch Intern Med 1981; 141:607-611.
9. Lynch HT, Fusaro RM, Roberts L et al. Muir-Torre syndrome in several members of a family with a variant of the cancer family syndrome. Br J Dermatol 1985; 113:295-301.
10. Warthin AS. Heredity with reference to carcinoma as shown by the cases examined in the pathological laboratory of the University of Michigan 1895-1913. Arch Intern Med 1913; 96:261-268.
11. Kolodner RD, Hall NR, Lipford J et al. Structure of the human MSH2 locus and analysis of two Muir-Torre kindreds for MSH2 mutations. Genomics 1994; 24:516-526.
12. Liu B, Parsons R, Papadopoulos N et al hMSH2 mutations in hereditary nonpolyposis colorectal cancer kindreds. Cancer Res 1994; 54:4590-4594.
13. Kruse R, Lamberti C, Wang Y et al. Is the mismatch repair deficient type of Muir-Torre syndrome confined to mutations in the hMSH2 gene? Hum Genet 1996; 98:747-750.
14. Bapat B, Xia L, Madlensky L et al. The genetic basis of Muir-Torre syndrome includes the hMLH1 locus. Am J Hum Genet 1996; 59:736-739.
15. Esche C, Kruse R, Lamberti C et al. Muir-Torre syndrome: clinical features and molecular genetic analysis. Br J Dermatol 1997; 136:913-917.
16. Kruse R, Rutten A, Lamberti C et al. Muir-Torre phenotype has a frequency of DNA mismatch-repair-gene mutations similar to that in hereditary nonpolyposis colorectal cancer families defined by the Amsterdarm criteria. Am J Hum Genet 1998; 63:63-70.
17. Southey MC, Young MA, Whitty J et al. Molecular pathologic analysis enhances the diagnosis and management of Muir-Torre syndrome and gives insight into its underlying molecular pathogenesis. Am J Surg Pathol 2001; 25:936-941.
18. Mercader P, García-Melgares ML, Roche E et al. Clinical follow-up and presence of visceral tumors in 12 patients with sebaceous gland tumours. Actas Dermosifiliogr 2008; 99:532-539.
19. Requena C, Requena L. Criterios histopatológicos de diferenciación sebácea. In: Requena L, editor. Neoplasias anexiales cutáneas. Madrid: Grupo Aula Médica 2004; 359-363.
20. Finan MC, Connolly SM. Sebaceous gland tumors and sistemic disease: a clinicipathologic análisis. Medicine 1984; 63:232-242.
21. Balzer F, Menetrier P. Report of a case of sebaceous adenoma. Arch Physiol 1885; 6:564-570.
22. Lever WF. Sebaceous adenoma: review of the literature and report of a case. Arch Dermatol 1948; 57:102-111.
23. Nussen S, Ackerman AB. Sebaceous "adenoma" is sebaceous carcinoma. Dermatopathol Pract Concep 1998; 4:5-14.
24. Lever WF. Histopathology of the skin. Philadelphia, J.B. Lippincott 1949:273.
25. Requena L, Requena C. Sebomatricoma. In: Requena L, ed. Neoplasias anexiales cutáneas. Madrid: Grupo Aula Médica 2004; 435-447.
26. Troy Jl, Ackerman AB. Sebaceoma. A distinctive benign neoplasm of adnexal epithelium differentiating towards sebaceous cells. Am J Dermatopathol 1984; 6:7-13.
27. Requena L, Requena C. Carcinoma basocelular de diferenciación sebácea. In: Requena L, ed. Neoplasias anexiales cutáneas. Madrid: Grupo Aula Médica; 2004; 459-465.
28. Sanchez-Yus E, Requena L, Simón P et al. Sebomatricoma: a unifying term that encompasses all benign neoplasm with sebaceous differentiation. Am J Dermatopathol 1995; 17:213-221.
29. Simon RS, de Eusebio E, Alvarez Viétieiz S et al. Proliferaciones con diferenciación sebácea. II: Hiperplasias y neoplasias. Actas Dermosifiliogr 1998; 89:651-664.
30. Allaire G. Contribution a l'etude du polyadénoma ou epitelioma intra-glandulaire. Thesis, Paris 1891.

31. Lazar AJF, McKee PH. Tumors and related lesions of the sebaceous glands. In: McKee Ph, Calonje E, Granter SR, eds. Pathology of the skin with clinical correlations. London: Elsevier 2005; 1565-1588.
32. Dores GM, Curtis RE, Toro JR et al. Incidence of cutaneous sebaceous carcinoma and risk of associated neoplasm: insight into Muir-Torre syndrome. Cancer 2008; 113:3372-3381.
33. Cohen PR, Kohn SR, Davis DA et al. Muir-Torre syndrome. Dermatol Clin 1995; 13:79-89.
34. Serleth HJ, Kisken WA. A Muir-Torre syndrome family. Am Surgeon 1998; 64:365-369.
35. Ponti G, Ponz de Leon M. Muir-Torre síndrome. Lancet Oncol 2005; 6:980-987.
36. Thornton CM, Hunt S. Sebaceous adenoma with a cutaneous horn. J Cutan Pathol 1995; 22:185-7.
37. Ferguson JW, Geary CP, Mac Alister AD. Sebaceous cell adenoma. Rare intra-oral occurrence of a tumor wich is a frequent marker of Torre's syndrome. Pathology 1987; 19:204-208.
38. Gnepp DR, Brannon R. Sebaceous neoplasm of salivary origin. Cancer 1984; 53:2155-2170.
39. Ni C, Sealr SS, Kuo PK et al. Sebaceous cell carcinomas of the ocular adnexa. Int Ophthalmol Clin 1982; 22:23-61.
40. Gurin DM, Rapini R. Aggressive sebaceous carcinoma of the eyelid: an elusive diagnosis. Cutis 1993; 52:40-42.
41. Requena L, Requena C, Sánchez M et al. Carcinoma sebáceo de párpado. Piel 1988; 3:6-10.
42. Brownstein S, Codere F, Jackson WB. Masquerade syndrome. Ophtalmology 1980; 87:259-262.
43. Gloor P, Ansari I, Sinard J. Sebaceous carcinoma presenting as unilateral papillary conjunctivitis. Am J Ophtalmol 1999; 127:458-459.
44. Requena L, Fernandez Herrera J. Carcinoma sebáceo. In: Requena L, editor. Neoplasias anexiales cutáneas. Madrid: Grupo Aula Médica 2004; 467-486.
45. Liu CJ, Chang KW, Chang RC. Sebaceous carcinoma of the buccal mucosa. Report of a case. Int J Oral Maxilofac Surg 1997; 26:293-294.
46. Ohara N, Taguchi K, Yamamoto M et al. Sebaceous carcinoma of the submandibular gland with high-grade malignancy: report of a case. Pathol Int 1998; 48:287-291.
47. Siriwardena BS, Tilakaratne WM, Rajapakshe RM. A case of sebaceous carcinoma of the parotid gland. J Oral Pathol Med 2003; 32:121-123.
48. Hisaoka M, Takamatsu Y, Hirano Y et al. Sebaceous carcinoma of the breast: case report and review of the literature. Virchows Arch 2006; 449:484-488.
49. Ribeiro-Silva A, Chang D, Bisson FW et al. Clinicopathological and immunohistochemical features of a sebaceous carcinoma arising within a benign dermoid cyst of the ovary. Virchows Arch 2003; 443:574-578.
50. Yen MT, Tse DT. Sebaceous cell carcinoma of the eyelid and the human immunodeficiency virus. Ophthal Plast Reconstr Surg 2000; 16:206-210.
51. Harwood CA, Swale VJ, Bataille VA et al. An association between sebaceous carcinoma and microsatellite instability in immunosuppressed organ transplant recipients. J Invest Dermatol 2001; 116:246-253.
52. Kivelä T, Asko-Seljavaara S, Pihkala U et al. Sebaceous carcinoma of the eyelid associated with retinoblastoma. Ophthalmology 2001; 108:1124-1128.
53. Rumelt S, Hogan NR, Rubin PA et al. Four-eyelid sebaceous cell carcinoma following irradiation. Arch Ophthalmol 1998; 116:1670-1672.
54. Schlernitzauer DA, Font RL. Sebaceous gland carcinoma of the eyelid. Following radiation therapy for cavernous hemangioma of the face. Arch Ophthslmol 1976; 94:1523-1525.
55. Lynch HT, Fusaro RM, Lynch PM. Sebaceous skin lesions as clues to hereditary nonpolyposis colorectal cancer. J Invest Dermatol 2006; 126:2158-159.
56. Ponti G, Losi L, Di Gegorio C et al. Identification of Muir-Torre síndrome among patients with sebaceous tumors and keratoacanthomas. Role of clinical features, microsatellite instability and immnuohistochemistry. Cancer 2005; 103:1018-1125.
57. Mangold E, Pagenstecher C, Leister M et al. A genotype-phenotype correlation in HNPCC: strong predominance of msh2 mutations in 41 patients with Muir-Torre syndrome. J Med Genet 2004; 41:567-572.
58. Rodenas JM, Herranz MT, Tercedor J et al. Muir-Torre síndrome associated with a family history of hyperlipidemia. J Am Acad Dermatol 1993; 28:285-288.
59. Yamamoto T, Katayama I, Nishioka K. A possible role of interleukin-8 in the induction of psoriasis-like lesions in Torre-Muir syndrome. Acta Derm Venereol 1996; 76:75-77.
60. Kruse R, Ruzicka T. DNA mismatch repair and the significance of a sebaceous skin tumor for visceral cancer prevention. Trends Mol Med 2004; 10:136-141.
61. Mangold E, Rahner N, Friedrichs N et al. MSH6 mutation in Muir-Torre syndrome: could this be a rare finding? British Journal of Dermatology 2007; 156:158-162.

62. Marazza G, Masouye I, Taylor S et al. An illustrative case of Muir-Torre syndrome. Contribution of immunohistochemical analysis in identifying indicator sebaceous lesions. Arch Dermatol 2006; 142:1039-1042.
63. Arnold A, Payne S, Fisher S et al. An individual with Muir-Torre syndrome found to have a pathogenic MSH6 gene mutation. Fam Cancer 2007; 6:317-321.
64. Chhibber V, Dresser K, Mahalingam M. MSH-6: extending the reliability of immunohistochemistry as a screening tool in Muir-Torre syndrome. Mod Pathol 2008; 21:159-64.
65. Ponti G, Ponz de Leon M, Maffei S et al. Attenuated familial adenomatous polyposis and Muir-Torre síndrome linked to compound biallelic constitutional MYH gene mutations. Clin Genet 2005; 68:442-447.

Chapter 18

Wilms' Tumor

Carlos H. Martínez, Sumit Dave and Jonathan Izawa*

Abstract

Wilms' tumor or nephroblastoma is the most frequent renal tumor in children and is associated with different congenital anomalies and syndromes. A significant number of studies in genetics and molecular biology have improved our understanding of this malignancy discovering as well how different genes play a critical role in the organogenesis process. Derived from these studies promising markers have emerged relating translational medicine with clinical prognosis, which has been traditionally defined by the histopathological changes and the tumoral stage.

During the last 40 years the therapeutic outcomes have improved after multi-centric and multidisciplinary efforts represented mainly by The National Wilms' Tumor Study Group (currently the Renal Tumor Committee of the Children's Oncology Group) from North America and the International Society of Paediatric Oncology from Europe and this has served as a role model for establishing similar trials for other pediatric tumors.

Our aim with this chapter of Wilms' tumor is to present the state of knowledge in translational and clinical areas in a balanced perspective.

Introduction

In his monograph in 1899, Max Wilms first described 7 children with nephroblastoma.[1] In 1877 Thomas Richard Jessop was the first to perform a successful nephrectomy in a child with Wilms' tumor, however the patient died 9 months after the surgery of recurrent disease.[2] Wilms' tumor was part of the original tumor group used for development of the "two-hit" model of tumor suppression by Knudson et al.[3] The exact etiology of nephroblastoma remains to be elucidated, yet important advances in our understanding of kidney development have been derived from its study.

Epidemiology

6.1% of the malignant tumors diagnosed in childhood (0-14 years) are renal tumors. Wilms' tumor is the most frequent renal tumor diagnosed in children of this group and is the second most frequent pediatric intra-abdominal solid organ tumor after neuroblastoma. In 2008, the Canadian Cancer Society/National Cancer Institute of Canada reported its incidence as 7.9 cases per million per year. It is more frequent in females, with a male to female ratio of 0.8.[4] The National Cancer Institute of Surveillance, Epidemiology and End Results (SEER) Program have reported an incidence of 7.3 cases per million per year between 2001 and 2005 in the United States.[5] The peak incidence of Wilms' tumor occurs between 0-4 years reaching 15.6-17.2 cases per million per year. The 5-year survival with appropriate therapy is 92%.[4]

*Corresponding Author: Jonathan Izawa— Departments of Surgery & Oncology, The University of Western Ontario, London Health Sciences Center-Victoria Hospital, 800 Commissioners Road East, London, Ontario, Canada, N6A 4G5. Email: jonathan.izawa@lhsc.on.ca

Diseases of DNA Repair, edited by Shamim I. Ahmad. ©2010 Landes Bioscience and Springer Science+Business Media.

Genetic Basis and the Molecular Mechanism of the Disease

WT1

This tumor suppressor gene is harbored in the 11p13 locus and was cloned in 1990. This region also contains the PAX6 gene related to the Wilms' tumor, aniridia, genitourinary abnormalities and mental retardation (WAGR) syndrome.[6] It encodes a 55 kDa transcription factor with 4 zinc fingers domains and a region rich in proline and glutamine and is involved in Wilms' tumor development.[7] The mechanism of how WT1 leads to Wilms' tumor formation remains unclear.

Isoforms

For WT1, 24 isoforms have been described and associated with different areas of the protein: exon 5 variants, KTS forms, AWT1 transcript with alternative exon 1, one isoform starting at the end of intron 5 and other alternative isoform with upstream CTG start codon, internal ATG start codon at the end of exon 1 and a residue in exon 6.[8] All these possible isoforms may have different roles. Some have different locations inside the cell[9] and the KTS isofoms are the only variations present in all vertebrates. Mice lacking anyone of these isoforms present with severe kidney defects and die soon after birth. However, mice lacking exon 5 are fertile and viable.[10] The complexity derived from isoforms for proteins encoded by the same gene and of course their different functions limit the studies, using cDNA and the complete understanding of the function of each of these isoforms.

Transcription Regulator

WT1 is a transcriptional activator and suppressor, depending of which isoform and/or cell line is analyzed.[11] The repressor domains are located in amino acid residues 71-180 and activator domains in residues 180-250. Another additional suppressor domain has been described at N-terminus of the repressor domain.[12] The variability in activation or repression may depend on the construct evaluated and may represent on the biological interactions with different molecules like CBP (CREB binding protein), TP73 (tumoral protein 73), CIAO1 (ciao 1 protein), UBE21 (ubiquitin conjugating enzyme E2I) and other proteins less characterized.[13]

Developmental Role

After demonstrating that WT1-null mice fail to develop normal kidneys,[14] several studies have emerged showing the WT1 presence in mature organs, tumors, developing tissues and adult tissues.[15] WT1 is expressed in the kidneys, gonads, spleen, mesothelial linings of thoracic and abdominal organs. It is likely that these tissues undergo a transition between mesenchymal and epithelial differentiation. During kidney formation WT1 can be detected in the metanephric mesenchyma and especially in the nephrogenic mesenchyma, which is closer to the mesenchymal area affecting the ureteral bud during the branching process and tubulogenesis.[13] After kidney formation is complete, WT1 expression is only detected in the podocytes.[16,17]

Tumorogenesis

It is not clear which role WT1 plays in Wilms' tumor formation as WT1 mutations occur in the germline of only 15% of sporadic Wilms' tumor patients.[18] However some possible mechanisms have been described; cancers, in which WT1 is expressed, generally are derived from epithelial cells. WT1 helps to maintain the mesenchymal-epithelial balance. In Wilms' tumor the cells are derived from mesenchymal tissue and WT1 forces the cells toward an epithelial state.[8] Tumor cells with WT1 mutations can escape the apoptotic process and be exposed to additional events, with respect to these mechanisms.[13]

WT2

This gene occupies the locus 11p15[19] and has classically been associated with overgrowth syndromes like Beckwith-Wiedemann syndrome.[20] However, recent studies have suggested the association of WT2 in all the individuals with Wilms' tumor after demonstrating constitutional defects in 11p15 as one of the most common causes.[21] This area of the chromosome is closely associated with the Insulin-like Growth Factor II (IGF-II) and IGF-II is a potential candidate for WT2.[18]

WT3

The gene 16q exhibits loss of heterozigosity in 17-20% of Wilms' tumors and has been associated with E-cadherin as the tumor suppressor derived from this region.[22] WT3 has also been associated with WT1 and WT2 with early genetic changes in the nephogenic rests, which may lead to nephroblastoma formation.[23]

WTX

This gene is harbored in chromosome X and is inactivated in 30% of the Wilms' tumors and seems to follow the "one hit hypothesis". In males inactivation of the single copy of X chromosome and in females the inactivation of the active copy follow a monoallelic pattern leading to increased tumorogenesis.[24]

p53 and Others Apoptosis Associated Markers

The p53 tumor suppressor gene is not involved in the pathogenesis of Wilms' tumor[25] and less than 5% of Wilms' tumors have a mutation in this protein. However, it is present in 75% of tumors with anaplasia and predicts a poor prognosis.[26] The role played by Bax, Bcl-2 and Bcl-X in Wilms' tumor remains unknown, but the ratio Bcl-2/Bax expression and the blastemal Bcl-X have been associated with survival outcomes and appear to have some prognostic significance.[27] However other studies have presented contrary findings.[28]

β-Catenin

Mutations in β-Catenin have been reported in 50-60% of pediatric tumors, 15% of adult tumors,[29] and were identified simultaneously in 19 of 20 patients with mutations in WT1. This significant association seems to follow different pathways.[30] Koesters et al showed a possible link between β-Catenin and the WNT signal transduction pathway through WNT 4.[31]

Growth Factors

A number of studies have reported the alterations of growth factors in patients with nephroblastoma. Insulin-like growth factors have been detected in Wilms' tumor and its implications are unknown. Expression of transforming growth factor α may play a role in the proliferation of Wilms' tumor, but the mechanism is unclear.[32] Vascular endothelial growth factor (VEGF) has been clearly associated with tumoral vascular proliferation and specifically VEGF-C has demonstrated in Wilms' tumor and its protein expression appears to have prognostic value.[33]

Cell Adhesion Molecules

Neural cell adhesion molecule (NCAM) has been suggested as a cancer stem cell marker. Cells positive for NCAM, using immunostaining, were localized in the blastema and are highly clonogenic and overexpress Wilms' tumor genes and TOP2A. In addition, treatment of these cells with Etoposide and Irinotecan showed down-regulation of NCAM.[34] CD44 is a transmembrane glycoprotein with different isoforms present in normal kidney tissue and Wilms' tumor cells; specifically the blastemal CD44v5 has been correlated with clinical progression and tumor related death.[35]

Other Markers

Multidrug resistant associated permeability glycoprotein (P-gp) expression in Wilms' tumor has been controversial due to methodological reasons and the results of a number of studies are inconclusive.[32] The presence of renin has been detected by different authors in nephroblastoma,[36,37] and more recently has been demonstrated that WT1 regulate the renin gene transcription,[17] which may explain in part, the hypertension and high levels of renin in Wilms' tumor patients.

DNA Repair

The literature in DNA repair in Wilms' tumor is scanty; however the presence of microsatellite instability (MSI) was reported, suggesting that DNA mismatch repair might play a role during the pathogenesis of Wilms' tumor.[38] In addition the expression of DNA mismatch repair proteins like hMSH2, hMLH1 and p21(waf1) have been studied on Wilms' tumor,[39] but unfortunately initial results showed no association with tumor treatment response.[39]

Familial Wilms' Tumor

1-2% of Wilms' tumor cases cluster within families and is inherited as an autosomal dominant trait with incomplete penetrance.[40] Usually the familial Wilms' tumor is diagnosed at more advanced age and stage than the sporadic cases. Three loci have been associated with most but not all the familial Wilms' tumors: WT1 (11p13), FWT1 (17q12-q21) and FWT2 (19q13).[41] However, secondary to the variability in terms of penetrance and genetics, the underlying cause remains unknown.

Wilms' Tumor Associated Syndromes

Beckwith-Wiedemann Syndrome

This is an overgrowth syndrome defined by the presence of macroglossia, macrosomia and abdominal wall defect. Due to the variable clinical manifestations, the diagnosis is established if three features are present, including the above criteria or the presence of hemihyperplasia, ear anomalies, visceromegaly, renal abnormalities, embryonal tumors, neonatal hypoglycemia, cleft palate and a positive family history.[42]

Children with this syndrome have increased risk of the development embryonic tumors, such as nephroblastoma, hepatoblastoma, neuroblastoma or rabdomyosarcoma. The reported frequency for such malignancies in Beckwith-Wiedemann syndrome vary from 4-21%.[43] Most tumors are detected before the age of 4 and rarely after 10 years of age. A meta-analysis reported by Rump et al stratified groups based on imprinting patterns and the relation with risk of tumor development. Although the study included all the embryonic tumors, 67% were Wilms' tumors and the overall risk was 5-10%. The higher risk of tumor development was represented by patients with loss of imprinting of H19 with a risk estimated at 35-45% and the group with loss of imprinting of LIT1 and H19 (uniparental dysomy 11p15) was 25-30%.[43] The results of this meta-analysis are encouraging several research analyses have been published showing epigenetics as a tool for the prediction of tumor risk.[44] Limitations in these studies do not allow use of the molecular markers to guide tumor surveillance. The actual recommendation for tumor surveillance for patients with Beckwith-Wiedemann syndrome consists of abdominal MRI at diagnosis and quarterly evaluation with abdominal ultrasound until the patient is 8 years of age and serum alpha fetoprotein until the 5 years of age, taking into account the higher levels of alpha fetoprotein in children.[42]

WAGR Syndrome

The syndrome involving Wilms' tumor, complete or partial aniridia, genitourinary anomalies (cryptochidism or ambiguous external genitalia) and mental retardation (WAGR) was initially described by Miller et al in 1964.[45] Cytogenetic analyses have associated WAGR syndrome with deletions at 11p13, specifically the contigous genes PAX6 and WT1.[6,7] However, it seems that patients with aniridia who have a deletion in WT, have a higher risk to develop Wilms' tumor than those without the deletion.[46] Patients with WAGR are at high risk of renal failure, reaching 36% according to the study combining the National Wilms' Tumor Study Group and the United States Renal Data System.[47]

Denys-Drash Syndrome

This syndrome is classically described with the triad of Wilms' tumor, nephropathy and genitourinary abnormalities, which are variable from mild alterations to pseudohermaphroditism.[48,49] The disease is associated with WT1 mutation in exon 9 (zinc finger III) and exon 8 (zinc finger II) affecting directly the recognition of DNA sequences.[50] In podocytes WT1 regulate the expression of key components of the cytoskeleton like filamentous actin leading to morphological changes,[51] and abnormal proliferation.[16] The nephropathy is diagnosed before the 18 months and the histological glomerular characteristic is diffuse mesangial sclerosis.[52]

Perlman Syndrome

This is a disorder characterized by neonatal macrosomia, polyhydramnios, nephromegaly, renal dysplasia, nephro-blastomatosis, distinctive facial appearance and predisposition to Wilms' tumor.[53,54] The clinical overlap with other overgrowth syndromes has been emphasized. The molecular bases are unknown. Perlman syndrome is considered an autosomal recessive condition and most of these patients die during the neonatal period.[55,56]

Histopathology

Macroscopic Features

Wilms' tumors cells are generally soft and friable and special care is required when sectioning to avoid transfer cells to the margins or other areas. It is recommended to collect samples for additional studies like electron microscopy in glutaraldehyde and frozen section for molecular genetic studies.[57] Recommendations of particularly valuable sections to demonstrate the relationship between tumor and extension have been described.[58] Nephroblastomas usually are spherical masses with grey or tan color. Areas with necrosis are more frequent when treated preoperatively. Extension to the renal vein has been described as well.

Microscopic Features

Wilms' tumor classically presents a triphasic appearance with blastemal, epithelial and stromal elements. The relative proportions of these components vary widely, making difficult the interpretation of biopsies.[59] The morphology of epithelial component is described as tubular, rosette-like and glomeruloid structures. The blastemal component manifest small round blue cell tumor with ovoid cells, high nuclear-cytoplasmic ratio, nuclear overlapping, closely packed nuclei and mitoses. Different patterns like diffuse, serpentine, nodular and basiloid for the blastemal component have been described. The stromal component may vary between hypocellular, loose areas of immature stellate cells within myxoid background to closely packed spindle cells recapitulating primitive mesenchyme. Occasionally components like fat, cartilage, glial and osteoid tissues are present.[59]

Nephroblastomas should be classified into favourable and unfavourable histological types and this is based on the presence or absence of anaplasia. The criteria for diagnosing anaplasia include the finding of atypical mitoses, nuclear enlargement (three times the size of adjacent cells minimum) and hyperchromasia. It is present in about 5% of tumors, is rare in children under 2 years and increases to up to 13% beyond 5 years of age.[60] Anaplasia is classified as focal or diffuse. Focal anaplasia occurs in only one region of the tumor which is completely resected and not present in vascular extensions or metastatic sites. On the other hand, diffuse anaplasia includes multifocal anaplasia, anaplasia at an extrarenal site, in a random biopsy and marked nuclear atypia. Studies based on these criteria showed that diffuse anaplasia has poorer prognosis than focal anaplasia.[60,61] Anaplasia has been associated with resistance to chemotherapy but seems not be generated or affected for it.[61] It is not yet known whether anaplasia is a marker for tumor aggressiveness or facilitates dissemination.

Two large scale groups have stratified the Wilms' tumor histology. The National Wilms' Tumor Study Group and Children's Oncology Group (NWTSG and COG) from North America uses this anaplasia classification, but the other group The International Society of Paediatric Oncology (SIOP) from Europe, analyze the risk as shown in Table 1.

Nephrogenic Rests and Nephroblastomatosis

Nephrogenic rests are defined as persistent metanephric tissue after 36 weeks of life and are considered a premalignant lesion. They are found in less than 1% of routine infant biopsies but

***Table 1. Revised SIOP pathological classification of renal tumors of childhood*[62]**

A. For Pretreated Cases
- a. Low-risk tumors
  - Mesoblastic nephroma
  - Cystic partially differentiated nephroblastoma
  - Completely necrotic nephroblastoma
- b. Intermediate-risk tumors
  - Nephroblastoma-epithelial type
  - Nephroblastoma-type
  - Nephroblastoma-mixed type
  - Nephroblastoma-regressive type
  - Nephroblastoma-focal anaplasia
- c. High-risk tumors
  - Nephroblastoma-blastemal type
  - Nephroblastoma-diffuse anaplasia
  - Clear cell sarcoma of the kidney
  - Rhabdoid tumor of the kidney

B. For Primary Nephrectomy Cases
- a. Low-risk tumors
  - Mesoblastic nephroma
  - Cystic partially differentiated nephroblastoma
- b. Intermediate-risk tumors
  - Non-anaplastic nephroblastoma and its variants
  - Nephroblastoma-focal anaplasia
- c. High-risk tumors
  - Nephroblastoma-diffuse anaplasia
  - Clear cell sarcoma of the kidney
  - Rhabdoid tumor of the kidney

are noted in at least a third of all sporadic tumors and at higher rates in syndromes predisposing to Wilms' tumor.[63] Nephrogenic rests are classified as perilobar, which surround a mature renal lobe; intralobar, which are inside the lobe itself; and combined, presence of perilobar and intralobar rests.[64] Nephroblastomatosis is defined like presence of diffuse and multifocal nephrogenic rests.[64,65] NWTS reported 41% of unilateral and 99% of bilateral Wilms' tumors were associated with nephrogenic rests.[63] The risk of metachronous Wilms' tumor in children with less than 1 year-old, diagnosed with unilateral tumor and nephrogenic rests is 15 times more than without rests, especially if the rests are perilobar.[66] Hyperplastic nephrogenic rests are very similar histologically and radiologically to Wilms' tumors and the one differentiating factor is the lack of a pseudocapsule in the former lesion.[67]

Clinical Features

Wilms' tumor usually presents as a painless abdominal mass detected by the caregiver, 30% may have hematuria and 25% hypertension.[68] Symptomatic patients can present bleeding or tumor rupture causing pain. Macroscopic hematuria may indicate collecting system or ureteral compromise.[69] Some patients may present with atypical symptoms secondary to compression of surrounding structures (10%) or invasion to vascular structures (4%). About 4% of tumors show vena caval or atrial involvement while renal vein involvement is seen in 11%. These symptoms include varicocele, hepatomegaly, ascites and heart failure.[70] Wilms' tumour can be associated with paraneoplastic syndromes, such as polycythemia, von Willebrand disease (8%),[71] factor VII deficiency[72] and may produce ACTH.[73] Nonspecific symptoms, such as fever, weight loss, malaise and anorexia may be related with Wilms' tumour. Syndromes associated with Wilms' tumor include hemihypertrophy, mental retardation, macrosomia, aniridia, genital malformations and nephropathy depending of which syndrome is present.

Imaging Studies

Ultrasonography

Abdominal or renal ultrasound is the first imaging test for Wilms' tumor and allows for accurate identification of the renal mass, compromise of adjacent organs and tumour thrombus with extension into the renal vein and/or inferior vena cava especially when using color Doppler.[74] Typically the tumor is large with solid hyperechoic masses and cystic areas. When associated with diffuse nephroblastomatosis, a thick hypoechoic band can be detected with the ultrasound.[75]

Despite the lack of evidence in terms of improved survival for Wilms' tumor screening in patients with genetic syndromes, this practice has been widespread. Considering the clear difficulties to obtain prospective trials, some recommendations have been proposed: (1) offer surveillance to children with more than 5% risk of Wilms' tumour; (2) patient reviewed by a clinical geneticist; (3) surveillance to be carried out until the age of 5 years except for Beckwith-Wiedeman syndrome,[76] Simpson-Golabi-Behmel syndrome and some familial tumors, in which case the surveillance should continue until the age 7; (4) surveillance should involve renal ultrasound every 4 months at the local centres by a radiologist with experience in paediatric ultrasonography; (5) detected lesions should be treated at a centre with experience and expertise in Wilms' tumor.[77]

Computed Tomography (CT) and Magnetic Resonance Imaging (MRI)

These imaging modalities of the abdomen and pelvis are necessary if delineation of tumour extension is considered relevant for treatment. Many European centers treat patients based on ultrasound findings alone. Usually CT is preferred over MRI, as the patients do not require sedation and CT is more widely available. On CT the tumor is heterogeneous and enhances less than the normal renal parenchyma. On MRI the tumor appears heterogeneous and hypointense on T1or hypo-isointense with respect to the renal cortex with hyperintense necrotic or cystic areas on T2. Contrast enhanced studies are necessary to assess the contralateral renal unit. Both studies are useful for extension of tumor thrombus.[74,75] Retrospective data suggests that patients with positive nodules on CT and negative chest X-ray can present higher recurrences rates and current protocols mandate radiographic imaging of the chest with a CT scan.[78,79] In addition, both CT and MRI can incorporate 3D imaging that allow the surgeon to visualize the spatial relationships of the tumor and surrounding organs and vasculature and hence avoid unexpected intraoperative injuries in cases with complex anatomy.[80] MRI can help in differentiating nephrogenic rests from Wilms' tumors and with the prevailing concept of ALARA (as low as reasonably achievable) radiation dose there is an increasing role of MRI in following bilateral tumors and assessing response to chemotherapy.[81,82]

Staging

Wilms' tumor is staged depending of the anatomical extension of the tumor and staging guides therapy. There are two staging systems SIOP and NWTSG and COG. (Tables 2 and 3).

Treatment

Critical improvements in the survival of Wilms' tumor patients have been achieved during the last century, with the initial 2-year survival being 0% and with contemporary therapy it now reaches almost 100%.[84]

Surgery

A proper transperitoneal radical nephrectomy plays the primary and critical role in curing patients with Wilms' tumor. A thorough exploratory laparotomy is carried out. Bloody peritoneal

Table 2. SIOP staging criteria for renal tumors of childhood after chemotherapy[62]

Stage	Criteria
I	Tumor limited to the kidney or surrounded with pseudocapsule (resection margins "clear")
	Tumor may protrude into collecting system but without infiltrating walls
	Renal vessels are not involved
	Intrarenal vessels may be involved
	Note: percutaneous core needle biopsy does not upstage
II	Tumor extends beyond the kidney into perirenal fat (resection margins "clear")
	Tumor infiltrates the renal sinus vessels but is completely resected
	Tumor infiltrate to adjacent organs or vena cava and is completely resected
III	Incomplete excision of the tumor (gross or microscopical margins positive)
	Any abdominal lymph nodes are involved
	Tumor rupture before or during surgery
	Tumoral peritoneal implants
	Tumor thrombi present at resection margins
	Tumor biopsy (open) prior surgery or chemotherapy
IV	Haematogenous metastases or lymph nodes positive outside abdomino-pelvic region
V	Bilateral renal tumors at diagnosis

Table 3. NWTSG and COG staging system for renal tumors before chemotherapy[83]

Stage	Criteria
I	Tumor confined to the kidney completely resected
II	Tumor extends beyond the kidney but is completely resected
	Penetration of capsule
	Invasion of renal sinus vessels
	Biopsy before removal
	Spillage of tumor locally during removal
III	Gross or microscopic residual tumor remains, include:
	Inoperable tumor
	Positive surgical margins
	Tumor spillage in peritoneal surfaces
	Transected thrombus
	Lymph node metastases
IV	Haematogenous metastases or lymph nodes metastases outside abdomen
V	Bilateral renal Wilms' tumors

fluid is considered a sign onf major spillage and upgrading to Stage III.[85] During the procedure special care is needed to avoid spilling tumor cells, which can affect recurrence rates. Preoperative chemotherapy can shrink the tumor and decrease the risk of tumor spill.[86] Studies suggest that contralateral exploration is not required if preoperative CT or MRI demonstrate a normal kidney. When feasible, preliminary renal artery and vein ligation is advisable. Resectability should be based on intraoperative findings and not on imaging, with the exception of suprahepatic caval thrombus.

Wilms' tumor frequently are adherent but not invading adjacent organs and major resection of adjacent organs should be avoided if possible in favour of preoperative chemotherapy if suspected intraoperatively. Clipping at the time of resection is recommended to mark areas of residual tumor and resection margins. Patients with suprahepatic extension of tumor thrombus should be biopsied and managed with prenephrectomy chemotherapy. Routine lymph node sampling from the hilum, peroicaval and paraaortic areas and removal of suspicious nodes should be performed, but a formal lymphadenectomy is not recommended.[87]

The importance of nephron sparing surgery in Wilms' tumor is unclear. Most of these patients present large and central masses with only 10% of them suitable for a safe procedure. After considering the probability of higher local recurrences, surgical morbidity in some series and equivalent results to radical nephrectomy in others, nephron sparing surgery should be reserved for patients with tumors in solitary kidney, bilateral disease and in those known to be at increased risk of metachronous tumors like BWS and aniridia. In patients with a unilateral tumor and a normal contralateral kidney, nephron sparing surgery remains controversial.[88,89]

National Wilms' Study Group

The NWTSG was set up in 1967 to study the biology and genetics of Wilms' tumor, to develop stratified treatment protocols based on disease extent and to study whether an intergroup mechanism was feasible. This group proposed radical nephrectomy to diagnose and stage the disease before chemotherapy. NWTS-1 demonstrated that the combination of actinomycin D and vincristine had better results than each medication alone in Stages II and III.[90] NWTS-2 demonstrated for Stage I disease 6 or 15 months of chemotherapy have similar outcomes and for Stage IV disease the addition of adriamycin to the chemotherapy regimen improved survival from 63 to 77%.[91] NWTS-3 showed that patients with Stage I disease, receiving either 10 weeks or 6 months of chemotherapy, have no statistically significant differences in survival. For patients Stage II disease radiotherapy did not add any survival advantage to chemotherapy. For Stage III disease no differences in survival was noted in patients receiving 1000 vs 2000 cGy and the addition of cyclophosphamide did not improve survival for Stage IV patients with unfavourable histology.[92] NWTS-4 demonstrated that actinomycin D and adriamycin can be safely administrated in pulse intensive regimens, reducing severe hematological toxicities and cost[85,93] NWTS-4 demonstrated the increased risk of recurrence if no lymph node sampling was done for suspicious lymph nodes and if tumor spill occurs.[87] Finally NWTS-5 concluded that loss of heterozygosity for chromosomes 1p and 16q identify a high risk group for relapse and death independent of stage and histology.[94] This last study also assessed nephrectomy alone without adjuvant chemotherapy for patients under 2 years of age with a tumor less than 550g and Stage 1 favorable histology. An overall survival of 98% was achieved but the COG is planning a similar trial to confirm these findings. The NWTSG was incorporated to the COG in 2001, which involves most of the oncological centers for children in North America. However there is an ongoing Late Effects Study which continues to collect data on long-term survivors. Regular follow-up post treatment and nephrectomy involves yearly blood pressure and urine protein assessments with a serum creatinine every 5 years. The presence of aniridia and intralobar nephrogenic rests is associated with a higher incidence of renal failure.[47]

International Society of Paediatric Oncology

The SIOP incorporates most of the paediatric oncology centres in Europe. The hallmark of this group is the neoadjuvant chemotherapy followed by nephrectomy. Moreover Stage IV tumors do not receive lung irradiation if there is a complete remission achieved with preoperative chemotherapy.[95] SIOP-1 reported no significant differences in survival between preoperative or postoperative radiotherapy and increased number of abdominal recurrences was associated with tumor rupture.[96] SIOP-2 was a nonrandomized study, which confirmed that tumor rupture during surgery occurred more often if the patients did not receive neoadjuvant chemotherapy.[97] SIOP-5 was a randomized trial which showed that chemotherapy alone was as good as radiotherapy to prevent tumor rupture.[98] SIOP-6 revealed that for Stage I Wilms' adjuvant vincristine and act inomycin D 17 weeks vs 38 weeks had similar survival and that therapy should be risk adapted.[95]

SIOP-9 showed that neoadjuvant chemotherapy of 4 vs 8 weeks is equivalent.[99] In addition, patients with total tumor necrosis after chemotherapy had excellent outcome.[100] SIOP 93-01 showed that shortening of postoperative chemotherapy from 18 weeks to 4 weeks in patients with Stage I disease and with anaplasia, intermediate risk did not affect the effectiveness and the reduced secondary chemotherapy effects.[101]

Analysis of NWTS and SIOP Protocols

The main issue for the treatment of patients with Wilms' tumor is the time when the nephrectomy should be done, after or before chemotherapy. The SIOP supports a delayed nephrectomy and the NWTSG and COG supports early intervention. Advantages of a delayed procedure are low risk of tumor rupture,[102] but has disadvantages, such as treating patients without a clear diagnosis and inaccurate staging.[103] Downstaging of the tumor with preoperative chemotherapy can reduce the overall treatment burden and toxicity and also allow treatment escalation in those with a poor clinical or histological response.[92,96] On the other hand, advocates of the NWTS approach site the higher relapse rate in SIOP-6 Stage II node negative patients who possibly had occult Stage III disease to begin with and did not receive adequate treatment due to the downstaging. Prechemotherapy nephrectomy increases the risk of tumor rupture,[104] however it seems to be more accurate for staging. Tumor spillage due to rupture is important in relapse free rates but has not shown any effect on overall outcome. Only one randomized trial has been published addressing the best time for surgery in this population and despite some methodological weaknesses, the timing of early or delayed surgical intervention did not significantly affect the 5-year overall survival (early 89.0% vs delayed 79.6%).[105] The error rate in SIOP-9 was 5% which included 1.6% with benign pathologies who received preoperative chemotherapy.[106] The UKW3 trial attempts to incorporate a prechemotherapy prenephrectomy core needle biopsy to avoid blind treatment but there have been concerns regarding bleeding and needle track recurrences.[107]

Bilateral Wilms' Tumor

The presence of bilateral and synchronous Wilms' tumor occurs in 4-6%.[108] For these patients the risk of renal failure is important and efforts to preserve the renal function must be carried out. Both the SIOP and NWTSG-COG have recommended initial chemotherapy. Partial nephrectomy should be done if feasible, without compromising resection margins. Patients with bilateral tumors treated with partial nephrectomy have an 8.2% recurrence rate vs 1.5-2.5% recurrence rate in unilateral tumors undergoing radical nephrectomy. However, local recurrences were not associated with positive margins.[108] Recently, it has been reported by the NWTSG that a group of patients with bilateral Wilms' tumor have been treated successfully with chemotherapy showing complete radiological response.[109]

Adult Wilms' Tumor

This is a rare tumor and usually the diagnosis is made histopathologically after surgery and no clinical or radiological differences with other tumors are noted. The literature is scanty and is mainly represented by case reports, however Izawa et al reported an analysis of 128 cases based on previously published cases and showed that adult Wilms' tumor has poorer prognosis than paediatric counterpart. The overall survival was 68% and more aggressive therapy than that recommended for the paediatric population is suggested, especially for those with unfavourable histology and better survival may be achieved by closely following pediatric therapy protocols.[110]

Conclusion

The NWTS-5 demonstrated the critical role that genetics changes play in Wilms' tumor, not only in relation with the origin, but with the survival and prognosis.[94] Considering these findings, further research evaluating treatments based on genetic markers may direct individualized therapy for each patient rather than grouping patients for therapy based upon risk factors like stage and pathology alone.

References

1. Wilms M. Die Mischgeschwulste. I. Die Mischgeschwulste der Niere, A. Georgi, Leipzig. 1899.
2. Willetts IE. Jessop and the Wilms' tumor. J Pediatr Surg 2003; 38:1496-1498.
3. Knudson AG Jr, Strong LC. Mutation and cancer: a model for Wilms' tumor of the kidney. J Natl Cancer Inst 1972; 48:313-324.
4. The Canadian Cancer Society/National Cancer Institute of Canada. Canadian Cancer Statistics 2008. Chidhood cancer in ages 0 to 14 2008; 2009.
5. National Cancer Institute. SEER Cancer Statistics Review 1975-2005. ICCC. 04/09/08; 2009.
6. Ton CC, Hirvonen H, Miwa H et al. Positional cloning and characterization of a paired box- and homeobox-containing gene from the aniridia region. Cell 1991; 67:1059-1074.
7. Call KM, Glaser T, Ito CY et al. Isolation and characterization of a zinc finger polypeptide gene at the human chromosome 11 Wilms' tumor locus. Cell 1990; 60:509-520.
8. Hohenstein P, Hastie ND. The many facets of the Wilms' tumour gene, WT1. Hum Mol Genet 2006; 15 Spec No 2:R196-201.
9. Dutton JR, Lahiri D, Ward A. Different isoforms of the Wilms' tumour protein WT1 have distinct patterns of distribution and trafficking within the nucleus. Cell Prolif 2006; 39:519-535.
10. Natoli TA, McDonald A, Alberta JA et al. A mammal-specific exon of WT1 is not required for development or fertility. Mol Cell Biol 2002; 22:4433-4438.
11. Scharnhorst V, van der Eb AJ, Jochemsen AG. WT1 proteins: functions in growth and differentiation. Gene 2001; 273:141-161.
12. Wagner KJ, Roberts SG. Transcriptional regulation by the Wilms' tumour suppressor protein WT1. Biochem Soc Trans 2004; 32:932-935.
13. Rivera MN, Haber DA. Wilms' tumour: connecting tumorigenesis and organ development in the kidney. Nat Rev Cancer 2005; 5:699-712.
14. Kreidberg JA, Sariola H, Loring JM et al. WT-1 is required for early kidney development. Cell 1993; 74:679-691.
15. Kreidberg JA, Hartwig S. Wilms' tumor-1: a riddle wrapped in a mystery, inside a kidney. Kidney Int 2008; 74:411-412.
16. Morrison AA, Viney RL, Saleem MA et al. New insights into the function of the Wilms tumor suppressor gene WT1 in podocytes. Am J Physiol Renal Physiol 2008; 295:F12-7.
17. Steege A, Fahling M, Paliege A et al. Wilms' tumor protein (-KTS) modulates renin gene transcription. Kidney Int 2008; 74:458-466.
18. Ehrlich PF. Wilms tumor: progress and considerations for the surgeon. Surg Oncol 2007; 16:157-171.
19. Karnik P, Chen P, Paris M et al. Loss of heterozygosity at chromosome 11p15 in Wilms tumors: identification of two independent regions. Oncogene 1998; 17:237-240.
20. Sparago A, Russo S, Cerrato F et al. Mechanisms causing imprinting defects in familial Beckwith-Wiedemann syndrome with Wilms' tumour. Hum Mol Genet 2007; 16:254-264.
21. Scott RH, Douglas J, Baskcomb L et al. Constitutional 11p15 abnormalities, including heritable imprinting center mutations, cause nonsyndromic Wilms tumor. Nat Genet 2008; 40:1329-1334.
22. Safford SD, Goyeau D, Freemerman AJ et al. Fine mapping of Wilms' tumors with 16q loss of heterozygosity localizes the putative tumor suppressor gene to a region of 6.7 megabases. Ann Surg Oncol 2003; 10:136-143.
23. Guertl B, Ratschek M, Harms D et al. Clonality and loss of heterozygosity of WT genes are early events in the pathogenesis of nephroblastomas. Hum Pathol 2003; 34:278-281.
24. Rivera MN, Kim WJ, Wells J et al. An X chromosome gene, WTX, is commonly inactivated in Wilms tumor. Science 2007; 315:642-645.
25. Waber PG, Chen J, Nisen PD. Infrequency of ras, p53, WT1, or RB gene alterations in Wilms tumors. Cancer 1993; 72:3732-3738.
26. Bardeesy N, Falkoff D, Petruzzi MJ et al. Anaplastic Wilms' tumour, a subtype displaying poor prognosis, harbours p53 gene mutations. Nat Genet 1994; 7:91-97.
27. Ghanem MA, Van der Kwast TH, Den Hollander JC et al. The prognostic significance of apoptosis-associated proteins BCL-2, BAX and BCL-X in clinical nephroblastoma. Br J Cancer 2001; 85:1557-1563.
28. Re GG, Hazen-Martin DJ, El Bahtimi R et al. Prognostic significance of Bcl-2 in Wilms' tumor and oncogenic potential of Bcl-X(L) in rare tumor cases. Int J Cancer 1999; 84:192-200.
29. Su MC, Huang WC, Lien HC. Beta-catenin expression and mutation in adult and pediatric Wilms' tumors. APMIS 2008; 116:771-778.
30. Maiti S, Alam R, Amos CI et al. Frequent association of beta-catenin and WT1 mutations in Wilms tumors. Cancer Res 2000; 60:6288-6292.
31. Koesters R, Ridder R, Kopp-Schneider A et al. Mutational activation of the beta-catenin proto-oncogene is a common event in the development of Wilms' tumors. Cancer Res 1999; 59:3880-3882.

32. Ghanem MA, van Steenbrugge GJ, Nijman RJ et al. Prognostic markers in nephroblastoma (Wilms' tumor). Urology 2005; 65:1047-1054.
33. Nowicki M, Ostalska-Nowicka D, Kaczmarek M et al. The significance of VEGF-C/VEGFR-2 interaction in the neovascularization and prognosis of nephroblastoma (Wilms' tumour). Histopathology 2007; 50:358-364.
34. Pode-Shakked N, Metsuyanim S, Rom-Gross E et al. Developmental Tumorigenesis: NCAM as a putative marker for the malignant renal stem/progenitor cell population. J Cell Mol Med 2008; .
35. Ghanem MA, Van Steenbrugge GJ, Van Der Kwast TH et al. Expression and prognostic value of CD44 isoforms in nephroblastoma (Wilms tumor). J Urol 2002; 168:681-686.
36. Lindop GB, Duncan K, Millan DW et al. Renin gene expression in nephroblastoma. J Pathol 1990; 161:93-97.
37. McKenzie KJ, Ferrier RK, Howatson AG et al. Demonstration of renin gene expression in nephroblastoma by in situ hybridization. J Pathol 1996; 180:71-73.
38. Mason JE, Goodfellow PJ, Grundy PE et al 16q loss of heterozygosity and microsatellite instability in Wilms' tumor. J Pediatr Surg 2000; 35:891-6; discussion 896-7.
39. Idikio HA. Expression of DNA mismatch repair proteins hMSH2 and hMLH1 and the cyclin G1 inhibitor, p21(waf1/cip1) in pediatric tumors: correlation with response to therapy. Oncol Rep 2001; 8:965-971.
40. Breslow NE, Olson J, Moksness J et al. Familial Wilms' tumor: a descriptive study. Med Pediatr Oncol 1996; 27:398-403.
41. Rapley EA, Barfoot R, Bonaiti-Pellie C et al. Evidence for susceptibility genes to familial Wilms tumour in addition to WT1, FWT1 and FWT2. Br J Cancer 2000; 83:177-183.
42. Weksberg R, Shuman C, Smith AC. Beckwith-Wiedemann syndrome. Am J Med Genet C Semin Med Genet 2005; 137C:12-23.
43. Rump P, Zeegers MP, van Essen AJ. Tumor risk in Beckwith-Wiedemann syndrome: A review and meta-analysis. Am J Med Genet A 2005; 136:95-104.
44. Bliek J, Gicquel C, Maas S et al. Epigenotyping as a tool for the prediction of tumor risk and tumor type in patients with Beckwith-Wiedemann syndrome (BWS). J Pediatr 2004; 145:796-799.
45. Miller RW, Fraumeni JF,Jr, Manning MD. Association of Wilms's Tumor with Aniridia, Hemihypertrophy and Other Congenital Malformations. N Engl J Med 1964; 270:922-927.
46. Muto R, Yamamori S, Ohashi H et al. Prediction by FISH analysis of the occurrence of Wilms tumor in aniridia patients. Am J Med Genet 2002; 108:285-289.
47. Breslow NE, Collins AJ, Ritchey ML et al. End stage renal disease in patients with Wilms tumor: results from the National Wilms Tumor Study Group and the United States Renal Data System. J Urol 2005; 174:1972-1975.
48. Denys P, Malvaux P, Van Den Berghe H et al. Association of an anatomo-pathological syndrome of male pseudohermaphroditism, Wilms' tumor, parenchymatous nephropathy and XX/XY mosaicism. Arch Fr Pediatr 1967; 24:729-739.
49. Drash A, Sherman F, Hartmann WH et al. A syndrome of pseudohermaphroditism, Wilms' tumor, hypertension and degenerative renal disease. J Pediatr 1970; 76:585-593.
50. Pelletier J, Bruening W, Kashtan CE et al. Germline mutations in the Wilms' tumor suppressor gene are associated with abnormal urogenital development in Denys-Drash syndrome. Cell 1991; 67:437-447.
51. Viney RL, Morrison AA, van den Heuvel LP et al. A proteomic investigation of glomerular podocytes from a Denys-Drash syndrome patient with a mutation in the Wilms tumour suppressor gene WT1. Proteomics 2007; 7:804-815.
52. Habib R, Loirat C, Gubler MC et al. The nephropathy associated with male pseudohermaphroditism and Wilms' tumor (Drash syndrome): a distinctive glomerular lesion—report of 10 cases. Clin Nephrol 1985; 24:269-278.
53. Perlman M, Goldberg GM, Bar-Ziv J et al. Renal hamartomas and nephroblastomatosis with fetal gigantism: a familial syndrome. J Pediatr 1973; 83:414-418.
54. Perlman M, Levin M, Wittels B. Syndrome of fetal gigantism, renal hamartomas and nephroblastomatosis with Wilms' tumor. Cancer 1975; 35:1212-1217.
55. Piccione M, Cecconi M, Giuffre M et al. Perlman syndrome: clinical report and nine-year follow-up. Am J Med Genet A 2005; 139A:131-135.
56. Alessandri JL, Cuillier F, Ramful D et al. Perlman syndrome: report, prenatal findings and review. Am J Med Genet A 2008; 146A:2532-2537.
57. Govender D. The Pathology of Nephroblastoma. Current Diagnostic Pathology. 2000; 6:45-54.
58. Perlman EJ. Pediatric renal tumors: practical updates for the pathologist. Pediatr Dev Pathol 2005; 8:320-338.
59. Sebire NJ, Vujanic GM. Paediatric renal tumours: recent developments, new entities and pathological features. Histopathology 2009; 54:516-528.

60. Faria P, Beckwith JB, Mishra K et al. Focal versus diffuse anaplasia in Wilms tumor—new definitions with prognostic significance: a report from the National Wilms Tumor Study Group. Am J Surg Pathol 1996; 20:909-920.
61. Vujanic GM, Harms D, Sandstedt B et al. New definitions of focal and diffuse anaplasia in Wilms tumor: the International Society of Paediatric Oncology (SIOP) experience. Med Pediatr Oncol 1999; 32:317-323.
62. Vujanic GM, Sandstedt B, Harms D et al. Revised International Society of Paediatric Oncology (SIOP) working classification of renal tumors of childhood. Med Pediatr Oncol 2002; 38:79-82.
63. Beckwith JB, Kiviat NB, Bonadio JF. Nephrogenic rests, nephroblastomatosis and the pathogenesis of Wilms' tumor. Pediatr Pathol 1990; 10:1-36.
64. Beckwith JB. Nephrogenic rests and the pathogenesis of Wilms tumor: developmental and clinical considerations. Am J Med Genet 1998; 79:268-273.
65. Hennigar RA, O'Shea PA, Grattan-Smith JD. Clinicopathologic features of nephrogenic rests and nephroblastomatosis. Adv Anat Pathol 2001; 8:276-289.
66. Coppes MJ, Arnold M, Beckwith JB et al. Factors affecting the risk of contralateral Wilms tumor development: a report from the National Wilms Tumor Study Group. Cancer 1999; 85:1616-1625.
67. Perlman EJ, Faria P, Soares A et al. Hyperplastic perilobar nephroblastomatosis: long-term survival of 52 patients. Pediatr Blood Cancer 2006; 46:203-221.
68. Varan A. Wilms' tumor in children: an overview. Nephron Clin Pract 2008; 108:c83-90.
69. Ritchey M, Daley S, Shamberger RC et al. Ureteral extension in Wilms' tumor: a report from the National Wilms' Tumor Study Group (NWTSG). J Pediatr Surg 2008; 43:1625-1629.
70. Ritchey ML, Kelalis PP, Breslow N et al. Intracaval and atrial involvement with nephroblastoma: review of National Wilms Tumor Study-3. J Urol 1988; 140:1113-1118.
71. Baxter PA, Nuchtern JG, Guillerman RP et al. Acquired von Willebrand syndrome and Wilms tumor: not always benign. Pediatr Blood Cancer 2009; 52:392-394.
72. Granger J, Gidvani VK. Acquired factor VII deficiency associated with Wilms tumor. Pediatr Blood Cancer 2009; 52:394-395.
73. Wang J, Zhang G. Paraneoplastic Cushing syndrome because of corticotrophin-releasing hormone-secreting Wilms' tumor. J Pediatr Surg 2008; 43:2099-2101.
74. Riccabona M. Imaging of renal tumours in infancy and childhood. Eur Radiol 2003; 13(Suppl 4):L116-29.
75. McHugh K. Renal and adrenal tumours in children. Cancer Imaging 2007; 7:41-51.
76. Tan TY, Amor DJ. Tumour surveillance in Beckwith-Wiedemann syndrome and hemihyperplasia: a critical review of the evidence and suggested guidelines for local practice. J Paediatr Child Health 2006; 42:486-490.
77. Scott RH, Walker L, Olsen OE et al. Surveillance for Wilms tumour in at-risk children: pragmatic recommendations for best practice. Arch Dis Child 2006; 91:995-999.
78. Meisel JA, Guthrie KA, Breslow NE et al. Significance and management of computed tomography detected pulmonary nodules: a report from the National Wilms Tumor Study Group. Int J Radiat Oncol Biol Phys 1999; 44:579-585.
79. Owens CM, Veys PA, Pritchard J et al. Role of chest computed tomography at diagnosis in the management of Wilms' tumor: a study by the United Kingdom Children's Cancer Study Group. J Clin Oncol 2002; 20:2768-2773.
80. Martinez CH, Chalasani V, Knudsen B et al. Virtual Reality May Improve Training of Renal Surgeons. SPIE Newsroom 2009.
81. Gylys-Morin V, Hoffer FA, Kozakewich H et al. Wilms tumor and nephroblastomatosis: imaging characteristics at gadolinium-enhanced MR imaging. Radiology 1993; 188:517-521.
82. Shah NB, Platt SL. ALARA: is there a cause for alarm? Reducing radiation risks from computed tomography scanning in children. Curr Opin Pediatr 2008; 20:243-247.
83. Kalapurakal JA, Dome JS, Perlman EJ et al. Management of Wilms' tumour: current practice and future goals. Lancet Oncol 2004; 5:37-46.
84. D'Angio GJ. The National Wilms Tumor Study: a 40 year perspective. Lifetime Data Anal 2007; 13:463-470.
85. Green DM, Breslow NE, Beckwith JB et al. Treatment with nephrectomy only for small, stage I/favorable histology Wilms' tumor: a report from the National Wilms' Tumor Study Group. J Clin Oncol 2001; 19:3719-3724.
86. Burger D, Moorman-Voestermans CG, Mildenberger H et al. The advantages of preoperative therapy in Wilms' tumour. A summarised report on clinical trials conducted by the International Society of Paediatric Oncology (SIOP). Z Kinderchir 1985; 40:170-175.
87. Shamberger RC, Guthrie KA, Ritchey ML et al. Surgery-related factors and local recurrence of Wilms tumor in National Wilms Tumor Study 4. Ann Surg 1999; 229:292-297.

88. Moorman-Voestermans CG, Aronson DC, Staalman CR et al. Is partial nephrectomy appropriate treatment for unilateral Wilms' tumor? J Pediatr Surg 1998; 33:165-170.
89. Zani A, Schiavetti A, Gambino M et al. Long-term outcome of nephron sparing surgery and simple nephrectomy for unilateral localized Wilms tumor. J Urol 2005; 173:946-8; discussion 948.
90. D'Angio GJ, Evans AE, Breslow N et al. The treatment of Wilms' tumor: Results of the national Wilms' tumor study. Cancer 1976; 38:633-646.
91. D'Angio GJ, Evans A, Breslow N et al. The treatment of Wilms' tumor: results of the Second National Wilms' Tumor Study. Cancer 1981; 47:2302-2311.
92. D'Angio GJ, Breslow N, Beckwith JB et al. Treatment of Wilms' tumor. Results of the Third National Wilms' Tumor Study. Cancer 1989; 64:349-360.
93. Green DM, Breslow NE, Beckwith JB et al. Comparison between single-dose and divided-dose administration of dactinomycin and doxorubicin for patients with Wilms' tumor: a report from the National Wilms' Tumor Study Group. J Clin Oncol 1998; 16:237-245.
94. Grundy PE, Breslow NE, Li S et al. Loss of heterozygosity for chromosomes 1p and 16q is an adverse prognostic factor in favorable-histology Wilms tumor: a report from the National Wilms Tumor Study Group. J Clin Oncol 2005; 23:7312-7321.
95. Tournade MF, Com-Nougue C, Voute PA et al. Results of the Sixth International Society of Pediatric Oncology Wilms' Tumor Trial and Study: a risk-adapted therapeutic approach in Wilms' tumor. J Clin Oncol 1993; 11:1014-1023.
96. Lemerle J, Voute PA, Tournade MF et al. Preoperative versus postoperative radiotherapy, single versus multiple courses of actinomycin D, in the treatment of Wilms' tumor. Preliminary results of a controlled clinical trial conducted by the International Society of Paediatric Oncology (S.I.O.P.). Cancer 1976; 38:647-654.
97. Graf N, Tournade MF, de Kraker J. The role of preoperative chemotherapy in the management of Wilms' tumor. The SIOP studies. International Society of Pediatric Oncology. Urol Clin North Am 2000; 27:443-454.
98. Lemerle J, Voute PA, Tournade MF et al. Effectiveness of preoperative chemotherapy in Wilms' tumor: results of an International Society of Paediatric Oncology (SIOP) clinical trial. J Clin Oncol 1983; 1:604-609.
99. Tournade MF, Com-Nougue C, de Kraker J et al. Optimal duration of preoperative therapy in unilateral and nonmetastatic Wilms' tumor in children older than 6 months: results of the Ninth International Society of Pediatric Oncology Wilms' Tumor Trial and Study. J Clin Oncol. 2001; 19:488-500.
100. Boccon-Gibod L, Rey A, Sandstedt B et al. Complete necrosis induced by preoperative chemotherapy in Wilms tumor as an indicator of low risk: report of the international society of paediatric oncology (SIOP) nephroblastoma trial and study 9. Med Pediatr Oncol 2000; 34:183-190.
101. de Kraker J, Graf N, van Tinteren H et al. Reduction of postoperative chemotherapy in children with stage I intermediate-risk and anaplastic Wilms' tumour (SIOP 93-01 trial): a randomised controlled trial. Lancet 2004; 364:1229-1235.
102. Godzinski J, Tournade MF, deKraker J et al. Rarity of surgical complications after postchemotherapy nephrectomy for nephroblastoma. Experience of the International Society of Paediatric Oncology-Trial and Study "SIOP-9". International Society of Paediatric Oncology Nephroblastoma Trial and Study Committee. Eur J Pediatr Surg 1998; 8:83-86.
103. de Kraker J, Delemarre JF, Lilien MR et al. Misstaging in nephroblastoma. Causes and consequences. A report of the Sixth Nephroblastoma Trial and Study of the International Society of Paediatric Oncology. Eur J Pediatr Surg 1999; 9:153-157.
104. Leape LL, Breslow NE, Bishop HC. The surgical treatment of Wilms' tumor: results of the National Wilms' Tumor Study. Ann Surg 1978; 187:351-356.
105. Mitchell C, Pritchard-Jones K, Shannon R et al. Immediate nephrectomy versus preoperative chemotherapy in the management of nonmetastatic Wilms' tumour: results of a randomised trial (UKW3) by the UK Children's Cancer Study Group. Eur J Cancer 2006; 42:2554-2562.
106. Fuchs J, Kienecker K, Furtwangler R et al. Surgical aspects in the treatment of patients with unilateral wilms tumor: a report from the SIOP 93-01/German Society of Pediatric Oncology and Hematology. Ann Surg 2009; 249:666-671.
107. Vujanic GM, Kelsey A, Mitchell C et al. The role of biopsy in the diagnosis of renal tumors of childhood: Results of the UKCCSG Wilms tumor study 3. Med Pediatr Oncol 2003; 40:18-22.
108. Horwitz JR, Ritchey ML, Moksness J et al. Renal salvage procedures in patients with synchronous bilateral Wilms' tumors: a report from the National Wilms' Tumor Study Group. J Pediatr Surg 1996; 31:1020-1025.
109. Hamilton TE, Ritchey ML, Argani P et al. Synchronous bilateral Wilm's tumor with complete radiographic response managed without surgical resection: a report from the National Wilm's Tumor Study 4. J Pediatr Surg 2008; 43:1982-1984.
110. Izawa JI, Al-Omar M, Winquist E et al. Prognostic variables in adult Wilms tumour. Can J Surg 2008; 51:252-256.

Chapter 19

Cerebro-Oculo-Facio-Skeletal Syndrome

Hiroshi Suzumura* and Osamu Arisaka

Abstract

Cerebro-oculo-facio-skeletal (COFS) syndrome is an autosomal recessive inherited disorder characterized by congenital microcephaly, congenital cataracts and/or microphthalmia, arthrogryposis, severe developmental delay, severe postnatal growth failure and facial dysmorphism with prominent nasal root and/or overhanging upper lip. This syndrome is now recognized as a disorder belonging to the spectrum of inherited defects in Nucleotide Excision Repair (NER) resulting in profound photosensitivity. In COFS syndrome, as in Cockayne syndrome, DNA repair is impaired in the transcription-coupled NER pathway, but not in the global genome NER pathway. Fourteen cases so far described as COFS syndrome have been studied at the molecular levels. All mutations have been found in Cockayne syndrome gene, *CSB*, xeroderma pigmentosum genes, *XPD* and *XPG* and *ERCC1* gene involved in the transcription-coupled NER pathway.

Introduction

Cerebro-oculo-facio-skeletal (COFS) syndrome is an autosomal recessive inherited disorder that was initially described by Pena and Shokeir.[1] This syndrome is now recognized as belonging to the spectrum of inherited defects in Nucleotide Excision Repair (NER) resulting in profound photosensitivity.

Clinical Features

COFS syndrome is characterized by the following features:

Head and Face: Microcephaly (95%), prominent root of the nose (100%), sloping forehead, large pinnae, overlapping lower lip, micrognathia, small mouth, cleft palate, high-arched palate, short neck.

Eyes: Deep-set eyes with blephalophimosis or microphthalmia (90%), nystagmus, cataracts and hypertelorism.

Limbs: Flexion contractures (especially involving the elbows and knees), camptodactyly, syndactyly, single palmar crease (30%), longitudinal groove in the soles (70%), rocker bottom feet due to vertical talus (70%), osteoporosis.

Brain and Neurological: Progressive demyelination of the brain, reduced white matter of the brain with gray mottling, hypogenesis or agenesis of the corpus callosum, cerebellar hypoplasia, intracranial calcification, generalized hypotonia, hyporeflexia or areflexia, mental retardation, sensorineural hearing loss.

*Corresponding Author: Hiroshi Suzumura—Department of Pediatrics, Dokkyo Medical University, 880 Kitakobayashi, Mibu, Shimotsuga-gun, Tochigi, Japan.
Email: suzumura@dokkyomed.ac.jp

Diseases of DNA Repair, edited by Shamim I. Ahmad. ©2010 Landes Bioscience and Springer Science+Business Media.

Others: Small for gestational age (50%), widely set nipples (50%), shallow thorax, congenital heart disease (patent ductus arteriosus, ventricular septal defect etc.), kyphosis or scoliosis (70%), acetabular hypoplasia, coxa valga, short umbilical cord, genitourinary hypoplasia, cryptorchidism.

Recently, new diagnostic criteria have been proposed by Laugel et al.[2]

Clinical Criteria: Congenital microcephaly, congenital cataracts and/or microphthalmia, arthrogryposis, severe developmental delay, severe postnatal growth failure and facial dysmorphism with prominent nasal root and/or overhanging upper lip.

Genetic Criteria: DNA repair defect in the transcription-coupled NER (TC-NER) pathway.

Prognosis

Infants with COFS syndrome usually need oxygen supplementation for respiratory distress during the neonatal period. Tube feeding is required for feeding trouble. Severe failure to thrive and developmental delays are seen and most infants die by 4-5 years old due to the recurrent respiratory infection.[1]

Differential Diagnosis

Neu-Laxova Syndrome

The main features of Neu-Laxova syndrome are microcephaly, micrognathia, arthrogryposis, overlapping fingers, edema of the skin, brain malformations and lethal prognosis. These symptoms are closely related to those of COFS syndrome. Severe kind of skin problems are common in Neu-Laxova syndrome, but is usually absent in COFS syndrome.[3] However, one case of COFS syndrome, complicated by congenital ichthyosis, has been described.[4] Silengo suggested that Neu-Laxova syndrome and COFS syndrome may represent different degrees of clinical expression for the same autosomal recessive mutation.[3]

Micro Syndrome (Warburg Micro Syndrome)

This syndrome was first reported by Warburg as autosomal recessive syndrome manifested with microcephaly, microcornea, cataracts, mental retardation, optic atrophy and hypogenitalism. Graham reported that NER analysis of cultured fibroblasts was a useful tool to differentiate Micro syndrome from other similar disorders.[5] Patients with Cockayne syndrome (CS) and COFS syndrome, both show hypersensitivity of fibroblast cells to ultraviolet (UV) irradiation due to defect in NER, but patients with Micro syndrome show normal NER. Graham also demonstrated that Micro syndrome should be distinguished from other similar clinical disorders by the presence of significant visual impairment and cortical blindness, frontal polymicrogyria, thin corpus callosum and cortical atrophy.[5]

Cockayne Syndrome

This autosomal recessively inherited disorder is characterized by low to normal birth weight, severe growth retardation, microcephaly, brain demyelination with calcium deposits, cutaneous photosensitivity, pigmentary retinopathy, cataracts and sensorineural hearing loss. Cells from patients with CS display increased sensitivity to cell death from UV radiation, with delayed recovery of DNA synthesis after UV. CS is caused by mutations in either the *CSA* or *CSB* genes, resulting in a selective defect of TC-NER.[6] Global genome NER (GG-NER) is typically normal in CS patients.[7]

Distinguishing between COFS syndrome and early-onset or Type 2 CS[8] is some time difficult.[2,9] Cutaneous photosensitivity is noted in patients with CS, but not in patients with COFS. COFS syndrome is associated with eye defects, whereas CS is not. Progressive demyelination with brain calcification in the severe infantile form of CS is quite similar to that is seen in COFS syndrome.[10] Laugel et al consider that COFS syndrome represents a different and even more severe part of the CS spectrum than early-onset CS.[2]

Cataract, Microcephaly, Failure to Thrive, Kyphoscoliosis (CAMFAK) Syndrome

CAMFAK syndrome was first reported in patients characterized by cataracts, microcephaly, failure to thrive and kyphoscoliosis. Associations with arthrogryposis, brain calcification and mental retardation have also been reported.[11] This syndrome might represent early-onset CS (Type 2 CS) or COFS syndrome.

Martsolf Syndrome

This syndrome is characterized by severe mental retardation, cataracts, short stature and hypogonadism.[12] Feeding problems, microcephaly, low nasal bridge, short philtrum and micrognathia are also evident. Patients with Martsolf syndrome do not show complications of severe facial dysmorphism including microphthalmia. No clinical photosensitivity is present and NER analysis has not yet been described. Inheritance is autosomal recessive.

Genetic Basis and the Molecular Mechanism

To date, 14 cases of COFS syndrome have been diagnosed and have also been described at the molecular level.[2] All these COFS patients showed an identical DNA repair defect in the TC-NER subpathway. All mutations have been found either in *CSB*, in xeroderma pigmentosum, *XPD or XPG*, or *ERCC1* genes involved in the TC-NER pathway. Laugel et al have suggested that the term COFS syndrome should be reserved for those patients with similar DNA repair defects.[2]

COFS1: ERCC6 Gene (CSB) at 10q11

Meira et al studied 3 COFS patients from the Manitoba aboriginal population. None of these patients showed clinical photosensitivity. Fibroblast cell lines, derived from two patients, revealed features indistinguishable from those of CS.[13] Cell lines from these patients were more sensitive to killing by UV radiation than cells from a control individual. Furthermore, normal levels of GG-NER were present. Fibroblasts from these COFS patients manifested delayed recovery of RNA synthesis after exposure to UV radiation, indicating that sensitivity to killing after UV radiation resulted from defective TC-NER. Sensitivity to killing by UV radiation was abolished by transfection with wild-type *CSB* cDNA. They also detected identical deletion mutations in the CS group B (*CSB*) gene in 3 patients with typical COFS syndrome phenotype. Of 3 patients, two were dizygotic twins and identical mutation was observed in one *CSB* allele in the parents of 1 patient. This mutation was also detected in DNA obtained from tissues of three other COFS syndrome patients. On the basis of these findings, they suggested that COFS syndrome represents an allelic, clinically severe form of CS.

Laugel et al reported 3 patients with genetically confirmed COFS syndrome who showed mutation in the *CSB* gene.[2] One of the three patients showed clinical photosensitivity with widespread facial erythema. DNA repair studies on cultured fibroblasts from all 3 patients showed severe decrease in recovery of RNA synthesis after UV irradiation, indicating a defect in TC-NER pathway. UV sensitivity studies showed increased sensitivity of fibroblasts to UV irradiation. Unscheduled DNA synthesis (UDS) assay revealed that GG-NER was normal in all 3 patients. Mutational analysis showed that mutation in *CSB* gene was present in all 3 patients (homozygous, n = 1; heterozygous, n = 2).

COFS2: ERCC2 Gene (XPD) at 19q13.2-q13.3

Graham et al reported a COFS syndrome patient with mutation in a third DNA-repair gene, *XPD*.[9] Cutaneous photosensitivity was noted in this patient. Fibroblasts from the patient showed a 10-fold increase in UV sensitivity, comparable to that in patients with severe xeroderma pigmentosa (mutation in *XPB* or *XPG*). The overall rate of NER (both GG-NER and TC-NER) activities were severely impaired and reduced to 1-2%, comparable to that in the most severe cases of XP (mutation in *XPB* or *XPG* gene) as restoration of NER activity occurred after fusion with cells from XPB and XPG, but not after fusion with cells from XPD patients.

Two heterozygous missense base transition mutations in the *XPD* gene were revealed in a patient suffering from COFS. In prenatal diagnosis of a subsequent pregnancy, Graham et al showed that a deficiency of DNA-repair synthesis in chorionic villus cells existed. They concluded that UV-sensitivity tests in cases with suspected COFS syndrome are important for the initial diagnosis and proposed that patients with UV-sensitive COFS syndrome should be included within the spectrum of impaired NER disorders.

COFS3: ERCC5 Gene (XPG) at 13q33

Hamel et al reported a male infant with severe early-onset CS (with typical features of COFS syndrome) who was small for gestational age.[7] He showed hypersensitive skin reaction to sun exposure. Other complications included fibroblasts from the patient total failure to recover to the rate with which RNA synthesis after UV irradiation occurs. DNA replication after UV exposure was also abnormally depressed. UV sensitivity by cell survival assay was more severe than usually observed in cells from CS patients. UV-induced UDS was <4% of that in normal cells, indicating a strong impairment of NER characteristic of XP, but not of CS. When cells from the patients of various XP complementation groups were fused to the cells from the patient, UDS was restored to normal levels after fusion with cells from XP group B and D, but not with XPG cells. This result shows the involvement of the *XPG* (*ERCC5*) gene in incision function in the NER process. Hamel et al emphasized the importance of performing extensive DNA repair studies in cases of suspected early onset CS or COFS syndrome.[7]

COFS4: ERCC1 Gene at 19q13.2-q13.3

Jaspers et al reported the first case of human inherited ERCC1 deficiency with clinical features compatible with a diagnosis of COFS syndrome.[14] Although the clinical features were very severe, cells showed only moderate hypersensitivity to UV and relatively mild impairment of NER, compared with cells from patients with *CS-B* or *XP-A*. UV-induced UDS was reduced to 15% of normal and RNA synthesis recovery after UV exposure was reduced to 13% of normal. Cells were four fold more sensitive to UV than were normal fibroblasts. They also showed two point mutations on the coding region of *ERCC1* and both parents were heterozygous for the respective mutations.

Laugel et al emphasized that COFS patients with mutations in the *XPD* or *XPG* gene are characterized by particularly pronounced cutaneous photosensitivity and by repair defects in both TC-NER and GG-NER.[2] They also analyzed 5 cases of suspected COFS syndrome showing normal DNA repair capacities and all 5 patients did not fulfill at least one of the clinical criteria they proposed, with some patients whose DNA repair capacities were normal showing additional features of cardiac or renal malformations, or muscular dystrophy. They suggest that testing of NER be undertaken in newborns or infants suspected of falling within the COFS-CS spectrum, whenever microcephaly, cataracts, postnatal growth failure and notably reduced development are present. Where TC-NER is deficient and GG-NER is normal, analysis of the *CSB* gene should be undertaken first. Where TC-NER and GG-NER are both deficient, *XPD*, *XPG*, or *ERCC1* genes should be analyzed.

Conclusion

COFS syndrome belongs to the spectrum of inherited defects in NER. In this syndrome the NER is impaired in the TC-NER pathway, but not in the GG-NER. Mutations have been found in *CSB*, *XPD*, *XPG* and *ERCC1* genes involved in the TC-NER pathway. Genetic analysis and biochemical testing of proteins of NER pathway should be carried in newborns or infants suspected of falling within the COFS-CS disease.

References

1. Pena SD, Shokeir MHK. Autosomal recessive Cerebro-Oculo-Facio-Skeletal (COFS) syndrome. Clin Genet 1974; 5:285-293.
2. Laugel V, Dalloz C, Tobias ES et al. Cerebro-oculo-facio-skeletal syndrome: three additional cases with CSB mutations, new diagnostic criteria and a approach to investigation. J Med Genet 2008; 45:564-571.
3. Silengo MC, Davi G, Biango R et al. The Neu-COFS (cerebro-oculo-facio-skeletal) syndrome: report of a case. Clin Genet 1984; 25:201-204.
4. Suzumura H, Nitta A, Arisaka O. Cerebro-oculo-facio-skeletal syndrome complicated with congenital ichthyosis. Clin Dysmorphol 2006; 15:39-40.
5. Graham JM, Hennekam R, Dobyns WB et al. MICRO syndrome: An entity distinct from COFS syndrome. Am J Med Genet 2004; 128A:235-245.
6. Cooper PK, Nouspikel T, Clarkson SG et al. Defective transcription-coupled repair of oxidative base damage in Cockayne syndrome patients from XP group G. Science 1997; 275:990-993.
7. Hamel BCJ, Raams A, Scuitema-Dijkstra AR et al. Xeroderma pigmentosum-Cockayne syndrome complex: a further case. J Med Genet 1996; 33:607-610.
8. Ahmad SI (ed.), Molecular Mecahnisms of Cockayne syndrome. Austin: Landes Bioscience, 2009.
9. Graham JM, Anyane-Yeboa K, Raams A et al. Cerebro-Oculo-Facio-Skeletal syndrome with a nucleotide excision-repair defect and a mutated XPD gene, with prenatal diagnosis in a triplet pregnancy. Am J Med Genet 2001; 69:291-300.
10. Nance MA, Berry SA. Cockayne syndrome: review of 140 cases. Am J Med Genet 1992; 42:68-84.
11. Scott-Emuakpor AB, Heffelfinger J, Higgins JV. A syndrome of microcephaly and cataracts in four siblings: a new genetic syndrome? Am J Dis Child 1977; 131:167-169.
12. Martsolf JT, Hunter AGW, Haworth JC. Severe mental retardation, cataracts, short stature and primary hypogonadism in two brothers. Am J Med Genet 1978; 1:291-299.
13. Meira LB, Graham JM, Greenberg CR et al. Manitoba aboriginal kindred with original Cerebro-oculo-facio-skeletal syndrome has a mutation in the Cockayne syndrome group B (CSB) gene. Am J Med Genet 2000; 66:1221-1228.
14. Jaspers NGJ, Raams A, Silengo MC et al. First reported patient with human ERCC1 deficiency has Cerebro-oculo-facio-skeletal syndrome with a mild defect in nucleotide excision repair and severe developmental failure. Am J Hum Genet 2007; 80:457-466.

Chapter 20

Dyskeratosis Congenita

Vineeta Gupta* and Akash Kumar

Abstract

Dyskeratosis congenita (DC) is an inheritable bone marrow failure syndrome characterized by reticulated hyperpigmentation, dystrophic nails and oral leukoplakia. Another name for the condition is Zinsser-Cole-Engman syndrome. Hematologic manifestations usually do not appear in childhood but later in early adulthood. Patients are also prone to carcinomas, particularly of the head and neck. The disease has X-linked or autosomal dominant/recessive inheritance. Early childhood variants (Hoyeraal-Hreidarsson syndrome) are associated with immunological abnormalities in the form of low T- and B-cell numbers. Four genes, namely DKC1 (codes for dyskerin), TERC and TERT (code for telomerase) and NOP10, have been implicated in the pathogenesis; the short telomeres provide a marker for genetic linkage studies. Androgens, with or without granulocyte colony stimulating factor, have been tried in the treatment of the conditions with variable results. Stem cell transplantation from matched sibling donor is currently the treatment of choice. It requires modified nonmyeloablative conditioning protocols, since the patients with DC are prone to pulmonary and hepatic complications.

Introduction

Bone marrow failure is characterized by its inability to produce adequate number of mature erythrocytes, granulocytes and platelets causing peripheral blood pancytopenia. This in turn is manifested with anemia, bleeding or infection leading to significant morbidity and mortality. In some cases, only one or two cell lines may be affected. In majority of these cases (70%), the primary etiology remains unknown and they are classified as 'idiopathic'. In some other cases, either virus or drug has been identified and suggested as the precipitating factor for bone marrow failure (15% cases). In other 10-20% cases, the disease is familial/inherited and presents with one or more somatic features. Another term which has been used for this group is 'constitutional aplastic anemia' and has been defined as "chronic bone marrow failure associated with other features, such as congenital anomalies, a familial incidence, or thrombocytopenia at birth".[1] However, the hematologic components may not always be evident at birth. Bone marrow failure may present at variable time after birth including adulthood in some cases. In the last one decade significant insight has been gained into the pathophysiology of these conditions. It has helped in better understanding of the genetic mechanisms involved in the disease process.

Some studies have reported the incidence of the inherited conditions in patients presenting with bone marrow failure/aplastic anemia. It was found to be approximately 30% in two studies.[2,3] In a study conducted at our centre, the incidence was approximately 10% in 55 children presenting with aplastic anemia in the age group of 1-14 years.[4]

Fanconi anemia is one of the best recognized inherited bone marrow failure syndrome. Some other conditions include, DC, Schwachman-Diamond syndrome, amegakryocytic

*Corresponding Author: Vineeta Gupta—Department of Pediatrics, Institute of Medical Sciences, Banaras Hindu University, Varanasi—221005, India. Email: vineetaguptabhu@gmail.com

Diseases of DNA Repair, edited by Shamim I. Ahmad. ©2010 Landes Bioscience and Springer Science+Business Media.

thrombocytopenia etc.[5] In this chapter, the inheritance, clinical features, molecular biology, treatment and outcome of DC will be discussed.

Clinical Features

DC is characterized by a triad of reticulated hyperpigmentation of face, neck and shoulders, dystrophic nails and mucosal leukoplakia. A number of nonmucocutaneous abnormalities of dental, gastrointestinal, genitourinary, neurological, ophthalmic, pulmonary and skeletal systems have been described. The condition is also known as Zinsser-Cole-Engman syndrome. The clinical features may appear during childhood with skin pigmentation and nail changes appearing first. The median reported age at diagnosis is around 15 years. Hematologic manifestations normally do not appear in childhood. Half of the patients develop aplastic anemia usually by the age of 20 years. Majority of the patients develop bone marrow failure by the age of 30 years. In a registry of DC patients, the cumulative incidence of marrow failure was 94% by the age of 40 years.[6] However, in some patients (silent carriers in DC families) no hematologic manifestations may appear.

Early childhood variants of DC are the Hoyeraal-Hreidarsson syndrome characterized by intrauterine growth retardation, developmental delay, microcephaly, cerebellar hypoplasia, immunodeficiency and bone marrow failure and Revesz syndrome which has exudative retinopathy in addition to above features.[6] In a classic case, a 5 year-old boy presented with reticulated pigmentation over chest, dystrophic nails and oral leukoplakia with aplastic anemia.[7]

Patients may have ocular abnormalities in the form of epiphora (excessive tearing), blepharitis, cataracts, conjunctivitis, ectropion, glaucoma, strabismus and loss of eyelashes. Dental caries and loss of teeth at an early age have been reported. Intracranial calcifications have also been noted. Osteporosis, fractures, scoliosis and aseptic necrosis may also be present. Urinary tract disorders include phimosis, hypospadias, penile leukoplakia and horseshoe kidney. Gastrointestinal anomalies are esophageal strictures, bifid uvula, umbilical hernia and anal leukoplakia. Pulmonary complications are interstitial pneumonitis and pulmonary fibrosis. Hypoplastic testes in males and vulvar leukoplakia in females have also been reported.

Genetics of DC

DC appears to have an X-linked inheritance and the male to female ratio is 4.5:1. Based on this observation it is supposed to have three modes of inheritance:

1. X-linked recessive: appears to be the most likely mode of inheritance as the disease has been reported in single-male cases, male siblings and males with maternal male cousins.
2. Autosomal recessive: Females with sporadic inheritance, consanguineous families and brother-sister sets have been reported.
3. Autosomal dominant: Few families have been described where males and females were affected after two to three generations with disease passage through both sexes.

X-linked inheritance can be identified with Xq28 restriction fragment length polymorphism and mutation analysis. Patients in the dominant group are of older ages at diagnosis and may have milder form of disease with fewer complications like aplastic anemia and cancer.

Genetic Defects

Lymphocytes from patients with DC do not show significant difference in the chromosomal breakage as compared to normal controls in response to clastogenic agents like Mitomycin C, bleomycin and gamma-irradiation. They are also prone to develop spontaneous chromosomal rearrangements like translocations, dicentrics and tricentrics in the absence of above compounds. In this respect the condition is similar to Fanconi anemia and can be considered a chromosomal instability disorder.[8]

The diagnosis of DC can easily be made when the patient has the diagnostic triad but no hematologic manifestations. The difficulty arises when patient has hematologic features in the form of thrombocytopenia or pancytopenia without physical findings. In such situations the diagnosis is facilitated by examination of the length of telomeres in leucocytes using flow-cytometry with fluorescence in situ hybridization (flow-FISH). Patients with acquired aplastic anemia may have short telomeres in granulocytes but patients with DC have short telomeres in all subsets of leucocytes. This test has both sensitivity and specificity of 90% in distinguishing DC from unaffected relatives and other bone marrow failure syndromes and may be considered the equivalent of chromosomal breakage test for Fanconi anemia.

Four genes have been identified in DC; one is X-linked DKC1, which codes for dyskerin protein. This protein is responsible for ribosomal RNA production, ribosomal assembly and maintenance of telomere length.[9,10] It is located close to nucleoli gene and has a role in the survival of cells that are highly proliferative like skin and bone marrow, the major tissues that are involved in the disease process.

The other genes involved are two autosomal dominant, TERC and TERT, which code for the mRNA for telomeres and telomerase enzyme respectively and one autosomal recessive, NOP10. These gene products are part of the telomere maintenance pathway by which the ends of chromosomes are prevented from shortening substantially with each cell replication. It has been suggested that DC may be a disease of the telomere maintenance rather than ribosomal biogenesis. Very short telomeres have provided a marker for linkage analysis studies. Bone marrow failure may be associated to premature shortening of telomeres, reducing the proliferative potential of the hematopoetic stem cells. Approximately 60% of the patients with DC lack mutation in either of the four known genes.[11]

Immunological Abnormalities

The cell mediated immunity has been found to be abnormal as evidenced by the absence or delayed hypersensitivity to skin test antigens and impaired response to mitogenic stimulation involving in vitro tests.[12] In another study markedly depressed CD4:CD8 ratio (0.38) and decrease in absolute CD4 positive cells were found.[13] Interstitial pneumonia, due to *Pneumocystis carinii* infection, has also been reported. A wide range of immunoglobulin abnormalities have also been reported including diffuse hypoglobulinemia, decreased IgG and IgA with slightly decreased IgM and increased IgG.[13] One study demonstrated severe B lymphopenia, decreased immunoglobulin M (IgM) levels and reduction in T-cells in members of a large family.[14] In rare instances, it can also present as Severe Combined Immunodeficiency (SCID) with complete absence of B and natural killer cells and may be a part of Hoyeraal-Hreidarsson syndrome. An infant with T+B-NK-SCID with marrow failure was reported who underwent sibling bone marrow transplantation.[15] In another study in our laboratory a child suffering from DC had normal immunoglobulin levels and no evidence of infection.[16]

Treatment

Androgens treatment is recommended for patients suffering from significant cytopenias. The treatment guidelines are similar to recommendations for Fanconi anemia.(see Chapter 10 of ref. 8). Oxymetholone can produce improvement in hematopoetic function in some patients. The treatment has to continue for long time in patients who respond to androgens. However, failure may occur after a prolonged treatment. Response to granulocyte colony stimulating factor (G-CSF), granulocyte-macrophage colony stimulating factor (GM-CSF) and erythropoietin has also been reported in some patients. However, combination of androgens and G-CSF has been reported to cause splenic peliosis and rupture.[17] Immunosuppression with antilymphocyte globulin and cyclosporine has been tried in patients without HLA-matched sibling and some improvement in hemogram noted.[18] Six months later the patient died of respiratory complications. Another choice of treatment is stem cell transplantation (SCT) from a HLA-matched sibling. As the siblings may be silent carriers, examination of telomeres in leucocyte subsets is highly recommended to avoid the use of a donor whose stem cells may fail to engraft. The rate of complications, such as pulmonary and vascular complications, has been high following allogenic SCT in DC patients probably because of preexisting pulmonary disease in a proportion of patients with DC. The preparative regimens may require modifications and agents causing pulmonary toxicity (radiotherapy, busulphan etc.)

may not be used. Non myeloablative Fludarabine based protocols have given encouraging results.[19] The follow-up is short to evaluate long-term toxicity and the natural course of the disease. It may be that correction of the telomerase defect is essential for the treatment of the disease, therefore, SCT may be performed on carefully selected patients. Supportive care with blood products and antibiotics may be provided to patients with aplastic anemia. Є-aminocaproic acid (0.1 g/kg) may be given every 6 hourly for symptomatic bleeding.

Prognosis and Outcome

The prognosis of patients with DC is not promising. Patients with autosomal dominant disease have milder form of disease and consequently a better outcome. Those with X-linked disease have a more severe course. The cause of death is usually complications of aplastic anemia, or failure of bone marrow transplantation or development of malignancies.

Approximately 15% of the patients of DC studied developed cancer. Majority of the cases had squamous cell carcinoma which included cancers of the oropharyngeal and gastrointestinal tracts. Hodgkin's lymphoma, adenocarcinoma of the pancreas and skin cancers have also been reported.[6] The cancers developed at a median age of 30 years whereas the median age of development of aplastic anemia was 11 years. The patients who developed aplastic anemia were younger as compared to those who developed cancers. However, leukemia and myelodysplastic syndrome have been reported rarely in patients with DC.

Conclusion

Dyskeratosis congenita (DC) is an inherited bone marrow failure syndrome characterized by a triad of reticulated hyperpigmentation, dystrophic nails and oral leukoplakia. Inheritance is autosomal recessive/dominant or X linked recessive. Four genes namely, DKC1, TERC, TERT and NOP10 have been implicated in the pathogenesis. Clinical features may appear during childhood but hematologic manifestations appear late usually in second decade. Some variants may have associated immunological abnormalities. Oxymethalone has been used in the treatment with variable results. Immunosuppressive treatment and stem cell transplantation (SCT) from matched sibling donor have also been tried, although rate of complications following SCT have been high. Prognosis of patients with DC is not very promising. Cause of death is bone marrow failure or development of malignancies.

References

1. Alter BP. Bone marrow failure syndrome in children. Pediatr Clin North Am 2002; 49:973-988.
2. Alter BP, Potter NU, Li FP. Classification and etiology of the aplastic anaemias. Clin Haematol 1978; 7:431-465.
3. Windass B, Vowels MR, O'Gorman Hughes D. Aplastic anemia in childhood: Prognosis and approach to therapy. Med J Aust 1987; 146:15-19.
4. Gupta V, Tripathi S, Singh TB et al. A study of bone marrow failure syndrome in children. Indian J Med Sci 2008; 62:13-18.
5. Tamary H, Alter BP. Current diagnosis of inherited bone marrow failure syndrome. Pediatr Hematol Oncol 2007; 24:87-99.
6. Alter BP. Diagnosis, genetics and management of inherited bone marrow failure syndrome. Hematology Am Soc Hematol Educ Program 2007;29-39.
7. Gupta V, Bhatia BD. Dyskeratosis congenita Indian Pediatr 2008; 45:936.
8. Ahmad, SI, Kirk, SH, eds. Molecular Mechanisms of Fanconi anemia. Austin/New York: Landes Bioscience/Springer Science/Business Media Inc., 2006.
9. Mitchell JR, Wood E, Collins K. A telomere component is defective in the human disease: Dyskeratosis Congenita. Nature 1999; 402:551-555.
10. Vulliamy TJ, Marrone A, Dokal I et al. Association between aplastic anemia and mutations in telomerase RNA. Lancet 2002; 359:2168-2170.
11. Vulliamy TJ, Marrone A, Knight SW et al. Mutations in dyskeratosis congenita: their impact on telomere length and the diversity of clinical presentation. Blood 2006; 107:2680-2685.
12. Ortega JA, Swanson VL, Landig BH et al. Congenital Dyskeratosis: Zimmer-Engman-Cole syndrome with thymic dysplasia and aplastic anemia. Am J Dis Child 1972; 124:701-704.

13. Lee BW, Yap HK, Quah TC et al. T-cell immunodeficiency in Dyskeratosis Congenita. Arch Dis Child 1992; 67:524-526.
14. Knudson M, Kulharni S, Ballas ZK et al. Association of immune abnormalities with telomere shortening in autosomal dominant DC. Blood 2005; 105:682-688.
15. Cossu F, Vulliamy TJ, Marrone A. A novel DKCI mutation, severe combined immunodeficiency (T+B-NK-SCID) and BHT in and infant with Hoyeraal- Hreidarsson syndrome. Br J Haematol 2002; 119:765-768.
16. Gupta V, Bhatia BD. Immunological abnormalities in Dyskeratosis congenita. Indian Pediatr 2009; 46:448.
17. Giri N, Pitel N, Green D. Splenic peliosis and rupture in patients with Dyskeratosis congenita on androgens and G-CSF. Br J Haematol 2007; 138:815-817.
18. Connoli P, Basso S, Huanga GC. Intensive immunosuppression therapy for aplastic anemia associated with Dyskeratosis congenita: report of a case. Int J Hematol 2006; 83:275-276.
19. de La Fuente J, Dokal I. Dyskeratosis congenita: advances in the understanding of the telomerase defect and the role of stem cell transplantation. Pediatr Transplant 2007; 11:584-594.

Chapter 21

Retinoblastoma

Dietmar Lohmann*

Abstract

Retinoblastoma (Rb) is a malignant tumor that originates from developing retina. Diagnosis is based on clinical signs and symptoms and is usually made in children under the age of five years. Mutations in both alleles of the RB1 gene are a prerequisite for this tumor to develop. In most patients with sporadic unilateral Rb, both RB1 gene mutations occur in somatic cells and are not passed over to offspring (nonhereditary Rb). Almost all patients with sporadic bilateral and virtually all patients with familial Rb are heterozygous for RB1 gene mutations that cause predisposition to Rb (hereditary Rb). In families, Rb predisposition is transmitted as an autosomal dominant trait (familial Rb). In addition to Rb, patients with hereditary disease also have an increased risk of tumors outside the eye (second cancer). This risk is enhanced in patients who have received external beam radiotherapy. Analysis of genotype-phenotype associations has shown that the mean number of tumor foci that develop in carriers of mutant RB1 alleles is variable depending on which functions of the normal allele are retained and to what extent. Moreover, phenotypic expression of hereditary retinoblastoma is subject to genetic modification. Identification of the genetic factors that underlie these effects will not only help to arrive at a more precise prognosis but may also point to mechanisms that can be used to reduce the risk of tumor development.

Clinical Aspects

Epidemiology of Retinoblastoma

Retinoblastoma (OMIM# 180200) is a malignant tumor of the eye that originates from cone precursors cells of the developing retina.[1] The estimated incidence is between 1 in 15,000 to 20,000 live birth children.[2,3] It is not yet clear if the incidence of retinoblastoma is the same worldwide.[4] Environmental factors, which may account for different incidence figures, have no known role in the etiology of Rb.[5] Specifically, neither HPV nor any other pRb-inactivating human DNA tumor viruses play a role in the development of this tumor.[6] In most children, diagnosis is made under the age of five years. In adults, Rb is very rare. It has been suggested that tumors diagnosed after childhood age developed from preexisting progenitor lesions (retinoma, see below).[7,8]

Diagnosis of Rb

Diagnosis of Rb is based on clinical signs and symptoms. Most often, the first presenting sign is a white pupillary reflex (leukocoria). Strabismus is the second most common sign and may accompany or precede leukocoria. Usually, diagnosis of Rb is established by examination of the fundus of the eye using indirect ophthalmoscopy. Additional diagnostic tools such as computer tomography scan (CTS), magnetic resonance imaging (MRI) and ultrasonography may be required for differential diagnosis and staging. If tumor material was obtained, histopathology can confirm Rb.

*Dietmar Lohmann—Institut fur Humangenetik, Universitatsklinikum Essen, Hufelandstrasse 55, D-45122 Essen, Germany. Email: dietmar.lohmann@uni-due.de

Diseases of DNA Repair, edited by Shamim I. Ahmad. ©2010 Landes Bioscience and Springer Science+Business Media.

Presentation and Family History

Most patients (60%) have Rb in one eye only (unilateral Rb). Occasionally, multiple tumor foci can be found (unilateral multifocal Rb). In about 40% of the cases, both eyes are affected (bilateral Rb) usually with more than one focus per eye (bilateral multifocal Rb). In children with bilateral Rb, diagnosis is made earlier than in children with unilateral Rb (median age at diagnosis 11 and 22 months, respectively). A few children, most with bilateral Rb, also develop pinealoblastoma (trilateral Rb). With respect to histological appearance and age at diagnosis, this brain tumour is similar to Rb and, therefore, pinealoblastoma is often not regarded as a second tumour (see below).

Patients with unilateral Rb most often have sporadic disease, i.e., no other case of Rb has been noted in their family. About 75% of patients with bilateral Rb are also sporadic, in the remainder (25%) there is a positive family history (familial Rb). Examination of the fundus of the eye in all first degree relatives of children with Rb is required to identify retinal scars or quiescent tumors (retinomas). The presence of such lesions in a relative indicates familial disease.[9]

Therapy and Prognosis

Compared to other solid neoplasms, Rbs are small at the time of initial diagnosis. Treatment depends on tumor stage, the number of tumor foci (unifocal, unilateral multifocal or bilateral disease), localization and size of the tumor(s) within the eye, presence of vitreous seeding and the age of the child. Treatment options include enucleation, external-beam radiation, cryotherapy, photocoagulation, brachytherapy with episcleral plaques. The latter are effective treatment options when the tumor size is limited (<5 mm) and the tumor is not complicated by extensive exudative retinal detachment, recurrences or widespread vitreous seeding. Novel treatment options include systemic chemotherapy combined with local therapy. Following successful treatment, children require frequent follow-up examinations for early detection of new intraocular tumors. If the tumor(s) have not invaded extraocular tissues, treatment is usually successful. In spite of recent improvements, metastasized Rb is often fatal.

Second Cancers in Patients with Hereditary Rb

Incidence of Second Cancers

Epidemiological studies have shown that patients with hereditary Rb have a lifelong increased risk for developing neoplasms outside of the eye (second non-ocular cancers). In larger series with a sufficient number of older survivors of Rb the incidence rate per year was estimated at 0.5 to 1%.[10] According to the largest study published so far the cumulative incidence for new malignancies 50 years after the diagnosis of Rb is 36% (95% CI 31-41%).

Spectrum of Second Cancers

Most second cancers are ostosarcomas that arise from the skull or long bones (ratio of observed to expected incidence [standardized incidence ratio] = 406 [95% CI = 318-511]). Sarcoma of the soft tissues and malignant melanomas are also frequent (SIR = 140, bzw. 30).[11] The spectrum of soft tissue sarcomas in patients with Rb is dominated by leiomyosarcoma, a histopathological subtype that is normally rare.[12] Patients who received radiation often show tumours of the nasal cavities and the orbits. In general, the spectrum of second cancers is broad and, according to recent studies, also includes tumours of epithelial origin (lung SIR = 3,2; bladder SIR = 7,9; colon SIR = 8.1).

Subcutaneous lipoma are benign neoplasms that are not infrequent in otherwise healthy adults. The frequency of these tumours is increased in adult survivors of hereditary Rb (about 4% of patients).[13] Moreover, Rb patients with lipoma most often develop several of these tumours.[14] Although lipomas do not progress further to malignant tumours, observations suggest that lipomas are more frequent in patients who develop sarcoma. This implies that the risk of second cancers is variable among patients with Rb, an idea that is also backed by the observation of familial clustering of patients with second cancers (see genetic risk factors below).

Development of Further Tumours after a Second Tumour

Abramson et al[15] have studied the incidence, timing, pattern and survival as the result of a third, fourth and fifth nonocular tumors in survivors of Rb after diagnosis of a second cancer. They showed that the incidence rate of tumours in patients with multiple neoplasms is increased and that the latency between diagnoses decreases with the number of tumours. Moreover, the kind and localization of the second cancers in these patients was linked to the kind and localization of third, fourth and fifth nonocular tumors. These observations indicate that variation of risk of second tumours is site-specific.

Ionising Radiation as a Risk Factor

Exposing children to ionising radiation increases the risk of cancer later in life. The New York Group has first shown in 1993 that external beam radiotherapy (EBRT) dramatically increases the risk of radiation induced malignant tumors in the radiation field for patients with hereditary Rb. According to recent studies the relative risk is about 3.1 compared to patients not exposed to external beam radiation; thus indicating a synergistic interaction.[16] However, for the individual patient the risk attributable to radiation is variable and is likely to be influenced by the radiotherapy technique used, radiation doses and field and, possibly, age at therapy.[17,18]

Chemotherapy as a Risk Factor

The risk of malignant neoplasms is increased in children who received chemotherapy for the treatment of cancer. There are a few cases of acute myelogenous leukemia (AML) in children with Rb who had received chemotherapy.[19] However, risk of second cancers in Rb patients treated with chemotherapy has not been studied systematically yet.

Genetic Risk Factors

The spectrum of oncogenic mutations that cause hereditary predisposition to Rb is heterogeneous. With respect to the manifestation of Rb, distinct genotype-phenotype associations have been identified. On average, the number of Rb foci in a patient is associated with the type of oncogenic germ line mutation: alterations that create premature stop codons almost invariably cause bilateral multifocal Rb (complete penetrance). In contrast, some missense, in-frame and splice mutations are associated with fewer tumor foci (often unilateral Rb only) or mutation carriers do not develop Rb (incomplete penetrance).[20] Familial clustering of second neoplasms in only some families with Rb suggests that the risk of second tumours is dependent on genetic factor(s).

Molecular Genetics

The importance of hereditary factors has been recognized since the 19th century. Familial aggregation of Rb was noted as early as 1821.[21] After treatment was improved, more patients reached adult age and had children with Rb. Thus it became evident that familial Rb follows an autosomal dominant mode of inheritance. Initially all cases were regarded as hereditary but later it was recognized that in a significant proportion of patients, with sporadic Rb, the etiology is nonhereditary.[22]

Hereditary and Nonhereditary Rb is Caused by Two Mutations

In 1971 Knudson proposed a model to explain the genetic mechanisms underlying hereditary and nonhereditary Rb.[23] According to his hypothesis, both the hereditary and nonhereditary form of Rb are caused by two mutations (two-hit hypothesis):

- In hereditary Rb, the first mutation is inherited via germinal cells. Tumor foci are caused by second mutations that occur in somatic cells.
- In the nonhereditary form of Rb, the two mutations occur in somatic cells.

Identification of the RB1 Gene

Molecular analyses showed that two mutations required for Rb to occur target the tow alleles at one gene locus on chromosome 13q14.[24,25] The second mutation, which precedes tumor formation, often results in loss of one allele via mitotic recombination or other chromosomal mechanisms that lead to loss of heterozygosity (LOH) at polymorphic loci on chromosome 13q14. The cDNA of the RB1 gene was reported in 1986.[26] The RB1 gene (accession number L11910) consists of 27 exons that are scattered over 183 kb of genomic sequence on chromosome 13q14. At its 5′-end, the RB1 gene has a CpG-island, which is normally unmethylated. Another CpG island, which is located in intron 2 and is derived from a processed pseudogene, shows parent of origin specific methylation. The promoter region at the 5′-end contains binding motifs for transcription factors Sp1 and ATF but no TATA or CAAT-elements. In tissues, investigated so far, the gene is transcribed into a 4.7 kb mRNA. The open reading frame, which starts in exon 1 and is terminated in exon 27, spans over 2.7 kb and is followed by a 2 kb untranslated region. Another mRNA, which is transcribed from the paternal allele only, starts with the alternative exon 2B with a possible translation start site in exon 3. Homologs of the human RB1 gene have been identified in a wide variety of organisms and show a high level of sequence similarity in translated regions. The part of the gene that encodes the domains for the A/B pocket (see below) has a homolog also in higher plants (mat3).

The RB1 Gene Encodes a Pocket Protein

The complete ORF encodes pRb, a 928 amino acid nuclear phosphoprotein migrating at 110 kD in SDS-PAGE when hypophosphorylated. pRb is a member of the pocket protein family that also includes p107 and p130 (for recent reviews see 27-29). These proteins share significant sequence similarity in two discontinuous areas that constitute the A/B pocket. Conditional on the phosphorylation status at multiple serine and threonine residues in other regions of the protein, the A/B pocket can bind to members the E2F-family of transcription factors and other nuclear proteins that contain the LxCxE peptide motif (such as histone deacetylases 1 and 2). The C-terminal region of pRB contains a nuclear localization signal and a cyclin-cdk interaction motif that enables it to be recognized and phosphorylated by cyclin-cdk complexes. The C-terminal region can also bind to the nuclear c-Abl tyrosine kinase and to MDM2, which are proteins with oncogenic properties.

The role of pRb, that is understood best, is its function as a gatekeeper that negatively regulates progression through G1 phase of the cell cycle. During the G1 phase of the cell cycle pRb is hypo-phosphorylated. When hypo-phosphorylated, the pocket proteins act as transcriptional cofactors and, by recruiting chromatin remodeling enzymes, repress the proliferation-promoting activities of different sets of E2F transcription factors. Beginning in late G1 and continuing to the M phase pRB is phosphorylated by G1 cyclin-dependent kinases. Phosphorylation results in derepression and activation of the respective subsets of E2F dependent gene promoters. In addition to control the G1-S cell cycle transition, pRb can induce apoptosis, in response to genotoxic stress and has important roles in directing differentiation and cell fate specification during embryogenesis.

Tumor Genetics of Rb

The development of a Rb focus in a heterozygous patient appears not to be an immediate consequence of a mutation of the second allele. This view is supported by the finding that, in addition to mutations of the RB1 gene, Rbs often show alterations at other loci. This has led to the suggestion that additional mutations (third mutations, M3) are required for development of this tumor.[30] Molecular studies have identified several candidate genes that may be the functional targets of the recurrent genomic copy number changes that are observed in Rbs.[30]

Phenotypic Consequences of RB1 Gene Mutations

Mutation of RB1/rb1, but not of the other pocket protein genes is consistently associated with tumor predisposition in both humans and mice. RB1 gene mutations have been identified in more than 2000 patients with Rb (RB1-LDSB). Almost all types of mutations have been identified including translocations, deletions, insertions, point mutations and epigenetic mutations (hypermethylation of the CpG-island in the promoter region at the 5′-end of the gene).

Genotype-Phenotype Associations

Heterozygous carriers of oncogenic RB1 gene mutations show variable phenotypic expression.[20] Patients may develop tumors in both eyes or in one eye only (variable expressivity). Some carriers show no Rb at all (incomplete penetrance). According to the two-hit hypothesis, variation of phenotypic expression is to be expected because the development of an individual tumor focus depends on the chance occurrence of a second somatic mutation. However, a quantitative analysis of phenotypic variation in families with retinoblastoma shows that stochastic effects can account for only a part of the observed differences. It is now well established that penetrance and expressivity of hereditary retinoblastoma depend on the nature of the predisposing mutation.

The majority of mutations that have been identified in patients with hereditary retinoblastoma are nonsense or frameshift mutations. These mutations are located in exons 1 to 25 of the RB1 gene. With rare exceptions, nonsense or frameshift mutations in internal exons (2 to 25) are associated bilateral retinoblastoma. Occasionally, such a mutation is identified in a patient with isolated unilateral retinoblastoma or in a unilaterally affected parent of a child with bilateral retinoblastoma. However, in some of these patients the mutation is present in a mosaic state. This parallels the findings in several other disorders with dominant inheritance where, compared to the phenotype in heterozygous mutation carriers, mosaicism is associated with milder phenotypic expression. In some genetic diseases, e.g., adenomatosis polyposis coli, mutant alleles with nonsense and frameshift mutations are associated with distinct phenotypic expression depending on the location of premature stop codon within the causative gene. Genotype-phenotype of this kind have not been observed in hereditary retinoblastoma where the site of the internal premature stop codon within the RB1 gene seems to have little or no effect on phenotypic expression. Possibly, such a genotype-phenotype association is not to be expected because available data suggest that transcripts of RB1 alleles with internal premature stop codons are recognized by nonsense-mediated decay, which is a posttranscriptional surveillance mechanism that causes degradation of mutant transcripts. Consequently, in cells heterozygous for a mutation that triggers nonsense mediated only transcripts of the normal allele are left and this results in a haploinsufficient situation. Interestingly, no mutation has been identified in the two terminal exons of the RB1 gene (exons 26 and 27) although this region contains two CGA-codons, which are potential hot spots of nonsense mutations. However, according to what is known from other genes, premature stop codons in these regions will not trigger nonsense mediated decay. This suggests that mutant pRBs with late C-terminal truncation may have sufficient tumor suppressive activity to prevent development of retinoblastoma. Interestingly, families with deletions or insertions in exon 1 that result in a frameshift can show incomplete penetrance.[31]

Premature termination codons can also result from splicing errors that are caused by point mutations in intronic or exonic sequences. As is expected from the genotype-phenotype associations outlined above, splice mutations that result in out-of-frame exon skipping typically are associated with complete penetrance. However, this is only valid for point mutations that alter invariable splice signals. Mutations that affect splice signal in exons or less conserved intronic splice signals can be associated with milder expressivity and incomplete penetrance.[32]

Carriers of missense and small in frame length alterations are expected to express mutant transcripts that are not recognized by posttranscriptional surveillance mechanisms. However, it is important to investigate the effect of supposed missense and inframe mutations on the RNA level because mutations in exons may result in altered splicing. For most reported missense and in frame mutations in the RB1 gene, such data are not available. Although the effect of individual mutations may be uncertain, the phenotypic expression that is associated with missense and in frame mutations is well distinguished from that of alleles with premature termination codons. Typically, heterozygous carriers of missense and in frame mutations show incomplete penetrance or milder expressivity. The amino acids that are substituted, deleted, or inserted in consequence of these mutations are most often part of the A/B pocket of the pRB. A few in frame mutations have been identified outside of regions that code for the A/B pocket, including deletions of exon 4 and of exon 24-25.[33,34] Functional studies have shown that mutant pRB expressed from alleles that are associated with incomplete penetrance show only a partial loss of normal function.[34-36] As noted above it is plausible that mutation of the second RB1 allele in a heterozygous patient is not sufficient for development of an Rb focus. Its conceivable that a retinoblastoma precursor cell that has mutations in both alleles of the RB1 may have various fates. The probability of progression towards retinoblastoma (vs apoptosis/differentiation) may reduce if residual pRB function is left because of weak mutations.

Other Factors That Modify Phenotypic Expression

Most of the phenotypic variation that is observed between Rb families can be explained by the different functional consequences of the various RB1 gene mutations. However, between families with identical mutations nonrandom differences of phenotypic expression have been observed (interfamilial variation). For example, there are two unrelated retinoblastoma families that both segregate an identical splice mutation, which results in skipping of exon 13[37,38] and that both show incomplete penetrance of retinoblastoma. However, only in one family mutation carriers have developed multiple subcutaneous lipoma. Lipoma are benign neoplasms of adipose tissue that, compared to the general population, occur more frequently in adult patients with hereditary retinoblastoma.[13] In the family with the lipoma/retinoblastoma, penetrance of lipma is almost complete whereas in the other family none of the carriers of the identical mutation has developed lipoma. This shows that predisposition to lipoma is not caused by the splice mutation per se but is due to a heritable modifying effect. As the lipoma phenotype was found to be linked to the mutant RB1 allele the modifying effect is most likely due to a genetic factor in cis relative to the mutant RB1 allele. It is conceivable that predisposition to second cancers, which are an important health problem in patients with retinoblastoma, is also subject to modifier effects. This would be consistent with the observation that second tumors are not associated with specific RB1 gene mutations.

Another modifier effect has been identified in two families that have the same base substitution in intron 6 of the RB1 gene. This mutation results in a premature termination codon because of exon skipping. Contrary to what is expected from the genotype-phenotype associations outlined both families show incomplete penetrance. Intriguingly, most mutation carriers that have received the mutant allele via the maternal germ line are unaffected whereas almost all mutation carriers that have received the mutant allele via the paternal germ line have developed retinoblastoma. It is not known yet if this parent-of-origin effect is mechanistically connected to the imprinted CpG-island in intron 2 of the RB1 gene.

Endophenotypes Observed in Rb Patients

Cells from patients with diverse hereditary cancer predisposition syndromes show hypersensitivity to ionizing radiation in vitro. Cellular hyper-radiosensitivity responses are most pronounced in those cancer syndromes where the causative genetic alteration affects DNA repair and cell cycle regulatory pathways.[39] As patients with hereditary Rb are functionally hemizygous for a gene that codes an important cell cycle regulator that has a role in mitotic fidelity and DNA repair, nonnormal response to ionizing radiation is to be expected. In fact, at the organismal level, patients with hereditary Rb that received external beam radiation show an increased risk of second cancers (see above). Various in vitro cellular responses have been documented in heterozygous carriers of RB1 gene mutations.[39] However, an enhanced sensitivity for cell killing and G1-phase arrest after acute gamma irradiation was also observed in primary fibroblasts from unaffected (RB1 homozygous normal) parents of Rb patients. At the molecular level, unirradiated fibroblast cultures from these family members showed distinctive gene expression profiles.[40] It was hypothesized that the aberrant responses in RB1 homozygous normal relatives are due to genetic variation at other loci.[41,42]

Conclusion

Retinoblastoma (Rb), a chidlhood malignant tumor of the eye, is the paradigm of hereditary tumor predisposion syndromes. Hereditary predisposition to retinoblastoma is an autosomal dominant trait that is caused by mutations in one allele of the retinoblastoma (RB1) gene. Development of individual tumor foci is initiated by mutational loss of the second allele (Knudson's two-hit hypothesis). The different functional consequences of the various predisposing (first) RB1 gene mutations are associated with penetrance and expressivity of hereditary retinoblastoma. In addition to retinoblastoma, patients heterozygous for RB1 mutations have a lifelong increased risk for cancer outside of the eye most notably ostosarcomas, sarcoma of the soft tissues, and malignant melanomas. Risk of second tumor is increased in pateints who received ionizing radiation. Findings in vitro show aberrant cellular responses following acute irradiation in patients with retinoblastoma. So far, no association of second cancer and specific RB1 mutation has been detected.

References

1. Xu XL, Fang Y, Lee TC et al. Retinoblastoma has properties of a cone precursor tumor and depends upon cone-specific MDM2 signaling. Cell 2009; 137:1018-1031.
2. Broaddus E, Topham A, Singh AD. Incidence of retinoblastoma in the USA: 1975-2004. Br J Ophthalmol 2009; 93:21-23.
3. Suckling RD, Fitzgerald PH, Stewart J et al. The incidence and epidemiology of retinoblastoma in New Zealand: A 30-year survey. Br J Cancer 1982; 46:729-736.
4. Olisa EG, Chandra R, Jackson MA et al. Malignant tumors in American black and Nigerian children: a comparative study. J Natl Cancer Inst 1975; 55:281-284.
5. Buckley JD. The aetiology of cancer in the very young. Br J Cancer 1992; (Suppl 18):S8-12.
6. Gillison ML, Chen R, Goshu E et al. Human retinoblastoma is not caused by known pRb-inactivating human DNA tumor viruses. Int J Cancer 2007; 120:1482-1490.
7. Odashiro AN, Pereira PR, de Souza Filho JP et al. Retinoblastoma in an adult: case report and literature review. Can J Ophthalmol 2005; 40:188-191.
8. Dimaras H, Khetan V, Halliday W et al. Loss of RB1 induces nonproliferative retinoma; increasing genomic instability correlates with progression to retinoblastoma. Hum Mol Genet 2008; 17:1363-1372.
9. Gallie BL, Ellsworth RM, Abramson DH et al. Retinoma: spontaneous regression of retinoblastoma or benign manifestation of the mutation? Br J Cancer 1982; 45:513-521.
10. Acquaviva A, Ciccolallo L, Rondelli R et al. Mortality from second tumour among long-term survivors of retinoblastoma: a retrospective analysis of the Italian retinoblastoma registry. Oncogene 2006; 25:5350-5357.
11. Kleinerman RA, Tucker MA, Tarone RE et al. Risk of new cancers after radiotherapy in long-term survivors of retinoblastoma: an extended follow-up. J Clin Oncol 2005; 23:2272-2279.
12. Kleinerman RA, Tucker MA, Abramson DH et al. Risk of soft tissue sarcomas by individual subtype in survivors of hereditary retinoblastoma. J Natl Cancer Inst 2007; 99:24-31.
13. Li FP, Abramson DH, Tarone RE et al. Hereditary retinoblastoma, lipoma and second primary cancers. J Natl Cancer Inst 1997; 89:83-84.
14. Rieder H, Lohmann D, Poensgen B et al. Loss of heterozygosity of the retinoblastoma (RB1) gene in lipomas from a retinoblastoma patient. J Natl Cancer Inst 1998; 90:324-326.
15. Abramson DH, Melson MR, Dunkel IJ et al. Third (fourth and fifth) nonocular tumors in survivors of retinoblastoma. Ophthalmology 2001; 108:1868-1876.
16. Yu CL, Tucker MA, Abramson DH et al. Cause-specific mortality in long-term survivors of retinoblastoma. J Natl Cancer Inst 2009; 101:581-591.
17. Abramson DH, Frank CM. Second nonocular tumors in survivors of bilateral retinoblastoma: a possible age effect on radiation-related risk. Ophthalmology 1998; 105:573-9; discussion 579-80.
18. Chauveinc L, Mosseri V, Quintana E et al. Osteosarcoma following retinoblastoma: age at onset and latency period. Ophthalmic Genet 2001; 22:77-88.
19. Gombos DS, Hungerford J, Abramson DH et al. Secondary acute myelogenous leukemia in patients with retinoblastoma: is chemotherapy a factor? Ophthalmology 2007; 114:1378-1383.
20. Lohmann DR, Gallie BL. Retinoblastoma: revisiting the model prototype of inherited cancer. Am J Med Genet C Semin Med Genet 2004; 129:23-28.
21. Kaelin A. Statistische Prüf- und Schätzverfahren für die relative Häufigkeit von Merkmalsträgern in Geschwisterreihen bei einem der Auslese unterworfenen Material mit Anwendung auf das Retinoblastom. Archiv der Julius Klaus-Stiftung 1955; 30:442-485.

22. Vogel F. (New studies on the genetics of retinoblastoma; glioma retinae.). Z Mensch Vererb Konstitutionsl 1957; 34:205-236.
23. Knudson AG. Mutation and cancer: statistical study of retinoblastoma. Proc Natl Acad Sci USA 1971; 68:820-823.
24. Godbout R, Dryja TP, Squire J et al. Somatic inactivation of genes on chromosome 13 is a common event in retinoblastoma. Nature 1983; 304:451-453.
25. Cavenee WK, Dryja TP, Phillips RA et al. Expression of recessive alleles by chromosomal mechanisms in retinoblastoma. Nature 1983; 305:779-784.
26. Friend SH, Bernards R, Rogelj S et al. A human DNA segment with properties of the gene that predisposes to retinoblastoma and osteosarcoma. Nature 1986; 323:643-646.
27. Genovese C, Trani D, Caputi M et al. Cell cycle control and beyond: emerging roles for the retinoblastoma gene family. Oncogene 2006; 25:5201-5209.
28. Stengel KR, Thangavel C, Solomon DA et al. RB/p107/130 pocket proteins: Protein dynamics and interactions with target gene promoters. J Biol Chem 2009.
29. Goodrich DW. The retinoblastoma tumor-suppressor gene, the exception that proves the rule. Oncogene 2006; 25:5233-5243.
30. Corson TW, Gallie BL. One hit, two hits, three hits, more? Genomic changes in the development of retinoblastoma. Genes Chromosomes Cancer 2007; 46:617-634.
31. Sánchez-Sánchez F, Ramírez-Castillejo C, Weekes DB et al. Attenuation of disease phenotype through alternative translation initiation in low-penetrance retinoblastoma. Hum Mutat 2007; 28:159-167.
32. Zhang K, Nowak I, Rushlow D et al. Patterns of missplicing caused by RB1 gene mutations in patients with retinoblastoma and association with phenotypic expression. Hum Mutat 2008; 29:475-484.
33. Dryja TP, Rapaport J, McGee TL et al. Molecular etiology of low-penetrance retinoblastoma in two pedigrees. Am J Hum Genet 1993; 52:1122-1128.
34. Bremner R, Du DC, Connolly-Wilson MJ et al. Deletion of RB exons 24 and 25 causes low-penetrance retinoblastoma. Am J Hum Genet 1997; 61:556-570.
35. Otterson GA, Modi S, Nguyen K et al. Temperature-sensitive RB mutations linked to incomplete penetrance of familial retinoblastoma in 12 families. Am J Hum Genet 1999; 65:1040-1046.
36. Otterson GA, Chen W, Coxon AB et al. Incomplete penetrance of familial retinoblastoma linked to germ-line mutations that result in partial loss of RB function. Proc Natl Acad Sci USA 1997; 94:12036-12040.
37. Scheffer H, Van Der Vlies P, Burton M et al. Two novel germline mutations of the retinoblastoma gene (RB1) that show incomplete penetrance, one splice site and one missense. J Med Genet 2000; 37:E6.
38. Genuardi M, Klutz M, Devriendt K et al. Multiple lipomas linked to an RB1 gene mutation in a large pedigree with low penetrance retinoblastoma. Eur J Hum Genet 2001; 9:690-694.
39. Deschavanne PJ, Fertil B. A review of human cell radiosensitivity in vitro. Int J Radiat Oncol Biol Phys 1996; 34:251-266.
40. Chuang EY, Chen X, Tsai MH et al. Abnormal gene expression profiles in unaffected parents of patients with hereditary-type retinoblastoma. Cancer Res 2006; 66:3428-3433.
41. Wilson PF, Nagasawa H, Warner CL et al. Radiation sensitivity of primary fibroblasts from hereditary retinoblastoma family members and some apparently normal controls: colony formation ability during continuous low-dose-rate gamma irradiation. Radiat Res 2008; 169:483-494.
42. Fitzek MM, Dahlberg WK, Nagasawa H et al. Unexpected sensitivity to radiation of fibroblasts from unaffected parents of children with hereditary retinoblastoma. Int J Cancer 2002; 99:764-768.

Chapter 22

Von Hippel Lindau Syndrome

Jenny J. Kim, Brian I. Rini and Donna E. Hansel*

Abstract

Von Hippel-Lindau syndrome (VHLS) is an autosomal dominant familial cancer syndrome arising from germ-line inactivation of the VHL gene on the short arm of chromosome 3. VHLS manifests in a myriad of hyper-vascular tumors of both benign and malignant nature. Incidence of VHLS is roughly 1 in 36,000 live births and has over 90% penetrance by the age of 65. Improved understanding of the natural history and biology of VHLS has led to the introduction of screening protocols, early interventions and improved treatments, all of which resulted in a substantially improved prognosis for this disease. Further details regarding variegated molecular pathways and mechanisms of VHLS are emerging with the subsequent advent of novel treatment protocols that are currently in clinical trials.

Genetic Basis and Molecular Mechanism of VHLS

Identification of the VHL Gene

Manifestations of the VHLS include hemangioblastomas of the CNS and retina, clear cell renal cell carcinoma (RCC), pheochromocytoma, endolymphatic sac tumors, cystadenomas and neuroendocrine tumors of the pancreas, papillary cystadenomas of epididymis and broad ligament (Table 1). The first patient with VHLS was reported by Collins in 1894. Subsequently, in early 1900s Eugene von Hippel described members of certain families who exhibited predisposition to development of angiomas (or hemangioblastomas) of the retina.[1,2] Later, Arvind Lindau reported that such individuals were also likely to develop histologically identical tumors in the brain and along the spinal cord.[3] In 1936, the term "von Hippel-Lindau disease" was first used clinically and since then several other characteristic tumors listed above have been associated with the disorder.[4-8] In the 1980s the cytogenetic studies and restriction fragment length polymorphism (RFLP) analysis of individuals with sporadic clear cell RCC consistently showed abnormalities involving the short arm of chromosome 3 (chromosome 3p), suggesting that similar abnormalities in chromosome 3p may also play a role in VHLS associated clear cell RCC.[9]

Genetic linkage analysis of multiple families with VHLS led to localization of *VHL* gene to chromosome 3p25-26.[10] It contains 3 exons,[11] with more than 300 germ-line mutations reported in the VHL syndrome. Recent studies have provided further insights into the genetics of VHLS, such as the concept of genotype-to-phenotype concordance in which specific types of *VHL* mutations are associated with specific tumor types in families with VHLS (Table 2).[12-14] Such information has further aided in risk-stratification and management of individuals and families with VHLS in recent years.

*Corresponding Author: Donna E. Hansel—Cleveland Clinic, Department of Anatomic Pathology, Glickman Urological and Kidney Institute, Taussig Cancer Institute and Genomic Medicine Institutue, 9500 Euclid Ave. L25, Cleveland, Ohio 44195, USA. Email: hanseld@ccf.org

Diseases of DNA Repair, edited by Shamim I. Ahmad. ©2010 Landes Bioscience and Springer Science+Business Media.

VHL Tumor Suppressor Gene

Similar to other hereditary cancer syndromes, cancers arising within the background of VHLS conform to the Knudson's 2-hit hypothesis[15,16] such that individuals with VHLS have a germline defect in one inherited *VHL* allele with subsequent somatic mutation of the remaining wild-type allele leading to disease manifestations. Investigators have also shown that inactivation of the *VHL* gene is a feature of sporadically occurring hemangioblastomas and clear cell RCCs.[17-24] In the latter case, promoter hypermethylation has also been implicated in the biallelic silencing of the *VHL* gene, in addition to gene mutation and deletion.[25]

Certain studies suggest possible role of protein product of *VHL* gene, pVHL in the mismatch repair mechanism. Tsuchiya et al, using immunoprecipitation assays, identified a novel protein, VHL binding protein 1 (VBP-1), which binds to the C terminus of pVHL with high affinity and is localized in the nucleus as a complex with pVHL protein.[26,27] Further studies of VBP-1 showed a consensus site for tyrosine phosphorylation, suggesting a possible protein kinase-mediated control.[28] It has been shown that VBP-1 plays an essential role in promoting the formation and assembly of functional α-, β- and γ-tubulin, suggesting a role in the early steps of the microtubule assembly pathway.[29-31] Also reported is the association of VBP-1 with the mismatch repair (MMR) genes, hMSH4 and hMSH4 splicing variant (hMSH4sv), that are critical in the meiotic processes. Inactivation of either Msh4 or Msh5 in mice results in testicular and ovarian degeneration and inactivation of both genes caused complete ovarian dysgenesis as a result of massive apoptosis.[32-34] The significance of hMSH4 (or hMSH4sv)-VBP1 interaction in humans is not currently clear, although the above observation suggests hMSH4 upon interaction with VBP-1 and possibly pVHL might perform additional functions during the meiotic process, including modification of the microtubule assembly to maintain proper chromosome positioning and to preserve chromosomal integrity.

The pVHL also plays a critical role in the ubiquitination of several key proteins within the cell.[35-41] Initially, the role of VHL gene and its protein product were unknown and pVHL did not appear to share homology with any known proteins. What was subsequently found was that the frequently mutated region of pVHL contained a binding domain for protein elongin C, which comprised a protein complex that included elongin B and cullin-2 (CUL2).[42,43] Elledge et al reported the resemblance of primary sequences of elongin C and CUL2 in human with the yeast proteins Skp1 and Cdc53, respectively, which bind an F-box protein to form an SCF (Skp1-Cdc53-F-box) ubiquitin-ligase complex in yeast.[44] The SCF ubiquitin-ligase complex in yeast catalyzes the polyubiquitination of proteins that ultimately leads to proteasome-mediated degradation.[45] Thus, pVHL appeared to complex with elongin B and cullin-2 and was hypothesized to function in polyubiquitination of protein targets for subsequent degradation.[43] Several studies followed which strengthened the above hypothesis: crystal structures of pVHL-elongin B-cullin-2 complex were similar to that of SCF ubiquitin-ligase complex[46] and immunoprecipitates of pVHL were shown to contain ubiquitin ligase function in the presence of ubiquitin-conjugating enzyme, E2.[47,48] Further, Rbx1, a protein responsible for ubiquitin ligase function of the SCF complex, was also found in the pVHL-elongin-CUL2 complexes.[49] All of the above evidence collectively illustrated that polyubiquitination function of pVHL is for subsequent proteasome-mediated degradation of certain proteins.

VHL-HIF Biology

pVHL is primarily a cytoplasmic protein, but can shuttle between the cytoplasm and nucleus.[50,51] There are 2 isoforms of pVHL with molecular weights of 30 and 19 kDa that result from differential splicing of the VHL gene.[52-54] The functions of both isotypes seem to be similar and are referred collectively as pVHL.

Several independent investigations, around the same time of cloning of the VHL gene, ultimately led to the discovery of hypoxia induced factor (HIF), a helix-loop-helix transcription factor, as a key regulator of cellular response to hypoxia.[55-57] HIF is a dimeric complex comprised of 2 subunits, HIFα and HIFβ. There are 3 HIFα genes and 3 HIFβ genes discovered.[58] HIFβ is constitutively synthesized whereas HIFα synthesis is tightly controlled via ubiquitination by pVHL and subsequent rapid proteasome-mediated degradation.[59]

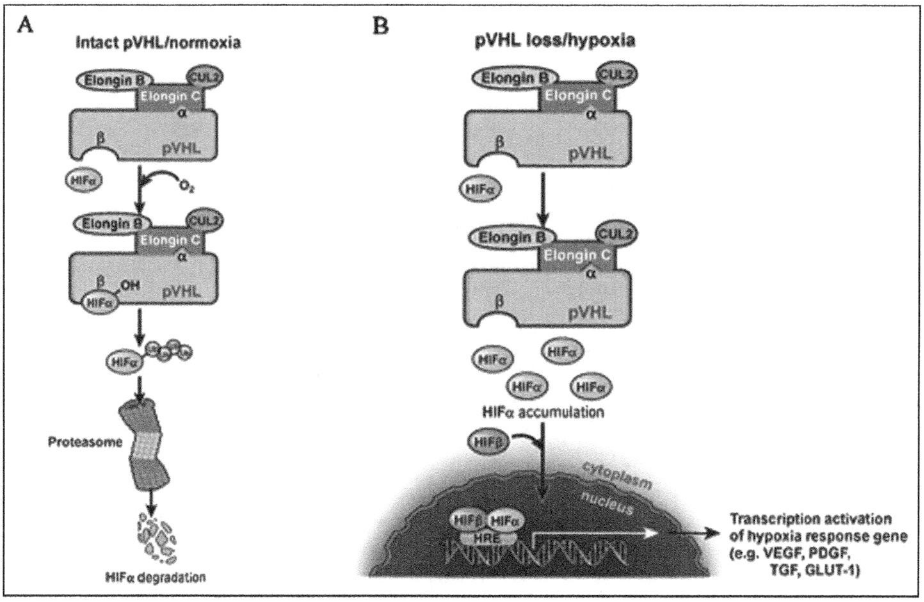

Figure 1. A) In state of normoxia and functional pVHL, HIFα is hydroxylated and interacts with pVHL, subsequently leading to ubiquitination and proteasome mediated degradation. B) In state of hypoxia and/or nonfunctioning pVHL, HIFα is not hydroxylated and therefore fails to interact with pVHL. HIFα is allowed to accumulate and dimerize with HIFβ which then translocates to nucleus. The heterodimer of HIFα-HIFβ binds to HRE (hypoxia response element of hypoxia inducible genes) thereby activating transcription of these genes.

The link between VHL and HIF was demonstrated by a landmark study by Maxwell et al, which illustrated that the pVHL plays an integral part in the regulation of the HIFα subunit (Fig. 1).[35] The β subunit of pVHL specifically binds to the HIFα subunit for polyubiquitination and subsequent rapid degradation of certain proteins. pVHL-HIFα/β binding requires hydroxylation of proline residues on HIFα subunits that occurs only in the presence of oxygen.[60-62] During a state of hypoxia, therefore, hydroxylation of HIFα fails to occur and pVHL-HIFα binding does not take place, leading to accumulation of HIFα. This excess HIFα heterodimerizes with constitutively secreted HIFβ and the HIFα/β complex translocates into the nucleus and binds to specific sequences known as hypoxia response elements (HRE) in the promoter regions of over 200 hypoxia inducible genes. Upregulation of hypoxia inducible genes is a cellular adaptive mechanism to hypoxia and many of these genes have been shown to have roles in tumorigenesis and cancer. Examples of hypoxia inducible genes include GLUt1, 6-phosphofructokinase 1 and pyruvate dehydrogenase kinase that are responsible for regulating glucose metabolism,[63,64] vascular endothelial growth factor (VEGF) and platelet derived growth factor (PDGF) that are implicated in angiogenesis,[59,65] matrix metalloproteinase 1(MMP1) and lysyl oxidase that affect extracellular matrix remodeling,[66,67] stromal cell derived factor (SDF1) and CXCR4, that mediate chemotaxis[68,69] and transforming growth factor α (TGFα) and epidermal growth factor receptor (EGFR) that are involved in cell proliferation and survival.[70,71]

In the absence of functional pVHL, similarly, constitutive up-regulation of HIFα and its transcription activation of hypoxia-inducible genes occur. Several studies have supported the causal role of defective pVHL and associated HIFα accumulation in tumorigenesis and furthermore VHL restoration in renal cell carcinoma cell lines with defective pVHL function led to cell-cycle exit and subsequent differentiation.[72]

Further insights into the different isoforms of HIF are emerging. HIF 1α and HIF 2α are structurally closely related and they are also subject to similar regulatory processes involving enzymatic hydroxylation and subsequent degradation via the VHL ubiquitination complex.[73] HIF1α and HIF2α also have considerably overlapping functions in the cellular response to hypoxia, including promotion of angiogenesis and extracellular matrix remodeling. However, more recent studies have illustrated that there are also distinct molecular effects unique to each HIF1α and HIF2α with respect to cellular metabolism, division and genetic stability.[74] For example, HIF1α, but not HIF2α, is able to stimulate glycolytic enzyme expression[75,76] while blocking anabolic biosynthesis of lipids and proteins by limiting mitochondrial pyruvate consumption.[77-79] Genetic studies have provided evidence for the nonredundant functions of HIF1α and HIF2α, where targeted inactivation of HIF1α and HIF2α has shown different patterns of response to hypoxia and low glucose stress.[80,81] Furthermore, different developmental defects are observed in mouse embryos mutated in HIF1α and HIF2α. HIF1α homozygous inactivation is lethal in embryonic development due to lack of vascular formation and global hypoxia, whereas HIF2α homozygous mutant embryos showed normal development of systemic vasculature but impaired lung maturation and adrenal function.[82,83] These results are likely due to the differential distribution of the two isoforms: HIF1α is ubiquitously expressed and HIF2α is detected most prominently in developing kidney, small vessels of the CNS, adrenal medulla, lung and intestine.[84] Furthermore, whereas HIF2α is undetected in most mature cells, it was shown to promote an undifferentiated phenotype in pleuripotent stem cells, suggesting silencing of HIF2α in many tissues after development.[85]

Additional studies have shown the distinct effects of HIF1α and HIF2α on the c-MYC oncoprotein and their subsequent differential effect on cell proliferation and genetic stability. While HIF1α has been shown to repress cell cycle progression by inhibiting the c-MYC oncoprotein,[86] HIF2α has been shown to have an opposite effect: HIF2α promotes cell cycle progression by enhancing c-MYC expression, which then leads to c-MYC-mediated activation of cyclin D2 and E2F1 and repression of p21 and p27.[87] Interestingly, during clear cell RCC development in VHLS, HIF1α expression was shown to gradually decrease whereas HIF2α expression gradually increased.[88,89] Furthermore, studies in clear cell RCC show that HIF 2α over-expression was primarily responsible for tumor development in pVHL-deficient xenograft models.[90,91] A predominance of the HIF1α isoform was shown in normal kidney and early cystic lesions whereas that of HIF2α isoform was seen in clear cell RCC.[92,93] Together, these observations provide an attractive hypothesis that a HIF2α effect on c-MYC oncogene may play a key role in the development of clear cell RCC in VHLS.

In addition to enhancing cellular proliferation, oncogenes can disrupt cell cycle controls and cellular metabolism, thereby contributing to genomic instability.[94] Cell cycle checkpoint inactivation allows DNA replication in aneuploid cells and may favor oncogenic genomic amplifications by repetitively triggering replication origins during a single S-phase.[95] Furthermore, many oncogenes alter cellular metabolism which results in the production of reactive oxygen species (ROS) that can directly damage DNA.[96] Consistent with these observations, high levels of the c-MYC oncogene have also been shown to result in accumulation of significant DNA damage.[99,100] Intermediate c-MYC expression levels, however, may actually promote genomic stability, which is supported by the observation that DNA damage is reduced when c-MYC is inhibited by HIF1α.[76,101]

Furthermore, interestingly, in hyperproliferative tumor cells, DNA damage from replication stress can actually trigger a signaling pathway including ATR, ATM, phospho-Chk1, phopho-Chk2 and phospho-histone H2AX (γH2AX)[97,98] which ultimately lead to cell cycle checkpoint activation/cell-intrinsic DNA damage response. What results is the enhanced ability to repair DNA damage through homologous recombination (HR) and nonhomologous end joining (NHEJ) to further promote cellular proliferation. Consistent with above results is the finding that HIF2α expression limits DNA damage accumulation.[102] HIF2α expressing tumors have been found to display enhanced expression of the homologous repair effectors BRCA1, BARD1 and XRCC2, enhanced expression of the spindle assembly checkpoint genes BUB1 and CENPE and lower levels of γH2AX, suggesting HIF2α association with increased DNA repair mechanism leading to increased genomic stability. On array-based Comparative Genomic Hybridization (CGH), a statistically significant 40% reduction in genomic aberrations was detected in clear cell RCC that expressed increased levels of HIF2α.[102] Together, the above observations suggest that HIF2α mediated up-regulation of c-MYC activity leads to higher rates of its proliferation that is partly enhanced by concurrent stimulation of homologous DNA repair mediators that allow efficient transit through S-phase and may ultimately be responsible for development of clear cell RCC in VHLS disease. Further studies are needed to validate this model.

HIF3α has been shown to be down-regulated with progression to cancer.[103] Three of the six splice variants are subject to similar pVHL mediated ubiquitination and proteasomal degradation.[104] HIF3α structurally resembles HIF1α and HIF2α but lacks transcription activation domain required for HIFα/HIFβ dimerization and nuclear translocation.[105] It appears to function as a dominant negative regulator of hypoxia-inducible gene expression in the human kidney.[105]

Non-HIF Related VHLS Pathways

In addition to the regulatory role of pVHL in the oxygen-dependent degradation of HIFα, pVHL has been shown to play a role in nonHIF related pathways including cell adhesion, ciliogenesis in microtubule assembly, extracellular matrix organization and other tumor suppressor pathways.

E-Cadherin and β-Catenin

Several recent studies have shown that diminished expression of E-cadherin occurs with decreased VHL function and with HIFα accumulation.[106-108] E-cadherin is a cell adhesion molecule that is associated with the normal development of mesenchymal cells into epithelial cells.[107] Loss of E-cadherin has been associated with reduced cell-cell adhesion and increased tumorigenic and metastatic potential of many tissue types.[107] Evans et al showed that patients with clear cell RCC exhibited significantly diminished expression of E-cadherin[107] and further studies by Krishnamachary et al and Esteban et al implicated accumulation of HIFα for the loss of E-cadherin expression in knockout mice.[107,108] Also it has been shown that subsequent down-regulation of HIFα is sufficient for the promotion of E-cadherin expression.[107,108]

Loss of pVHL, furthermore, may lead to the activation of β-catenin.[109-111] β-catenin serves as a structural link within the E-cadherin/β-catenin/α-catenin complex at the cell-cell junctions, which then associates with the actin cytoskeleton.[112-114] β-catenin also acts as a transcriptional activator in conjunction with the lymphoid enhancer factor/T-cell factor (LEF/TCF) DNA binding proteins.[115-117] Studies on β-catenin have shown that inactivated VHL leads to constitutive phosphorylation of cMET, which in turn leads to the disruption of cell adherence by causing dissociation of the E-cadherin/β-catenin complex at cell-cell junctions.[112] Dissociated β-catenin then accumulates and is subsequently translocated into the nucleus and acts upon transcription of genes responsible for proliferation, invasion and morphogenesis.[112]

Ciliogenesis

Evidence also suggests that pVHL, along with glycogen synthase kinase (GSK)3β, play a key role in the maintenance of primary cilia and microtubule assembly.[118,119] GSK3β is a protein kinase which has been shown to play a critical role in microtubule assembly, its stability and dynamics.[118] pVHL has recently been identified in the primary cilia of mouse embryo fibroblasts, as well in primary cilia of primary human renal proximal tubule epithelial cells.[118-121] Furthermore, studies have shown that pVHL is integral in the formation of primary cilia in cancer cell lines. When VHL function was restored in pVHL-deficient RCC cell lines, maintenance of primary cilia was completely rescued.[119-121] Collectively, the above data suggest that loss of the primary ciliary network may be linked to increased epithelial cell proliferation which may be associated with the development of clear cell RCC.

Extracellular Matrix Regulation

Recent studies suggest that pVHL binding of certain target proteins may play a critical role in the proper organization of extracelluar matrix molecules.[122-123] One example is the binding of pVHL to fibronectin.[123] Studies have also shown that pVHL binding to fibronectin is essential in proper fibronectin extracellular matrix assembly[123] and all oncogenic VHL mutations to harbour a defect in fibronectin binding.[124-125] This failure of pVHL-fibronectin binding results in growth of clear cell RCC in mouse xenograft models.[126] Furthermore, VHL mutant cell lines, lacking fibronectin binding, but retaining HIFα regulatory function, have been associated with clear cell RCC tumors of significantly higher microvessel density compared to cell lines with intact fibronectin binding.[127] These studies suggest that pVHL and fibronectin binding likely promote proper organization of extracellular matrix network that may hinder the formation of microvessel network during tumor-induced angiogenesis. pVHL also has been shown to bind to α2 chains of fibrillar collagen Type IV, an interaction which has been shown to be required in proper integration of collagen into the extracellular matrix network.[128-129] Collectively, the above studies suggest that proper organization of extracellular matrix molecules is a nonHIF related tumor suppressor function of pVHL.

p53 Tumor Suppressor Stabilization and Transactivation

pVHL also has been shown to positively regulate another tumor suppressor protein, p53.[130] p53 was found to specifically bind to the alpha domain of pVHL, which subsequently blocked the Mdm2-mediated ubiquitination and nuclear export of p53.[130-132] Furthermore, pVHL was shown to enhance the stabilization and transactivation of p53 upon DNA damage by forming a multimeric complex with p53 and ATM and p300, two enzymes that are known to promote p53 stabilization and activation.[131-133] Taken together, tumor suppressor function of pVHL may extend into a critical role in enhancing the tumor suppressor role of p53 upon DNA damage.

Clinical Features

Characteristic tumors of VHLS include hemangioblastomas of the CNS and retina, clear cell RCC, pheochromocytoma, endolymphatic sac tumors, cystadenomas and neuroendocrine tumors of the pancreas, papillary cystadenomas of epididymis and broad ligament (Table 1).[8,134]

VHLS is further divided into 2 types (Type I and II) depending on the likelihood of pheochromocytoma within the family; with Type I patients generally exhibiting a lower risk of pheochromocytoma development.[135-137] Type II disease is again subdivided, based upon the risk of development of RCC, into Type IIB in which patients exhibiting high risk of RCC and Type IIC, in which patients are affected by pheochromocytomas exclusively (Table 2). More recently, the concept of genotype-to-phenotype is emerging as a result of studies which link specific mutations to specific tumor(s) in the VHLS.[12-14,138] Pheochromocytomas have been almost always been associated with small missense mutations, whereas deletion mutations that would disrupt the pVHL function, are associated with a markedly reduced risk of pheochromocytomas. Although the reason for this is unclear, the development of pheochromocytoma may be due to a pVHL gain-of-function via nucleotide exchange or a partial loss, rather than a complete loss of pVHL.

Diagnosis of VHLS is often clinical.[139,140] With a positive family history, a patient with a CNS hemangioblastoma, pheochromocytoma or clear cell RCC is diagnosed with the disease. In the absence of a family history, however, 2 or more CNS hemangioblastomas or one CNS hemangioblastoma and a visceral tumor (excluding epididymal and renal cysts) meet the diagnostic criteria.

Table 1. **VHL Syndrome-Associated Tumors**

1.	Hemangioblastomas of the CNS
2.	Retinal angiomas
3.	Clear cell renal cell carcinoma (RCC)
4.	Pheochromocytomas
5.	Endolymphatic sac tumors of the middle ear
6.	Serous cystadenomas and neuroendocrine tumors of the pancreas
7.	Papillary cystadenomas of the epididymis and broad ligament

Table 2. Subtypes of VHL disease

Type	Characteristics	Types of Mutation
Type I	Hemangioblastomas	Large mutations
	Low risk of pheochromocytoma	
	Clear cell RCC	
Type IIA	Hemangioblastomas	Missense mutations
	High risk of Pheochromocytoma	
	Low risk of clear cell RCC	
Type IIB	Hemangioblastomas	Missense mutations
	High risk of Pheochromocytoma	
	High risk of clear cell RCC	
Type IIC	Pheochromocytoma **only**	Missense mutations

Table 3. Cambridge screening regimen for asymptomatic VHL individuals (134).

1.	Annual physical examination.
2.	Annual urine testing for cytology.
3.	Direct and indirect ophthalmologic exam from age 5.
4.	Fluorescein angiography or angiography from age 10 until age 60.
5.	MRI or CT brain scan every 3 years to age 50 and every 5 years thereafter.
6.	Annual renal ultrasound scan, with CT scan every 3 years.
7.	Annual 24 hr urine collection for VMA and metaneprine starting at age 5.

For patients affected with VHLS, diligent adherence to proposed surveillance protocol is crucial in order to decrease morbidity and improve overall prognosis (Table 3).[134]

Hemangioblastomas of the CNS

CNS hemangioblastoma is the most common tumor in VHLS, affecting 60-80% of the patients with the disease.[141-143] Hemangioblastomas are highly vascular, commonly multifocal benign tumors that are composed of vacuolated stromal cells harbouring VHL inactivation admixed with blood vessels. Average age at presentation is 33 years and these tumors are commonly associated with oedema and cysts along the craniospinal axis.[141] Lesions tend to be multifocal and evolve with enlargement and quiescent phases, alternating in an unpredictable manner.[141] The lesions ultimately lead to significant morbidity due to mass effect. Symptoms include headache, pain, emesis, hyper-reflexia, gait-ataxia, vertigo, weakness and dysmetria depending on the exact location of the tumor.[142]

Imaging study of choice for hemangioblastoma is contrast-enhanced T1-weighted MRI, although T2-weighted or FLAIR (fluid-attenuated inversion recovery) MRI is used to evaluate peritumoral edema and cysts (Figs. 2,3).[45,140,144] Screening examination for patients over 10 years of age is recommended (Table 3). Patients with known tumors are carefully followed clinically for onset of symptoms as well as recommended MRIs every 6-12 months, more frequently during phases of growth and less frequently during periods of dormancy.[141-144]

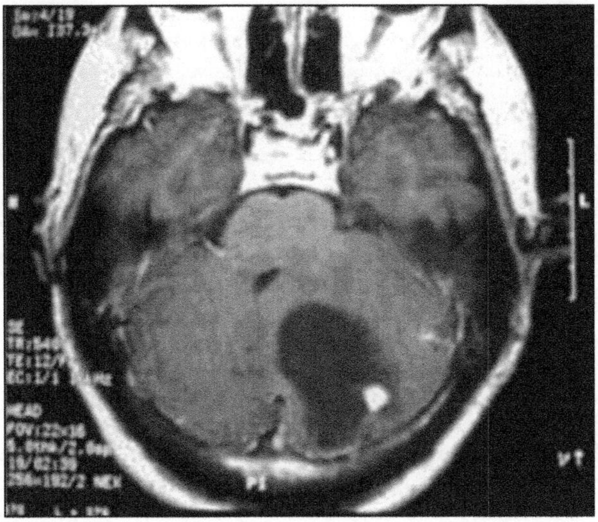

Figure 2. T1-weighted post contrast MRI of cystic hemangioblastoma with peripheral nodular enhancement in the left cerebellar hemisphere. Reproduced with permission from: Torreggiani WC et al. Clin Radiol 2002; 57:670-680;[145] ©2002 Elsevier.

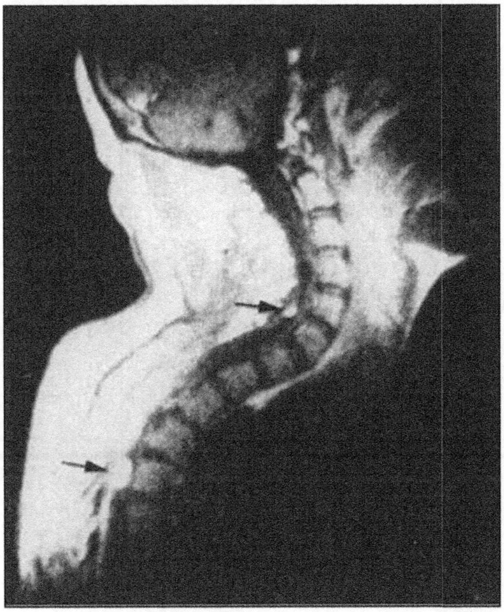

Figure 3. T1 weighted MRI illustrates multiple hemangioblastomas within the spinal canal (arrows). Reproduced with permission from: Filling-Katz MR et al. Neurology 1991; 41:41-46;[143] ©1991 BMJ Publishing Group Ltd.

Due to the multifocal nature of this lesion and the unpredictable evolution of these tumors, treatment is often deferred until development of symptoms.[141,146-148] Most lesions can be excised safely and effectively. Stereotactic radiation has also been employed in the treatment of CNS hemangioblastomas especially when they are not readily accessible by surgical means.[149-152] Notably, Chang and colleagues reported the largest case series to date which showed that for tumors less than 3 cm without mass effect stereotactic radiotherapy may be a reasonable alternative, with 3% rate of progression following treatment after a median follow-up of 43 months.[152] Due to the unpredictable nature of these tumors, which include long periods of dormancy, definitive relative safety and effectiveness of radiotherapy as compared to surgical intervention remain to be seen after a longer period of follow-up.

Retinal Hemangioblastoma

Retinal hemangioblastomas are seen in as many as 60% of patients with VHLS.[153,154] They are often multifocal and bilateral with a mean age of presentation of 25 years, although an estimated 5% of cases involve patients less than 10 years of age.[140] Although initially asymptomatic, these tumors can lead to significant morbidity including partial and total vision loss as they increase in size. The vision loss results from increased vascular permeability and subsequent accumulation of subretinal fluid, subsequently leading to the development of hard exudates at the macula that ultimately leads to retinal detachment and vision compromise (Fig. 4A,B).[140]

Small retinal hemangioblastomas may be managed by observation if the tumors are very small (<500 μm), are not associated with exudation or subretinal fluid and are of peripheral location and therefore not threatening of visual loss.[155,156] Otherwise, laser photocoagulation and cryotherapy comprise the mainstays of treatment in most cases.[156] Vitrectomy may be required in advanced cases that involve substantial tractional retinal detachment.[156] Rarely, enucleation may be performed if end stage disease such as glaucoma causes late complications including severe pain symptoms.[156] Early diagnosis and treatment are imperative in the prevention of serious late sequelae. For patients with known VHLS, routine, yearly ophthalmological screening starting at age 5 is recommended (Table 3).

Endolymphatic Sac Tumors

Endolymphatic sac tumors (ELSTs) are highly vascular papillary cystadenomas that arise from the posterior petrous bone which affects approximately 11-17% of patients with VHL disease (Fig. 5).[157,158] ELSTs are a common cause of hearing loss, tinnitus, vertigo and facial muscle weakness in patients with VHL disease. Although extremely rare, ELSTs can occur sporadically

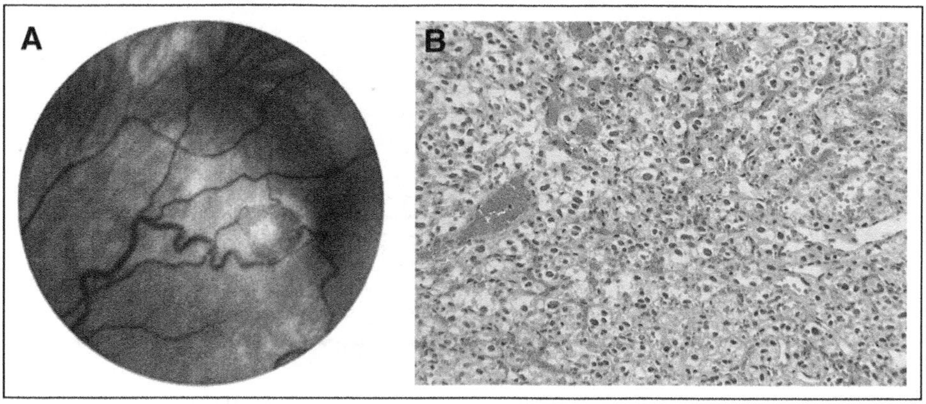

Figure 4. A) Retinal hemangioblastoma on ocular examination. Reproduced with permission from: Dollfus H et al. Invest Ophthalmol Vis Sci 2002; 43:3067-74.[153] B) Microscopic image of hemangioblastoma demonstrates pale vacuolated cells and associated blood vessels. Photomicrograph courtesy of Dr. Richard Prayson, Department of Anatomic Pathology, The Cleveland Clinic.

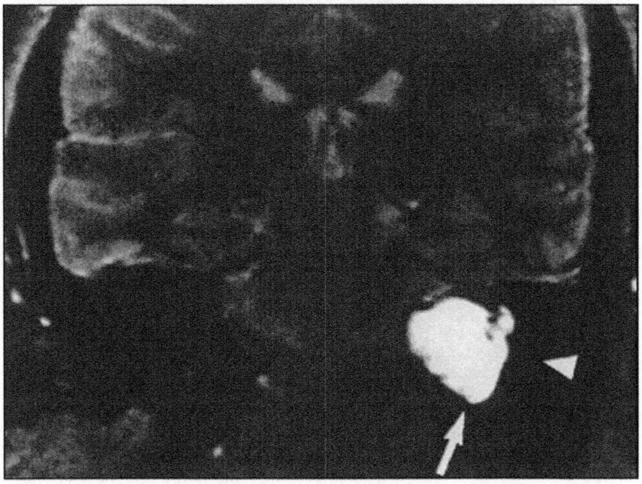

Figure 5. T2 weighted MRI shows endolymphatic sac tumor arising from petrous bone (arrow head) extending inferiorly to jugular fossa (arrow). Reproduced with permission from: Manski TJ et al. JAMA 1997; 277:1461-1466;[157] ©1997 American Medical Association.

in the general population. However, earlier age of onset as well as bilaterality tend to be more common in patients with VHLS.[157] Clinical symptoms are attributed to 3 mechanisms: (1) invasion of otic capsule resulting in destruction of the membranous labyrinth and disruption of endolymphatic flow; (2) intralabyrinthine hemorrhage; and (3) blockage of endolymphatic sac resorption of fluid.[157]

Treatment of ELSTs is mainly surgical which is curative with recovery of baseline hearing, when complete resection is feasible, provided that the tumors do not extensively invade the labyrinth.[156] Factors such as baseline hearing, severity of symptoms, possible surgical complications and the rate of tumor growth should be carefully considered prior to intervention.

Renal Cell Carcinomas and Renal Cysts

VHL disease is the most common cause of hereditary renal cell carcinoma and comprises about 2-3% of the total RCC incidence.[159] RCCs in this setting are invariably of the clear cell subtype.[160] RCCs tend to be multifocal and bilateral in the setting of VHL disease.[159] Symptoms of RCC include hematuria, flank pain, flank mass and nonspecific B symptoms such as fatigue and night sweats. Mean age of development of RCC is 35-40 years and up to 69% of patients with VHL disease have been shown to develop these tumors at some point in their lifetime.[134] RCC stands as the most lethal clinical manifestation in patients with VHLS, responsible for approximately 50% of mortality in affected patients.[161]

Renal lesions in VHLS patients span a spectrum of manifestations, from simple to complex cysts to renal cell carcinomas. Renal lesions are often multiple and bilateral as with other VHL associated tumors (Fig. 6A,B).[145] They tend to grow more slowly with low metastatic potential compared to their sporadic counterpart, especially when less than 3 cm.[160,162,163] Importantly however, in contrast to sporadic cases, simple cysts in setting of VHLS, have been shown to either transform into malignant lesions or already contain malignant cells in their walls, consequently warranting diligent monitoring with surveillance imaging studies every 6 to 12 months.[163] Growth and/or increasing complexity of a given cyst are features of tumor progression and are indications for surgical intervention.[163] Complex cysts have been shown to harbour malignant cells in approximately 25% of cases and require similar monitoring in this group of patients.[164] Yearly screening

A)

B)

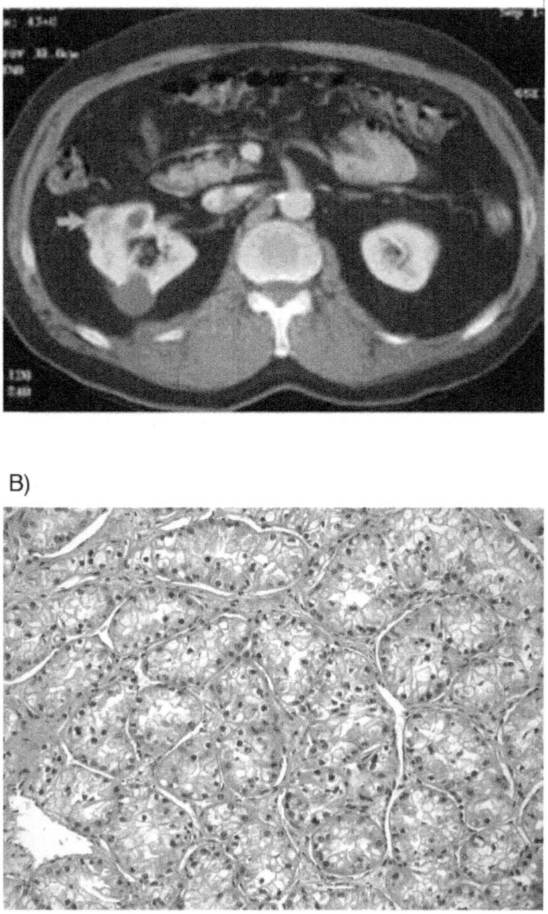

Figure 6. A) RCC and coexisting cystic disease on contrast enhanced CT.[145] B) H and E section of clear cell RCC demonstrating nest of pale tumor cells separated by a delicate vascular network. Reproduced with permission from: Torreggiani WC et al. Clin Radiol 2002; 57:670-680;[145] ©2002 Elsevier.

examination with ultrasound beginning at age 15 is recommended for individuals affected with VHL syndrome.[162] Patients with small, known renal lesions are monitored with serial computed tomography (CT) scans and/or magnetic resonance imaging (MRI) q6-12 months until the largest lesion reaches 3 cm at which point surgical resection is recommended.[163]

The mainstay of treatment for localized RCC is surgical resection. Nephron or renal-sparing surgeries have gained favour in recent years which provide adequate disease control as well as preservation of renal function especially in smaller tumors ($\leq$3 cm).[163] Walther and colleagues reported a 10-year investigation of renal-sparing surgery for tumors less than 3 cm.[163] Among 52 patients treated with renal-sparing surgery for tumors less than 3 cm, there was no evidence of metastases and no need for dialysis or kidney transplantation after a median follow-up of 60 months. Newer techniques such as laparoscopic cryoablation and radiofrequency ablation are under investigation in treatment of small, multiple tumors.[165-169] Following treatment for localized

RCC, continued surveillance is crucial in this patient population since a large portion of these individuals, approximately 30% at five years and up to 85% at 10 years, develop further tumors sometime in their lifetime.[170]

For advanced clear cell RCC, improved understanding of VHL tumor suppressor gene and its interplay with HIF biology have led to the development of molecular signalling inhibitors which have caused a paradigm shift in its treatment from cytokine based therapy to signalling inhibitor therapy. Although a comprehensive mechanism of action of these agents is not completely understood, they are inhibitors of angiogenesis, an end event of HIF accumulation and its transcription activation of hypoxia inducible genes. Some of the available agents include bevacizumab, a monoclonal antibody to VEGF, which has shown efficacy not only in RCC but other solid tumors including breast, lung and colon cancer.[171-173] Others include small molecule multikinase inhibitors of the VEGF receptor (VEGFR) and PDGF receptor (PDGFR), namely sunitinib and sorafenib, with other potent agents such as axitinib and pazopanib in advanced stages of development. Other relevant pathway that emerged as a therapeutic target is the mTOR pathway, which is a downstream molecule of the PI3K/AKT pathway and has been shown to contribute to the up-regulation of HIF and subsequent RCC tumor phenotype and angiogenesis.[174]

Anecdotal evidence that shows efficacy of sunitinib in retinal hemangioblastomas in VHL patients for rapid and durable recovery of visual function exists.[175] Currently, clinical trials involving sunitinib evaluating its safety and efficacy in patients with VHL associated tumors are on-going (NCT00330564, NCT00673816). Other targeted agents including vandetanib (VEGFR-2 inhibitor) and ranibizumab (VEGF monoclonal antibody fragment) are also undergoing investigation in renal tumors and retinal tumors respectively (NCT00566995, NCT00089765).

Pheochromocytomas

Pheochromocytomas arise from chromaffin cells of the adrenal gland. They can also arise as paragangliomas in extra-adrenal locations such as in the carotid body, glomus jugulare and periaortic tissues.[176] Overall incidence in setting of VHLS ranges from 7-20% with a mean age of approximately 30 years.[176] As with other manifestations of VHL disease, these tumors also tend to present in younger patients and can be multifocal and bilateral but rarely malignant (<10%).[176] Patients may be asymptomatic at the time of diagnosis, although a sudden release of catecholamines may prove life threatening. Common symptoms of pheochromocytoma include hypertensive crisis, intermittent hypertension, palpitations, flushing and headaches. Onset of disease may be prior to 10 years of age and initiation of early screening with 24 h urine or blood catecholamines prior to age 5 or 6 then yearly is recommended.[176]

Pheochromocytomas typically enhance after contrast media on CT scan (Fig. 7A,B).[145] CT scan, however, is less ideal for investigating ectopic sites where MRI is superior in that setting.[177] Pheochromocytomas demonstrate high signal intensity on T2 weighted MRI. Metaiodobenzylguanidine test (MIBG) is highly sensitive (95%) and specific (100%), but may not detect very small lesions.[178] When an adrenal lesion is discovered by CT scan, an MRI is usually performed. If the lesion is less than 3.5 cm, patient is asymptomatic and catecholamines are normal, the patient is generally followed with CT scans at yearly intervals.[144,179] If symptoms or elevated catecholamines develop, MIBG is performed prior to surgery to exclude additional sites. Further, with sudden death from unsuspected release of catecholamines is a major concern and screening before elective surgeries and/or prior to delivery in high risk individuals is imperative.[180,181]

In the past, open bilateral total adrenalectomies were treatment of choice for bilateral pheochromocytomas which led to significant complications due to adrenal deficiency with up to 23% of patients experiencing Addisonian crises.[182] More recently, however, keeping with the principle of organ preservation in VHL related tumors, combined with the fact that only 10% of adrenal gland is needed for maintenance of adequate adrenal function, "partial" or "cortical-sparing" adrenalectomy has gained favour in the surgical management of these tumors. Indications for surgery include tumors with abnormal function, meta-iodobenzylguanidine uptake or tumors larger than 3.5 cm.[180]

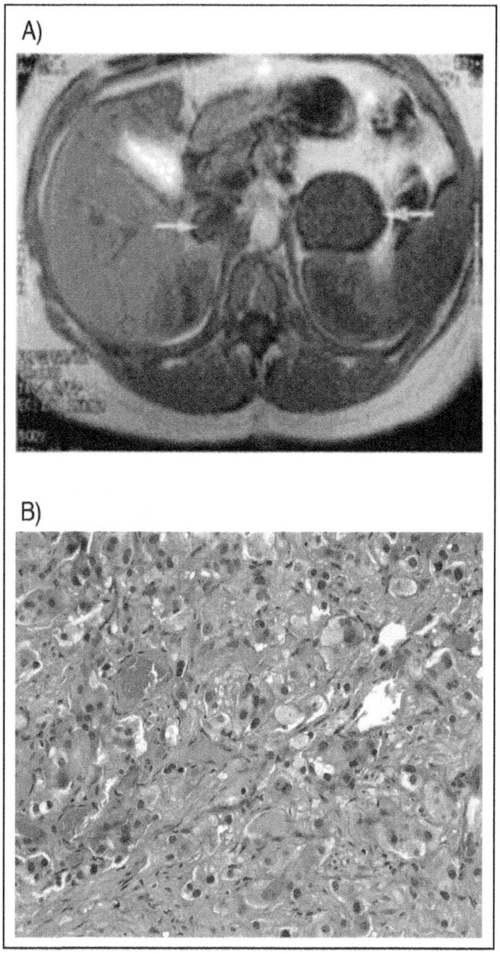

Figure 7. A) T1 weighted image of bilateral pheochromocytoma. Reproduced with permission from: Torreggiani WC et al. Clin Radiol 2002; 57:670-680;[145] ©2002 Elsevier. B) H and E section of pheochromocytoma demonstrating nest of tumor cells ("zellballen") surrounded by sustentacular cells in a background of hemorrhage.

Studies have shown that early surgical intervention with cortical-sparing approach is associated with low rates of recurrence and improved preservation of adrenal function.[183,184]

Pancreatic Neuroendocrine Tumors and Cysts

Pancreatic abnormalities are common in patients with VHLS with up to 77% of patients showing lesions on imaging studies.[185] Up to 70% of pancreatic lesions are cysts and 8-17% comprising neuroendocrine tumors.[185-1987] Mean age of presentation is 35 years. Pancreatic cysts are usually dormant and hence do not cause symptoms in patients with VHL disease and thus generally do not require treatment.

Pancreatic neuroendocrine tumors arise from pancreatic islet tumors and are also usually dormant.[174-176] They appear as a well-circumscribed, encapsulated enhancing mass on contrast CT which should be confirmed by an MRI for diagnosis (Fig. 8).[145] Most of the neuroendocrine

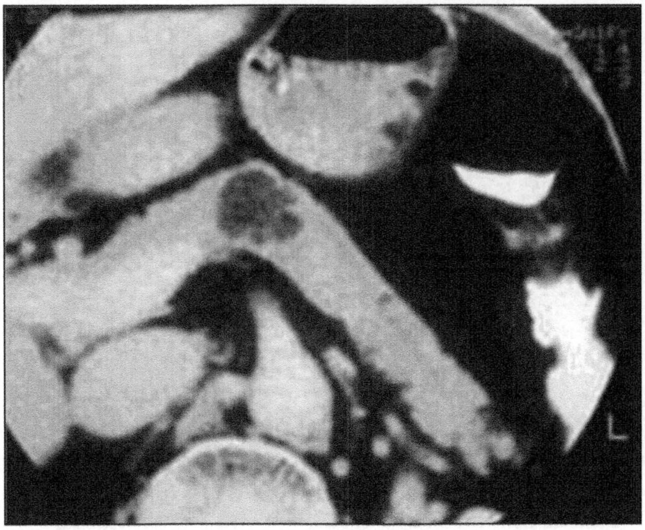

Figure 8. Cystadenoma of pancreas on contrast enhanced CT. Reproduced with permission from: Torreggiani WC et al. Clin Radiol 2002; 57:670-680;[145] ©2002 Elsevier.

tumors also do not cause symptoms and go undetected for many years.[187] In rare cases, however, active tumors can arise which may cause symptoms such as diarrhoea and hypoglycaemic episodes due to the secretion of vasoactive intestinal peptides and insulin.[185,186] Treatment of neuroendocrine tumors is mainly surgical (Table 4).[187]

Epididymal Cystadenomas

Epididymal cystadenomas occur in 25-60% of patients with VHLS.[187-189] These tumors usually manifest during the teenage period and generally do not show symptoms.[189] VHL disease-related cystadenomas tend to be multiple and bilateral as compared to sporadic cases which tend to be singular and unilateral. They originate from the epididymal duct, a derivative of the embryonic mesonephric duct.[189] Diagnosis is clinical and confirmed by ultrasonography (Fig. 9).[145] Sonographic criteria for diagnosis include: (1) solid tumor greater than 10 × 14 mm; (2) occurrence in setting of VHLS; and (3) slow growth.[189] As these tumors do not show symptoms in patients, no treatment is usually required and patients are managed conservatively with regular observation.

Broad Ligament Cystadenomas

Broad ligament cystadenomas are considered female counterpart lesions of epididymal cystadenomas. These tumors are quite rare and are believed to originate from the embryonic mesonephric

Table 4. Guidelines for surgical resection of pancreatic neuroendocrine tumors (Libutti et al, ref. 187)

1.	No evidence of metastatic disease
2.	Tumor size greater than 3 cm in the body or tail OR
3.	Tumor greater than 2cm in the head of the pancreas
4.	Patient undergoing laparotomy for other lesions

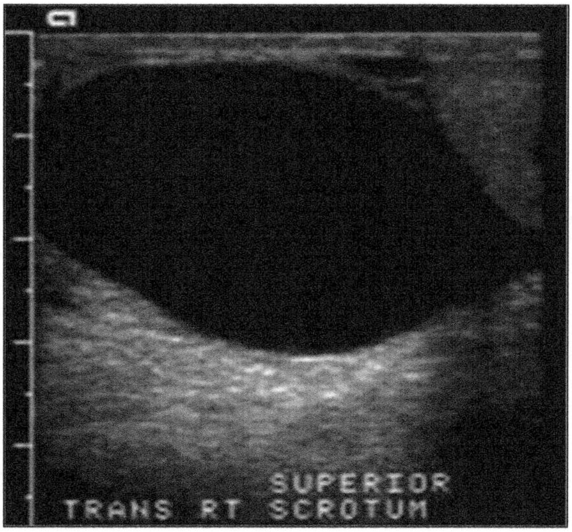

Figure 9. Hypoechoic lesion showing simple epididymal cyst on ultrasound. Reproduced with permission from: Torreggiani WC et al. Clin Radiol 2002; 57:670-680;[145] ©2002 Elsevier.

duct.[190] Patients are usually asymptomatic and therefore specific treatment is generally unnecessary. Conservative management with observation combined with CT or ultrasonography to monitor their size over time usually suffice.

Conclusion

Since the initial description of characteristic tumors in affected families over 100 years ago, great strides have been made in elucidating molecular pathways and mechanisms that underlie the VHLS. Improved understanding of the VHLS has subsequently led to timely screening and improves interventions which markedly improved clinical outcome in patients. Further analysis such as genotype-phenotype correlations and HIF related as well as nonHIF related pathways in the pathogenesis of VHLS are emerging, with applications of which are under active investigation for therapeutic purposes. A multidisciplinary approach including diligent adherence to lifelong surveillance, timely surgical/medical/radiological interventions of VHL manifestations as well as continued efforts in improvement of treatment modalities remain equally vital in the optimal management of individuals affected by this disease.

Acknowledgements

The Authors would like to thank Mr. David Schumick for his efforts with illustrations in the preparation of this manuscript.

References

1. Collins E. Intraocular growths (two cases, brother and sister, with peculiar vascular new growth, probably retinal, affecting both eyes). Trans Ophthalmol Soc UK 1894; 14:141-149.
2. Hippel Ev. Ueber eine sehr seltene erkrankung der nethat. Graefe Arch Ophthalmol 1904; 59:83-106.
3. Lindau A. Zur frage der angiomatosis retinai und ihrer hirncomplikation. Acta Ophthalmol 1927; 4:193-226.
4. Davison C, Brock S, Cyke CG. Retinal and central nervous hemangioblastomatosis with visceral changes (von Hippel-Lindau's disease). Bull Neurol Inst New York 1936; 5:72-93.
5. Glushien AS, Mansuy MM, Littman DS. Pheochromocytoma. Its relationship to the neurocutaneous syndrome. AM J Med 1953; 14:318-27.

6. Melmon KL, Rosen SW. Lindau's disease: review of the literature and study of a large kindred. Am J Med 1964; 36:595-617.
7. Manski TJ, Heffner DK, Glenn GM et al. Endolymphatic sac tumors. A source of morbid hearing loss in von Hippel Lindau disease. J Am Med Assoc 1997; 277:1461-1466.
8. Maher ER, Kaelin WG Jr. von Hippel-Lindau disease. Medicine 1997; 76:381-391.
9. Seizinger BR et al. Von Hippel-Lindau disease maps to the region of chromosome 3 associated with renal cell carcinoma. Nature 1988; 332:268-269.
10. Latif F et al. Identification of the von Hippel-Lindau disease tumor suppressor gene. Science 1993; 260:1317-1320.
11. Woodward E et al. Comparative sequence analysis of the VHL tumor suppressor gene. Genomics 2000; 65:253-265.
12. Chen F et al. Germline mutations in the von Hippel-Lindau disease tumor suppressor gene: correlations with phenotype. Hum Mutat 1995; 5:66-75.
13. Zbar B et al. Germline mutations in the von Hipple-Lindau (VHL) gene in families from North America, Europe and Japan. Hum Mutat 1996; 8:348-357.
14. Neumann H, Bender B. Genotype-phenotype correlations in von Hipple-Lindau disease. J Intern Med 1998; 243:541-545.
15. Knudson AGJ, Strong LC. Mutation and cancer: neuroblastoma and pheochromocytoma. Am J Hum Genet 1972; 24:514-532.
16. Knudson AGJ. Genetics of human cancer. Ann Rev Genet 1986; 20:231-251.
17. Kanno H, Kondo K, Ito S et al. Somatic mutations of the von Hippel-Lindau tumour suppressor gene in sporadic central nervous system hemangioblastomas. Cancer Res 1994; 54:4845-47.
18. Lee JY, Dong SM, Park WS et al. Loss of heterozygosity and somatic mutations of the VHL tumour suppressor gene in sporadic cerebellar hemangioblastomas. Cancer Res 1998; 58:504-08.
19. Oberstrass J, Reifenberger G, Reifenberger J et al. Mutation of the Von Hippel-Lindau tumour suppressor gene in capillary haemangioblastomas of the central nervous system. J Pathol 1996; 179:151-56.
20. Tse JY, Wong JH, Lo KW et al. Molecular genetic analysis of the von Hippel-Lindau disease tumour suppressor gene in familial and sporadic cerebellar hemangioblastomas. Am J Clin Pathol 1997; 107:459-66.
21. Gnarra JR, Tory K, Weng Y et al. Mutations of the VHL tumour suppressor gene in renal carcinoma. Nat Genet 1994; 7:85-90.
22. Shuin T et al. Frequent somatic mutations and loss of heterozygosity of the von Hippel-Lindau tumor suppressor gene in primary human renal cell carcinomas. Cancer Res 1994; 54:2852-2855.
23. Foster K et al. Somatic mutations of the von Hippel-Lindau disease tumor suppressor gene in nonfamilial clear cell renal carcinoma. Hum Mol Genet 1994; 3:2169-2173.
24. Kondo K, Yao M, Yoshida M et al. Comprehensive mutational analysis of the VHL gene in sporadic renal cell carcinoma: relationship to clinicopathological parameters. Genes Chromosomes Cancer 2002; 34:58-68.
25. Herman JG et al. Silencing of the VHL tumor-suppressor gene by DNA methylation in renal carcinoma. Proc Natl Acad Sci USA 1994; 91:9700-9704.
26. Her C, Wu X, Griswold MD et al. MutS Homologue MSH4 Physically Interacts with von Hippel-Lindau Tumor Suppressor-binding Protein 1. Cancer Research 2003; 63:865-872.
27. Tsuchiya H, Tokuhiro I, Hino O. Identification of a Novel Protein (VBP-1) Binding to the von Hippel-Lindau (VHL) Tumor Suppressor Gene Product. Cancer Research 1996; 56:2881-2885.
28. Brinke A, Green PM, Giannelli F. Characterization of the Gene (VBP1) and Transcript for the von Hippel-Lindau Binding Protein and Isolation of the Highly Conserved Murine Homologue. Genomics 1997; 45:105-112.
29. Geissler S, Siegers K, Schiebel E. A novel protein complex promoting formation of functional- and -tubulin. EMBO J 1998; 17:952-966.
30. Vainberg IE, Lewis SA, Rommelaere H et al. Prefoldin, a chaperone that delivers unfolded proteins to cytosolic chaperonin. Cell 1998; 93:863-873.
31. Alvarez P, Smith A, Fleming J et al. Modulation of tubulin polypeptide ratios by the yeast protein Pac10p. Genetics 1998; 149:857-864.
32. Kneitz B, Cohen PE, Avdievich E et al. MutS homolog 4 localization to meiotic chromosomes is required for chromosome pairing during meiosis in male and female mice. Genes Dev 2000; 14:1085-1097.
33. Edelmann W, Cohen PE, Kneitz B et al. Mammalian MutS homologue 5 is required for chromosome pairing in meiosis. Nat Genet 1999; 21:123-127.
34. De Vries SS, Baart EB, Dekker M et al. Mouse MutS-like protein Msh5 is required for proper chromosome synapsis in male and female meiosis. Genes Dev 1999; 13:523-531.

35. Maxwell PH, Wiesener MS, Chang GW et al. The tumour suppressor protein VHL targets hypoxia-inducible factors for oxygen-dependent proteolysis. Nature 1999; 399:271-275.
36. Tanimoto K, Makino Y, Pereira T et al. Mechanism of regulation of the hypoxia-inducible factor-1 alpha by the von Hippel-Lindau tumor suppressor protein. EMBO J 2000; 19:4298-4309.
37. Kamura T, Sato S, Iwai K et al. Activation of HIF1alpha ubiquitination by a reconstituted von Hippel-Lindau (VHL) tumor suppressor complex. Proc Natl Acad Sci USA 2000; 97:10430-10435.
38. Bardos JI, Ashcroft M. Hypoxia-inducible factor-1 and oncogenic signalling. Bioessays 2004; 26:262-269.
39. Gordan JD, Simon MC. Hypoxia-inducible factors: central regulators of the tumor phenotype. Curr Opin Genet Dev 2007; 17:71-77.
40. Hoffman MA, Ohh M, Yang H et al. von Hippel-Lindau protein mutants linked to type 2C VHL disease preserve the ability to downregulate HIF. Hum Mol Genet 2001; 10:1019-1027.
41. Knauth K, Bex C, Jemth P et al. Renal cell carcinoma risk in type 2 von Hippel-Lindau disease correlates with defects in pVHL stability and HIF-1alpha interactions. Oncogene 2006; 25:370-377.
42. Pause A et al. The von Hippel-Lindau tumor-suppressor gene product forms a stable complex with human CUL-2, a member of the Cdc53 family of proteins. Proc Natl Acad Sci USA 1997; 94:2156-2161.
43. Pause A, Peterson B, Schaffar G et al. Studying interactions of four proteins in the yeast two-hybrid system: structural resemblance of the pVHL/elongin BC/hCUL-2 complex with the ubiquitin ligase complex SKP1/cullin/F-box protein. Proc Natl Acad Sci USA 1999; 96:9533-9538.
44. Bai C et al. SKP1 connects cell cycle regulators to the ubiquitin proteolysis machinery through a novel motif, the F-box. Cell 1996; 86:263-274.
45. Deshaies R. SCF and Cullin/Ring H2-based ubiquitin ligases. Annu Rev Cell Dev Biol 1999; 15:435-467.
46. Stebbins CE, Kaelin WG, Pavletich NP. Structure of the VHL—elongin-C—elongin-B complex: implications for VHL tumor suppressor function. Science 1999; 284:455-461.
47. Lisztwan J, Imbert G, Wirbelauer C et al. The von Hippel-Lindau tumor suppressor protein is a component of an E3 ubiquitin—protein ligase activity. Genes Dev 1999; 13:1822-1833.
48. Iwai K et al. Identification of the von Hippel-Lindau tumor suppressor protein as part of an active E3 ubiquitin ligase complex. Proc Natl Acad Sci USA 1999; 96:12436-12441.
49. Kamura T et al. Rbx1, a component of the VHL tumor suppressor complex and SCF ubiquitin ligase. Science 1999; 284:657-661.
50. Los M et al. Expression pattern of the von Hippel-Lindau protein in human tissues. Lab Invest 1996; 75:231-238.
51. Lee S et al. Transcription-dependent nuclear-cytoplasmic trafficking is required for the function of the von Hippel-Lindau tumor suppressor protein. Mol Cell Biol 1999; 19:1486-1497.
52. Schoenfeld A, Davidowitz E, Burk R. A second major native von Hippel-Lindau gene product, initiated from an internal translation start site, functions as a tumor suppressor. Proc Natl Acad Sci USA 1998; 95:8817-8822.
53. Iliopoulos O, Ohh M, Kaelin W. pVHL19 is a biologically active product of the von Hippel-Lindau gene arising from internal translation initiation. Proc Natl Acad Sci USA 1998; 95:11661-11666.
54. Blankenship C, Naglich J, Whaley J et al. Alternate choice of initiation codon produces a biologically active product of the von Hippel-Lindau gene with tumor suppressor activity. Oncogene 1999; 18:1529-1535.
55. Semenza G. Regulation of mammalian O_2 homeostasis by hypoxia-inducible factor 1. Annu Rev Cell Dev Biol 1999; 15:551-578.
56. Stadler W. Chromosomes, hypoxia, angiogenesis and trial design: a brief history of renal cancer drug development. Clin Cancer Res 2007; 13:1630-1633.
57. Wang GL, Semenza GL. General involvement of hypoxia inducible factor 1 in transcriptional response to hypoxia. Proc Natl Acad Sci USA 1993; 90:4304-4308.
58. Semenza GL. HIF-1 and mechanisms of hypoxia sensing. Curr Opin Cell Biol 2001; 13:167-171.
59. Iliopoulos O, Levy AP, Jiang C et al. Negative regulation of hypoxia-inudcible genes by the von Hippel-Lindau protein. Proc Natl Acad Sci USA 1996; 93:10595-105599.
60. Ivan M, Kondo K, Yang H et al. HIF-alpha targeted for VHL-mediated destruction by praline hydroxylation: Implications for O_2 sensing. Science 2001; 292:464-468.
61. Jaakkola P, Mole DR, Tian YM et al. Targeting of HIF-alpha to the von Hippel-Lindau ubiquitylation complex by O_2-regulated prolyl hydroxylation. Science 2001; 292:468-472.
62. Yu F, White SB, Zhao Q et al. HIF-1alpha binding to VHL is regulated by stimulus-sensitive proline hydroxylation. Proc Natl Acad Sci USA 2001; 98:9630-9635.
63. Papandreou I, Cairns RA, Fontana L et al. HIF-1 mediates adaptation to hypoxia by actively downregulating mitochondrial oxygen consumption. Cell Metab 2006; 3:187-197.

64. Kim JW, Tchernyshyov I, Semenza GL et al. HIF-1 mediated expression of pyruvate dehydrogenase kinase: a metabolic switch required for cellular adpaptation to hypoxia. Cell Metab 2006; 3:177-185.
65. Levy AP, Levy NS, Iliopoulos O et al. Regulation of vascular endothelial growth factor by hypoxia and its modulation by the von Hippel-Lindau tumor suppressor gene. Kidney Int 1997; 51:575-578.
66. Petrella BL, Lohi J, Brinckerhoff CE. Identification of membrane type-1 matrix metalloproteinase as a taret of hypoxia-inducible factor-2α in von Hippel-Lindau renal cell carcinioma. Oncogene 2005; 24:1043-1052.
67. Erler JT et al. Lysyl oxidase is essential for hypoxia-induced metastasis. Nature 2006; 440:1222-1226.
68. Staller P et al. Chemokine receptor CXCR4 downregulated by von Hippel-Lindau tumour suppressor pVHL. Nature 2003; 425:307-311.
69. Zagzag D et al. Stromal cell-derived factor-1α and CXCR4 expression in hemangioblastoma and clear cell-renal cell carcinoma: von Hippel-Lindau loss-of-function induces expression of a ligand and its receptor. Cancer Res 2005; 65:6178-6188.
70. Knebelmann B, Ananth S, Cohen HT et al. Transforming growth factor alpha is a target for the von Hippel-Lindau tumor suppressor. Cancer Res 1998; 58:226-231.
71. Franovic A et al. Translational up-regulation of the EGFR by tumor hypoxia provides a nonmutational explanation for its expression in human cancer. Prox Natl Acad Sci USA 2007; 104:13092-13097.
72. Kaelin WG Jr. Molecular basis of the VHL hereditary cancer syndrome. Nat Rev Cancer 2002; 2:673-682.
72. Sowter HM, Raval R, Moore J et al. Predominant Role of Hypoxia-Inducible Transcription Factor (HIF)-1 versus HIF-2 in Regulation of the Transcriptional Response to Hypoxia1. Cancer Research 2003; 63:6130-6134.
73. Wiesener MS, Turley H, Allen WE et al. Induction of endothelial PAS domain protein-1 by hypoxia: characterization and comparison with hypoxia-inducible factor-1. Blood 1998; 92:2260-2268.
74. Gordan JD, Thompson CB, Simon MC. HIF and c-Myc: sibling rivals for control of cancer cell metabolism and proliferation. Cancer Cell 2007; 12(2):108-13.
75. Hu CJ, Wang LY, Chodosh LA et al. Differential roles of hypoxia-inducible factor 1alpha (HIF-1alpha) and HIF-2alpha in hypoxic gene regulation. Mol Cell Biol 2003; 23:9361-9374.
76. Huang LE, Bindra RS, Glazer PM et al. Hypoxia-induced genetic instability—a calculated mechanism underlying tumor progression. J Mol Med 2007; 85:139-148.
77. Kim JW, Tchernyshyov I, Semenza GL et al. HIF-1-mediated expression of pyruvate dehydrogenase kinase: a metabolic switch required for cellular adaptation to hypoxia. Cell Metab 2006; 3:177-185.
78. Lum JJ, Bui T, Gruber M et al. The transcription factor HIF-1alpha plays a critical role in the growth factor-dependent regulation of both aerobic and anaerobic glycolysis. Genes Dev 2007; 21:1037-1049.
79. Papandreou I, Cairns RA, Fontana L et al. HIF-1 mediates adaptation to hypoxia by actively down-regulating mitochondrial oxygen consumption. Cell Metab 2006; 3:187-197.
80. Ema M, Hirota K, Mimura J et al. Molecular mechanisms of transcription activation by HLF and HIF1 in response to hypoxia: their stabilisation and redox signal-induced interaction with CBP/p300. EMBO J 1999; 18:1905-1914.
81. Brusselmanns K, Bono F, Maxwell PH et al. Hypoxia-inducible factor-2 is involved in the apoptotic response to hypoglycemia but not to hypoxia. J Biol Chem 2001; 276:39192-39196.
82. Ryan HE, Lo J, Johnson RS. HIF-1 is required for solid tumor formation and embryonic vascularization. EMBO J 1998; 17(11):3005-15.
83. Tian H, Hammer RE, Matsumoto AM et al. The hypoxia-responsive transcription factor EPAS1 is essential for catecholamine homeostasis and protection against heart failure during embryonic development. Genes Dev 1998; 12(21):3320-4.
84. Smaldone MC, Maranchie JK. Clinical implications in hypoxia inducible factor in renal cell carcinoma. Urologic Oncology: Seminars and Original Investigations 2009; 27:238-245.
85. Holmquist L, Jogi A, Pahlman S. Phenotypic persistence after reoxygenation of hypoxic neuroblastoma cells. Int J Cancer 2005; 116(2):218-25.
86. Koshiji M, Kageyama Y, Pete EA et al. HIF-1alpha induces cell cycle arrest by functionally counteracting Myc. EMBO J 2004; 23:1949-1956.
87. Gordan JD, Bertout JA, Hu CJ et al. HIF-2alpha Promotes Hypoxic Cell Proliferation by Enhancing c-Myc Transcriptional Activity. Cancer Cell 2007a; 11:335-347.
88. Mandriota SJ, Turner KJ, Davies DR et al. HIF activation identifies early lesions in VHL kidneys: evidence for site-specific tumor suppressor function in the nephron. Cancer Cell 2002; 1:459-468.
89. Raval RR, Lau KW, Tran MG et al. Contrasting properties of hypoxia-inducible factor 1 (HIF-1) and HIF-2 in von Hippel-Lindau associated renal cell carcinoma. Mol Cell Biol 2005; 25:5675-5686.

90. Wiesener MS, Jurgensen JS, Rosenberger C et al. Widespread hypoxia-inducible expression of HIF-2 in distinct cell populations of different organs. FASEB J 2003; 17(2):271-3.
91. Mandriota SJ, Turner KJ, Davies DR et al. HIF activation identifies early lesions in VHL kidneys. Evidence for site-specific tumor suppressor function in the nephron. Cancer Cell 2002; 1(5):459-68.
92. Kim CM, Vocke C, Torres-Cabala C et al. Expression of hypoxia inducible factor-1 and 2 in genetically distinct early renal cortical tumors. J Urol 2006; 175(5):1908-14.
93. Kondo K, Klco J, Nakamura E et al. Inhibition of HIF is necessary for tumor suppression by the von Hippel-Lindau protein. Cancer Cell 2002; 1:237-246.
94. Halazonetis TD, Gorgoulis VG, Bartek J. An oncogene-induced DNA damage model for cancer development. Science 2008; 319:1352-1355.
95. Hook SS, Lin JJ, Dutta A. Mechanisms to control rereplication and implications for cancer. Curr Opin Cell Biol 2007; 19:663-671.
96. Lee AC, Fenster BE, Ito H et al. Ras proteins induce senescence by altering the intracellular levels of reactive oxygen species. J Biol Chem 1999; 274:7936-7940.
97. Bartkova J, Horejsi Z, Koed K et al. DNA damage response as a candidate anti-cancer barrier in early human tumorigenesis. Nature 2005; 434:864-87.
98. Gorgoulis VG, Vassiliou LV, Karakaidos P et al. Activation of the DNA damage checkpoint and genomic instability in human precancerous lesions. Nature 2005; 434:907-913.
99. Ray S, Atkuri KR, Deb-Basu D et al. MYC can induce DNA breaks in vivo and in vitro independent of reactive oxygen species. Cancer Res 2006; 66:6598-6605.
100. Vafa O, Wade M, Kern S et al. c-Myc can induce DNAvdamage, increase reactive oxygen species and mitigate p53 function: a mechanism for oncogeneinduced genetic instability. Mol Cell 2002; 9:1031-1044.
101. Koshiji M, To KK, Hammer S et al. HIF-1alpha induces genetic instability by transcriptionally downregulating MutSalpha expression. Mol Cell 2005; 17:793-803.
102. Gordan JD, Lal P, Dondeti VR et al. HIF-alpha effects on c-Myc distinguish two subtypes of sporadic VHL-deficient clear cell renal carcinoma. 2008; 14(6):435-46.
103. Maranchie JK et al. The contribution of VHL substrate binding and HIF1α to the phenotype of VHL Loss in renal cell carcinoma. Cancer Cell 2002; 1:247-255.
104. Gu YZ, Moran SM, Hogenesch JB et al. Molecular characterization and chromosomal localization of a third-class hypoxia inducible factor subunit, HIF3. Gene Expr 1998; 7(3):205-13.
105. Maynard MA, Qi H, Chung J et al. Multiple splice variants of the human HIF-3 locus are targets of the von Hippel-Lindau E3 ubiquitin ligase complex. J Biol Chem 2003; 278(13):11032-40.
106. Maynard MA, Evans AJ, Hosomi T et al. Human HIF-3-4 is a dominant-negative regulator of HIF-1 and is down-regulated in renal cell carcinoma. FASEB J 2005; 19(11):1396-406.
107. Esteban MA, Tran MG, Harten SK et al. Regulation of E-cadherin expression by VHL and hypoxia-inducible factor. Cancer Res 2006; 66:3567-3575.
108. Evans AJ, Russell RC, Roche O et al. VHL promotes E2 box-dependent E-cadherin transcription by HIF-mediated regulation of SIP1 and snail. Mol Cell Biol 2007; 27:157-169.
109. Krishnamachary B, Zagzag D, Nagasawa H et al. Hypoxia-inducible factor-1-dependent repression of E-cadherin in von Hippel-Lindau tumor suppressor null renal cell carcinoma mediated by TCF3, ZFHX1A and ZFHX1B. Cancer Res 2006; 66:2725-2731.
110. Nelson WJ, Nusse R. Convergence of Wnt, beta-catenin and cadherin pathways. Science 2004; 303:1483-1487.
111. Brembeck FH, Rosario M, Birchmeier W. Balancing cell adhesion and Wnt signaling, the key role of beta-catenin. Curr Opin Genet Dev 2006; 16:51-59.
112. Nakaigawa N, Yao M, Baba M et al. Inactivation of von Hippel-Lindau gene induces constitutive phosphorylation of MET protein in clear cell renal carcinoma. Cancer Res 2006; 66:3699-3705.
113. Peruzzi B, Athauda G, Bottaro DP. The von Hippel-Lindau tumor suppressor gene product represses oncogenic beta-catenin signaling in renal carcinoma cells. Proc Natl Acad Sci USA 2006; 103:14531-14536.
114. Drees F, Pokutta S, Yamada S et al. α-Catenin is a molecular switch that binds E-cadherin—β-catenin and regulates actin-filament assembly. Cell 2005; 123:903-915.
115. Yamada S, Pokutta S, Drees F et al. Deconstructing the cadherin-catenin-actin Complex. Cell 2005; 123:889-901.
116. Behrens J, von Kries JP, Kuhl M et al. Functional interaction of β-catenin with the transcription factor LEF-1. Nature 1996; 382:638-642.
117. Huber O, Korn R, McLaughlin J et al. Nuclear localization of β-catenin by interaction with transcription factor LEF-1. Mech Dev 1996; 59:3-10.
118. Thoma CR, Frew IJ, Hoerner CR et al. pVHL and GSK3-beta are components of a primary cilium-maintenance signalling network. Nat Cell Biol 2007; 9:588-595.

119. Schermer B, Ghenoiu C, Bartram M et al. The von Hippel-Lindau tumor suppressor protein controls ciliogenesis by orienting microtubule growth. J Cell Biol 2006; 175:547-554.
120. Esteban MA, Harten SK, Tran MG et al. Formation of primary cilia in the renal epithelium is regulated by the von Hippel-Lindau tumor suppressor protein. J Am Soc Nephrol 2006; 17:1801-1806.
121. Lutz MS, Burk RD. Primary cilium formation requires von hippel-lindau gene function in renal-derived cells. Cancer Res 2006; 66:6903-6907.
122. Ohh M, Yauch RL, Lonergan KM et al. The von Hippel-Lindau tumor suppressor protein is required for proper assembly of an extracellular fibronectin matrix. Mol Cell 1998; 1:959-968.
123. Tang N, Mack F, Haase VH et al. pVHL Function is essential for endothelial extracellular matrix deposition. Mol And Cell Biol 2006; 26(7):2519-2530.
124. Clifford SC, Cockman ME, Smallwood AC et al. Contrasting effects on HIF-1alpha regulation by disease-causing pVHL mutations correlate with patterns of tumourigenesis in von Hippel-Lindau disease. Hum Mol Genet 2001; 10:1029-1038.
125. Hoffman MA, Ohh M, Yang H et al. von Hippel-Lindau protein mutants linked to type 2C VHL disease preserve the ability to downregulate HIF. Hum Mol Genet 2001; 10:1019-1027.
126. Stickle NH, Chung J, Klco JM et al. pVHL modification by NEDD8 is required for fibronectin matrix assembly and suppression of tumor development. Mol Cell Biol 2004; 24:3251-3261.
127. Kurban G, Hudon V, Duplan E et al. Characterization of a von Hippel Lindau pathway involved in extracellular matrix remodeling, cell invasion and angiogenesis. Cancer Res 2006; 66:1313-1319.
128. Grosfeld A, Stolze IP, Cockman ME et al. Interaction of hydroxylated collagen IV with the von hippel-lindau tumor suppressor. J Biol Chem 2007; 282:13264-13269.
129. Kurban G, Duplan E, Ramlal N et al. Collagen matrix assembly is driven by the interaction of von Hippel-Lindau tumor suppressor protein with hydroxylated collagen IV alpha 2. Oncogene 2008; 27:1004-1012.
130. Roe JS, Kim HS, Lee SM et al. p53 stabilization and transactivation by von Hippel-Lindau protein. Mol Cell 2006; 22:395-405.
131. Brooks CL, Gu W. Ubiquitination, phosphorylation and acetylation: the molecular basis for p53 regulation. Curr Opin Cell Biol 2003; 15:164-71.
132. Kubbutat MH, Jones SN, Vousden KH. Regulation of p53 stability by Mdm2. Nature 1997; 387:299-303.
133. Sakaguchi K, Herrera JE, Saito S et al. DNA damage activates p53 through a phosphorylation-acetylation cascade. Genes Dev 1998; 12:2831-41.
134. Maher ER, Yates JR, Harries R et al. Clinical features and natural history of von Hippel-Lindau disease. Q J Med 1990; 77(283):1151-1163.
135. Couch V, Lindor MS, Noralane M et al. von Hippel-Lindau disease. Mayo Clin Proc 2000; 75:265-272.
136. Kaelin WG. The von Hippel-Lindau tumor suppressor protein and clear cell renal carcinoma. Clin Cancer Res 2007; 13:680s-684s.
137. Weirich G, Klein B, Wohl T et al. VHL2C phenotype in a German von Hippel-Lindau family with concurrent VHL germline mutations P81S and L188V. J Clin Endoc Metab 2002; 87:5241-5246.
138. Hes F, Zewald R, Peeters T et al. Genotype-phenotype correlations in families with deletions in the von Hippel-Lindau (VHL) gene. Hum Genet 2000; 106:425-31.
139. Lamiell JM, Salazar FG, Hsia YE. Von Hippl-Lindau disease affecting 43 members of a single kindred. Medicine 1989; 68:1-29.
140. Lonser RR, Glenn GM, Walther M et al. Von Hippel Lindau disease. Lancet 2003; 361:2059-2067.
141. Wanebo JE, Lonser RR, Glenn GM et al. The natural history of hemangioblastomas of the central nervous system in patients with von Hippel-Lindau disease. J Neurosurg 2003; 98:82-94.
142. Neumann HP, Eggert HR, Scheremet R et al. Central nervous system lesions in von Hippel-Lindau syndrome. J Neurol Neurosurg Psychiatry 1992; 55:898-901.
143. Filling-Katz MR, Choyke PL, Oldfield E et al. Central nervous system involvement in Von Hippel-Lindau disease. Neurology 1991; 41:41-46.
144. Choyke PL, Glenn GM, Walther MM et al. Von Hippel-Lindau disease: Genetic, clinical and imaging features. Radiology 1995; 194:629-642.
145. Torreggiani WC, Keogh C et al. Von Hippel-Lindau disease: A radiological essay. Clin Radiol 2002; 57:670-680.
146. Pavesi G, Feletti A, Berlucchi S et al. Neurosurgical treatment of von Hippel-Lindau-associated hemangioblastomas: benefits, risks and outcome. J Neurosurg Sci 2008; 52:29-36.
147. Lonser RR, Weil RJ, Wanebo JE et al. Surgical management of spinal cord hemangioblastomas in patients with von Hippel-Lindau disease. J Neurosurg 2003; 98:106-116.
148. Weil RJ, Lonser RR, DeVroom HL et al. Surgical management of brainstem hemangioblastomas in patients with von Hippel-Lindau disease. J Neurosurg 2003; 98:95-105.

149. Niemela M, Lim YJ, Soderman M et al. Gamma knife radiosurgery in 11 hemangioblastomas. J Neurosurg 1996; 85:591-96.
150. Page KA, Wayson K, Steinberg GK et al. Stereotaxic radiosurgical ablation: an alternative treatment for recurrent and multifocal hemangioblastomas: a report of four cases. Surg Neurol 1993; 40:424-28.
151. Patrice SJ, Sneed PK, Flickinger JC et al. Radiosurgery for hemangioblastoma: results of a multi-institutional experience. Int J Radiat Oncol Biol Phys 1996; 35:493-99.
152. Chang SD, Meisel JA, Hancock SL et al. Treatment of hemangioblastomas in von Hippel-Lindau disease with linear accelerator-based radiosurgery. Neurosurgery 1998; 43:28-34.
153. Dollfus H, Massin P, Taupin P et al. Retinal hemangioblastoma in von Hippel-Lindau disease: a clinical and molecular study. Invest Ophthalmol Vis Sci 2002; 43:3067-74.
154. Webster AR, Maher ER, Moore AT. Clinical characteristics of ocular angiomatosis in von Hippel-Lindau disease and correlation with germline mutation. Arch Ophthalmol 1999; 117:371-78.
155. Oosterhuis JA, Rubinstein K. Hemangioma at the optic disc. Ophthalmologica 1972; 164:362-374.
156. Singh AD, Nouri M, Shields CL et al. Treatment of retinal capillary hemangioma. Ophthalmology 2002; 109:1799-1806.
157. Manski TJ, Heffner DK, Glenn GM et al. Endolymphatic sac tumors. A source of morbid hearing loss in von Hippel-Lindau Disease. JAMA 1997; 277:1461-1466.
158. Luff DA, Simmons M, Malik T et al. Pathology in Focus. Endolymphatic sac tumours. J of Laryngology and Otology 2002; 116:398-401.
159. Meister M, Choyke P, Anderson C et al. Radiological evaluation, management and surveillance of renal masses in von Hippel-Lindau disease. Clinical Radiology 2009; 64:589-600.
160. Solomon D, Schwartz A. Renal pathology in von Hippel-Lindau disease. Hum Pathol 1988; 19:1072-1079.
161. Lamiell JM, Salazar FG, Hsia YE. von Hippel-Lindau disease affecting 43 members of a single kindred 1989; 68(1):1-29.
162. Neumann HP, Bender BU, Berger DP et al. Prevalence, morphology and biology of renal cell carcinoma in von Hippel-Lindau disease compared to sporadic renal cell carcinoma. J Urol 1995; 54:2010-2014.
163. Walther MM, Choyke PL, Glenn G et al. Renal cancer in families with hereditary renal cancer: Prospective analysis of a tumor size threshold for renal parenchymal sparing surgery. J Urol 2001; 161:1475-1479.
164. Poston CD, Jaffe GS, Lubensky IA et al. Characterization of the renal pathology of a familial form of renal cell carcinoma associated with von Hippel-Lindau disease: clinical and molecular genetic implications. J Urol 1995; 153:22-26.
165. Pavlovich CP, Walther MM, Choyke PL et al. Percutaneous radio frequency ablation of small renal tumours: initial results. J Urol 2002; 167:10-15.
166. Shingleton WB, Sewell PE Jr. Percutaneous renal tumour cryoablation with magnetic resonance imaging guidance. J Urol 2001; 165:773-776.
167. Gill IS, Novick AC, Meraney AM et al. Laparoscopic renal cryoablation in 32 patients. Urology 2000; 56:748-753.
168. Gill IS, Remer EM, Hasan WA et al. Renal cryoablation: outcome at 3 years. J Urol 2005; 173:1903-1907.
169. Hafron J, Kaouk JH. Ablative techniques for the management of kidney cancer. Nat Clin Pract 2007; 4:261-269.
170. Chauveau D, Duvic C, Chretien Y et al. Renal involvement in von Hippel-Lindau disease. Kidney Int 1996; 50:944-951.
171. Hurwitz H, Fehrenbacher L, Novotny W et al. Bevacizumab plus irinotecan, fluorouracil and leucovorin for metastatic colorectal cancer. N Engl J Med 2004; 350:2335-2342.
172. Sandler A, Gray R, Perry MC et al. Paclitaxel-carboplatin alone or with bevacizumab for nonsmall-cell lung cancer. N Engl J Med 2006; 355:2542-2550.
173. Miller K, Wang M, Gralow J et al. Paclitaxel plus bevacizumab versus paclitaxel alone for metastatic breast cancer. N Eng J Med 2007; 357:2666-2676.
174. Kim JJ, Rini BI. Recent advances in molecularly targeted therapy in advanced renal cell carcinoma. Therapy 2009; 6(3):309-320.
175. Aiello LP, George DJ, Cahill MT et al. Rapid durable recovery of visual function in a patient with von Hippel-Lindau syndrome after systemic therapy with vascular endothelial growth factor receptor inhibitor SU5416. Amer Acad Opth 2002; 109(9):1745-1751.
176. Walther MM, Reiter R, Keiser HR et al. Clinical and genetic characterization of pheochromocytoma in von Hippel-Lindau families: comparison with sporadic pheochromocytoma gives insight into natural history of pheochromocytoma. J Urol 1999; 162:659-664.
177. Weise M, Merke DP, Pacak K et al. Utility of plasma free metanephrines for detecting childhood pheochromocytoma. J Clin Endocrinol Metab 2002; 87:1955-60.

178. Neumann HP, Berger DP, Sigmund G. Pheochromocytomas, multiple endocrine neoplasia Type 2 and von Hippel Lindau disease. N Engl J Med 329:1531-1538.
179. Eisenhofer G, Lenders JW, Linehan WM et al. Plasma normetanephrine and metanephrine for detecting pheochromocytoma in von Hippel-Lindau disease and multiple endocrine neoplasia type 2. N Engl J Med 1999; 340:1872-79.
180. Mugawar M, Rajender Y, Purohit AK et al. Anesthetic management of von Hippel-Lindau syndrome for excision of cerebellar hemangioblastoma and pheochromocytoma surgery. Anesth Analg 1998; 86:673-674.
181. Harrington JL, Farley DR, van Heerden JA et al. Adrenal tumors and pregnancy. World J Surg 1999; 23:182-186.
182. Lairmore TC, Ball DW, Baylin SB et al. Management of pheochromocytomas in patients with multiple endocrine neoplasia type 2 syndromes. Ann Surg 1993; 217:595-601.
183. Pavlovich CP, Linehan WM, Walther MM. Partial adrenalectomy in patients with multiple adrenal tumours. Curr Urol Rep 2001; 2:19-23.
184. Baghai M, Thompson GB, Young WF Jr et al. Pheochromocytomas and paragangliomas in von Hippel-Lindau disease: a role for laparoscopic and cortical-sparing surgery. Arch Surg 2002; 137:682-88.
185. Hammel PR, Vilgrain V, Terris B et al. Pancreatic involvement in von Hippel-Lindau disease. The Groupe Francophone d'Etude de la Maladie de von Hippel-Lindau. Gastroenterology 2000; 119:1087-1095.
186. Binkovitz LA, Johnson CD, Stephens DH. Islet cell tumors in von Hippel-Lindau disease: increased prevalence and relationship to the multiple endocrine neoplasias. AJR Am J Roentgenol 1990; 155:501-505.
187. Libutti SK, Choyke PL, Bartlett DL et al. Pancreatic neuroendocrine tumors associated with von Hippel Lindau disease: Diagnostic and management recommendations. Surgery 1998; 124:1153-1159.
188. Gruber MB, Healey GB, Toguri AG et al. Papillary cystadenoma of epididymis: component of von Hippel-Lindau syndrome. Urology 1980; 16:305-306.
189. Choyke PL, Glenn GM, Wagner JP et al. Epididymal cystadenomas in von Hippel-Lindau disease. Urology 1997; 49:926-31.
190. Funk KC, Heiken JP. Papillary cystadenoma of the broad ligament in a patient with von Hippel-Lindau disease. AJR Am J Roentgenol 1989; 153:527-558.

Index

G

H

I

K

L

M

N

O

P

R

S

T

U

V

W

X

Z

GPSR Compliance

The European Union's (EU) General Product Safety Regulation (GPSR) is a set of rules that requires consumer products to be safe and our obligations to ensure this.

If you have any concerns about our products, you can contact us on ProductSafety@springernature.com

In case Publisher is established outside the EU, the EU authorized representative is:

Springer Nature Customer Service Center GmbH
Europaplatz 3
69115 Heidelberg, Germany

Batch number: 10762179

Printed by Printforce, the Netherlands